T0189017

Pseudo-Differential Operators

Volume 15

Associate Editors

Guozhen Lu, University of Connecticut, Storrs, CT, USA

Alberto Parmeggiani, Università di Bologna, Bologna, Italy

Luigi G. Rodino, Università di Torino, Torino, Italy

Bert-Wolfgang Schulze, Universität Potsdam, Potsdam, Germany

Johannes Sjöstrand, Université de Bourgogne, Dijon, France

Sundaram Thangavelu, Indian Institute of Science at Bangalore, Bangalore, India

Maciej Zworski, University of California at Berkeley, Berkeley, CA, USA

Managing Editor

M. W. Wong, York University, Toronto, Canada

Pseudo-Differential Operators: Theory and Applications is a series of moderately priced graduate-level textbooks and monographs appealing to students and experts alike. Pseudo-differential operators are understood in a very broad sense and include such topics as harmonic analysis, PDE, geometry, mathematical physics, microlocal analysis, time-frequency analysis, imaging and computations. Modern trends and novel applications in mathematics, natural sciences, medicine, scientific computing, and engineering are highlighted.

Boris Plamenevskii • Oleg Sarafanov

Solvable Algebras of Pseudodifferential Operators

Boris Plamenevskii
Department of Higher Mathematics
and Mathematical Physics
St. Petersburg State University
St. Petersburg, Russia

Oleg Sarafanov
Department of Higher Mathematics
and Mathematical Physics
St. Petersburg State University
St. Petersburg, Russia

ISSN 2297-0355 ISSN 2297-0363 (electronic)
Pseudo-Differential Operators
ISBN 978-3-031-28397-0 ISBN 978-3-031-28398-7 (eBook)
https://doi.org/10.1007/978-3-031-28398-7

© The Editor(s) (if applicable) and The Author(s), under exclusive license to Springer Nature Switzerland AG 2023
This work is subject to copyright. All rights are solely and exclusively licensed by the Publisher, whether the whole or part of the material is concerned, specifically the rights of reprinting, reuse of illustrations, recitation, broadcasting, reproduction on microfilms or in any other physical way, and transmission or information storage and retrieval, electronic adaptation, computer software, or by similar or dissimilar methodology now known or hereafter developed.
The use of general descriptive names, registered names, trademarks, service marks, etc. in this publication does not imply, even in the absence of a specific statement, that such names are exempt from the relevant protective laws and regulations and therefore free for general use.
The publisher, the authors, and the editors are safe to assume that the advice and information in this book are believed to be true and accurate at the date of publication. Neither the publisher nor the authors or the editors give a warranty, expressed or implied, with respect to the material contained herein or for any errors or omissions that may have been made. The publisher remains neutral with regard to jurisdictional claims in published maps and institutional affiliations.

This book is published under the imprint Birkhäuser, www.birkhauser-science.com by the registered company Springer Nature Switzerland AG
The registered company address is: Gewerbestrasse 11, 6330 Cham, Switzerland

Contents

Introduction

Pseudodifferential operators (ΨDO) are widely used in the theory of partial differential equations, mathematical physics, functional analysis, and topology. Several monographs [9, 15, 18, 39, 42, 43] are devoted to such operators in the smooth situation (i.e., on smooth manifolds and with smooth symbols). For a long time, "one-dimensional" singular integral operators with discontinuous coefficients on composite contours have been playing an important role in studying the boundary value problems on plain domains with non-smooth boundary. These operators have been discussed from different points of view in [10,12,20], etc. At present, there are also several monographs (for example, [17, 21, 35, 36]) dealing with the ΨDO on manifolds of dimension $n \geq 2$ with singularities. This monograph has a few intersections with [21]; it has no essential intersections with other books. To read this monograph, it is desirable to have some familiarity with the smooth theory of ΨDO and with C^*-algebras. However, in the first introductory chapter, some necessary preliminaries are given.

Now let us explain the subject of the book. In what follows, algebra and morphism mean C^*-algebra and $*$-morphism, $\mathcal{B}H$ denotes the algebra of all bounded operators in a Hilbert space H, and $\mathcal{K}H$ is the ideal of all compact operators in H. The spectrum $\widehat{\mathcal{A}}$ of algebra $\mathcal{A}$ is the set of all (equivalence classes of) irreducible representations of this algebra endowed with a natural topology (the so-called Jacobson topology). Let $\mathcal{M}$ be a smooth compact manifold without boundary and let $\mathcal{A}$ be the algebra generated in $L_2(\mathcal{M})$ by scalar ΨDO with smooth symbols. We assume that the principal symbols of such ΨDO are defined on the bundle of non-zero cotangent vectors, are homogeneous functions of degree zero on every fiber, and belong to the class $C^\infty(S^*(\mathcal{M}))$, where $S^*(\mathcal{M})$ is the cospherical bundle. As is known, the algebra $\mathcal{A}$ contains the ideal $\mathcal{K}L_2(\mathcal{M})$ and the quotient algebra $\mathcal{A}/\mathcal{K}L_2(\mathcal{M})$ is commutative. The spectrum $(\mathcal{A}/\mathcal{K}L_2(\mathcal{M}))^\wedge$ (i.e., the space of maximal ideals) can be identified with the bundle $S^*(\mathcal{M})$. If $P \in \mathcal{A}$ and Φ is the principal symbol of the operator P, then the map

$$(\mathcal{A}/\mathcal{K}L_2(\mathcal{M}))^\wedge \ni \pi \mapsto P(\pi) := \pi[P]$$

is implemented as the function $\pi \mapsto \Phi(\pi) \in \mathbb{C}$; here, $[P]$ is the equivalence class of the operator P in the algebra $\mathcal{A}/\mathcal{K}L_2(\mathcal{M})$. The operator P is Fredholm if and only if

$\pi[P] \neq 0$ for all $\pi \in (\mathcal{A}/\mathcal{K}L_2(\mathcal{M}))^\wedge$. (An operator $A \in \mathcal{B}H$ is called Fredholm if its range $\mathcal{R}(A)$ is closed and the spaces $\ker A$ and $\operatorname{coker} A$ are finite-dimensional.)

If the manifold $\mathcal{M}$ or the symbols of ΨDO have singularities, then the quotient algebra $\mathcal{A}/\mathcal{K}L_2(\mathcal{M})$ is non-commutative. Among its irreducible representations, there are infinite-dimensional ones (as $\dim \mathcal{M} > 1$). For $A \in \mathcal{A}$, the map

$$(\mathcal{A}/\mathcal{K}L_2(\mathcal{M}))^\wedge \ni \pi \mapsto A(\pi) := \pi[A]$$

is an analogue of the principal symbol. In contrast to the commutative case, not only can scalars serve as $\pi[A]$ but also non-trivial operators in infinite-dimensional Hilbert spaces. The "principal symbol" still gives a criterion for an operator to be Fredholm: $A \in \mathcal{A}$ is Fredholm if and only if any operator $A(\pi) = \pi[A]$ is invertible as $\pi \in (\mathcal{A}/\mathcal{K}L_2(\mathcal{M}))^\wedge$. Thus, to apply this criterion, one must list all (equivalence classes of) irreducible representations of the quotient algebra $\mathcal{A}/\mathcal{K}L_2(\mathcal{M})$ and to find implementations of these representations. One of this book's purposes is to describe the spectra of algebras generated by ΨDO with discontinuous symbols (coefficients) on piecewise smooth manifolds.

For the algebra generated by ΨDO with smooth symbols on a smooth manifold, the following relations hold:

$$\{0\} \subset \mathcal{K}L_2(\mathcal{M}) \subset \mathcal{A}, \qquad \mathcal{A}/\mathcal{K}L_2(\mathcal{M}) \simeq C(S^*(\mathcal{M})).$$

In the general situation, one can try to simplify the study of the (non-commutative) quotient algebra $\mathcal{A}/\mathcal{K}L_2(\mathcal{M})$ by taking the additional quotient modulo some ideal $J \supset \mathcal{K}L_2(\mathcal{M})$. The resulting loss of information is not large if the algebra $J/\mathcal{K}L_2(\mathcal{M})$ is comparatively simple.

Definition An algebra $\mathcal{L}$ is called solvable if there is a composition series $\{0\} = I_{-1} \subset I_0 \subset \cdots \subset I_N = \mathcal{L}$ of ideals I_j, such that the successive quotients I_j/I_{j-1} consist of continuous operator-functions with compact values; more precisely, if there is an isomorphism

$$I_j/I_{j-1} \simeq C_0(X_j) \otimes \mathcal{K}H_j, \qquad j = 0, \ldots, N < \infty,$$

where H_j is a Hilbert space, X_j is a locally compact Hausdorff space, and $C_0(X_j)$ is the algebra of continuous functions on X_j tending to zero at infinity. A composition series possessing such property is called solving, and the number N is called the length of the solving composition series.

Thus, in the smooth situation, the series $\{0\} \subset \mathcal{K}L_2(\mathcal{M}) \subset \mathcal{A}$ is solving, and its length is 1. Let, for example, the algebra $\mathcal{A}$ be generated by ΨDO with smooth symbols on a smooth manifold and by operators of multiplication by functions a continuous everywhere except at a fixed point x_0. At that point, the coefficients may have discontinuities "of the first kind" (there exists $\lim a(x)$ as $x \to x_0$ that depends on the direction in which x approaches x_0). The ideal $\operatorname{com} \mathcal{A}$ spanned by the commutators of the elements in $\mathcal{A}$ is larger than $\mathcal{K}L_2(\mathcal{M})$. The algebra $\mathcal{A}$ is solvable, and the (shortest) solving series is

$$\{0\} \subset \mathcal{K}L_2(\mathcal{M}) \subset \operatorname{com} \mathcal{A} \subset \mathcal{A},$$

while

$$\operatorname{com} \mathcal{A}/\mathcal{K}L_2(\mathcal{M}) \simeq C_0(\mathbb{R}) \otimes \mathcal{K}L_2(S^{m-1}),$$
$$\mathcal{A}/\operatorname{com} \mathcal{A} \simeq C(S^*(\mathfrak{C})),$$

where $\mathfrak{C}$ is obtained by gluing the boundary (i.e., a sphere) to the manifold $\mathcal{M} \setminus x_0$.

A solving composition series allows us to determine the collection of symbols $\sigma_1(A), \ldots, \sigma_N(A)$ for an operator A in a solvable algebra $\mathcal{L}$. The symbol $\sigma_j(A)$ is an operator-valued function on the space X_j. The invertibility of $\sigma_j(A), \ldots, \sigma_N(A)$ at each point is necessary and sufficient for the invertibility of A up to a summand in the ideal I_{j-1}. Composition series can be useful for studying the groups $K_*(\mathcal{L})$ of the operator K-theory related to the algebra $\mathcal{L}$. In this book, we present solving series for considered algebras; other questions mentioned in this paragraph are not discussed.

In the first part of the book, we study the algebras generated by pseudodifferential operators on a smooth manifold with continuous symbols and by operators of multiplication by discontinuous coefficients. Thus, the initial objects, i.e. the generators of the algebras, do not require special definitions in this situation.

The second part is devoted to operators on manifolds with (smooth non-intersecting) "edges" of arbitrary dimensions (conical points are edges of dimension 0). On such manifolds, we must begin with the definition of pseudodifferential operators. Here, we introduce ΨDO of arbitrary order and study their properties in detail, then we describe the spectrum of C^*-algebras generated by the operators of zero order. In [29, 30], this theory is developed for a wider class of "stratified" piecewise smooth manifolds (informally, manifolds with intersecting edges). The results of these papers are not included in the book.

Let us briefly describe the content of the chapters.

Chapter 1 is divided into three sections. Section 1.1 contains a standard introduction into the smooth theory of "classical" ΨDO. In Sect. 1.2, the necessary facts about special meromorphic ΨDO are listed. Here, we restrict ourselves by formulations, the proofs can be found in the monograph [21]. The meromorphic ΨDO are used everywhere in the book; they participate in the implementation of irreducible representations of the considered algebras. Section 1.3 contains some facts about C^*-algebras given without proofs (with references to the monograph [3]) and several results presented earlier only in papers; these results are accompanied by detailed proofs.

In Chap. 2, we consider the algebra $\mathcal{A}$ generated in the space $L_2(\mathcal{M})$ on a smooth compact n-dimensional manifold $\mathcal{M}$ by the operators of two classes. One of the classes comprises zero order pseudodifferential operators with smooth symbols. The other class consists of the operators of multiplication by functions ("coefficients") that may have discontinuities along a given submanifold $\mathcal{N}$. All the equivalence classes of irreducible representations of $\mathcal{A}$ are listed. If $0 < \dim \mathcal{L} < n - 1$, then the algebra $\mathcal{A}$ has, in addition to the one-dimensional representations, two series of infinite-dimensional irreducible representations; if $\dim \mathcal{L} = n - 1$, then the representations in one of these series become two-dimensional. The topology on the spectrum $\widehat{\mathcal{A}}$ is described, and the shortest solving series is constructed; it coincides with the series of maximal radicals. This chapter is the core of the book: although, in the next chapters, we consider more complicated algebras, which can possess many distinct series of infinite-dimensional irreducible representations, but any such series is in a sense analogous to one of the two mentioned series of representations of the algebra $\mathcal{A}$.

In Chap. 3, the study of algebras of ΨDO with discontinuous symbols on a smooth manifold is continued. As in Chap. 2, the algebra $\mathcal{A}$ is spanned by ΨDO with smooth symbols and by operators of multiplication on "coefficients" which now may have discontinuities along a given set of submanifolds (with boundary) of different dimensions; the submanifolds are allowed to have non-empty intersections. All equivalence classes of irreducible representations are listed, the topology on the spectrum is described, and a solving composition series is constructed. Chapter 3 is not used in Chaps. 4 and 5.

We start to study ΨDO on manifolds with smooth closed edges in Chap. 4. Here, we have to start with the definition of pseudodifferential operators. A class of manifolds is described, on which ΨDO of arbitrary order are introduced and general properties of these operators are discussed. Shortly, this chapter generalizes the theory from Sect. 1.1 for manifolds with edges.

In Chap. 5, the spectrum of C^*-algebras generated by zero order ΨDO is studied. Thus, in Chaps. 4 and 5, the results of Chap. 2 are generalized for manifolds with smooth closed edges.

In the body of the book, we restrict ourselves to the technical references. The detailed references are collected in the Bibliographical sketch at the end of the book.

1 Preliminaries

This chapter presents preliminary information that is of use in the book. Section 1.1 contains an elementary introduction (with proofs) to the theory of $\Psi DO's$ with smooth symbols. This section is addressed to the reader having no knowledge of pseudodifferential operators.

Section 1.2 is devoted to a special class of "meromorphic $\Psi DO's$" depending on complex parameter. Such operators are involved in the implementation of C^*-algebra irreducible representations in the remaining chapters (except Chap. 4). Here, we restrict ourselves to statements. The proofs can be found in [21].

In Sect. 1.3, a summary of some facts in the general theory of C^*-algebras has been given; the summary is accompanied by references to the monograph [3]. Moreover, Sects. 1.3.6–1.3.9 contain results (with proofs) before only published in papers.

1.1 Pseudodifferential Operators

For $x \in \mathbb{R}^n$, we introduce $\langle x \rangle = (1 + |x|^2)^{1/2}$ and denote by $\mathbb{Z}_+$ the set of non-negative integers and by $\mathbb{Z}_+^n$ the set of multi-indices $\alpha = (\alpha_1, \ldots, \alpha_n)$, where $\alpha_j \in \mathbb{Z}_+$. As it usually is, $\partial = \partial_x = (\partial/\partial x_1, \ldots, \partial/\partial x_n)$ and $D = D_x = -i\partial_x = (-i\partial/\partial x_1, \ldots, -i\partial/\partial x_n)$.

1.1.1 Amplitudes

Let Ω be an open set in $\mathbb{R}^n$, which we will call a domain for brevity, and let $\mu \in \mathbb{R}$.

© The Author(s), under exclusive license to Springer Nature Switzerland AG 2023

B. Plamenevskii, O. Sarafanov, *Solvable Algebras of Pseudodifferential Operators*, Pseudo-Differential Operators 15, https://doi.org/10.1007/978-3-031-28398-7_1

Definition 1.1.1 A function $a \in C^\infty(\Omega \times \Omega \times \mathbb{R}^n)$ is called an amplitude of order μ (in $\Omega \times \Omega$), if for any compact $K \subset \Omega \times \Omega$ and for all $\alpha, \beta, \gamma \in \mathbb{Z}^n_+$ there exists a constant $C = C(\alpha, \beta, \gamma, K)$ such that

$$|\partial_x^\alpha \partial_y^\beta \partial_\xi^\gamma a(x, y, \xi)| \le C\langle\xi\rangle^{\mu-|\gamma|} \text{ for } (x, y, \xi) \in K \times \mathbb{R}^n. \tag{1.1.1}$$

The collection of all μ order amplitudes is denoted by $S^\mu(\Omega, \Omega)$, or more simply, by S^μ. We also set $S^{-\infty}(\Omega, \Omega) = \bigcap_\mu S^\mu(\Omega, \Omega)$ (the amplitudes of order $-\infty$) and $S(\Omega, \Omega) = \bigcup_\mu S^\mu(\Omega, \Omega)$. The following amplitude properties are evident.

1. For each μ, the set S^μ is a complex linear space.
2. From $\mu_1 \le \mu_2$ it follows that $S^{\mu_1} \subset S^{\mu_2}$, so the amplitude order is not uniquely determined.
3. Given $a \in S^{\mu_1}$ and $b \in S^{\mu_2}$, $ab \in S^{\mu_1+\mu_2}$ holds.
4. Given $a \in S^\mu$, $\partial_x^\alpha \partial_y^\beta \partial_\xi^\gamma a \in S^{\mu-|\gamma|}$ holds.

Example 1 Assume that $a_\alpha \in C^\infty(\Omega \times \Omega)$. Then, $a(x, y, \xi) = \sum_{|\alpha|\le\mu} a_\alpha(x, y)\xi^\alpha$ is an amplitude of order μ. In particular, for $\mu = 0$, the function $(x, y, \xi) \mapsto a(x, y)$ with $a \in C^\infty(\Omega \times \Omega)$ is an amplitude of order 0.

Example 2 Let a function $a \in C^\infty(\Omega \times \Omega \times \mathbb{R}^n)$ be homogeneous of degree $\mu \in \mathbb{C}$ with respect to ξ for large $|\xi|$, that is, $a(x, y, t\xi) = t^\mu a(x, y, \xi)$, where $t \ge 1$ and $|\xi| \ge 1$. Then $a \in S^{\operatorname{Re}\mu}$.

Example 3 For $\mu \in \mathbb{R}$, the function $(x, y, \xi) \mapsto \langle\xi\rangle^\mu$ is the amplitude of order μ.

Every space $S^\mu(\Omega, \Omega)$ can be endowed with a locally compact topology. To this end, for $a \in S^\mu$ with $\mu > -\infty$, we set

$$p^{(\mu)}_{\alpha\beta\gamma K}(a) = \sup\{|\partial_x^\alpha \partial_y^\beta \partial_\xi^\gamma a(x, y, \xi)|\langle\xi\rangle^{|\gamma|-\mu}; (x, y, \xi) \in K \times \mathbb{R}^n\}, \tag{1.1.2}$$

where α, β, and γ are any multi-indices and $K \subset \Omega \times \Omega$ is an arbitrary compact. It is clear that $p^{(\mu)}_{\alpha\beta\gamma K}(a)$ is the minimal constant C satisfying (1.1.1). Formula (1.1.2) defines on S^μ the function $p^{(\mu)}_{\alpha\beta\gamma K}$ having all seminorm properties. The family of all such seminorms makes S^μ into a Frechet space, that is, into a complete metrizable locally compact space. The space $S^{-\infty}$ is provided with the projective limit topology so that

$$a_j \to a \text{ in } S^{-\infty} \Leftrightarrow a_j \to a \text{ in } S^\mu \ \forall\mu > -\infty.$$

In what follows, we make use of the next simple assertions.

(i) For any $\mu_1, \mu_2 \in [-\infty, +\infty)$ there is a continuous bilinear mapping

$$S^{\mu_1} \times S^{\mu_2} \ni (a, b) \mapsto ab \in S^{\mu_1+\mu_2}.$$

(ii) For any α, β, and $\gamma \in \mathbb{Z}^n_+$ and for $\mu \in [-\infty, +\infty)$ there is a continuous mapping

$$S^{\mu} \ni a \mapsto \partial_x^{\alpha} \partial_y^{\beta} \partial_{\xi}^{\gamma} a \in S^{\mu-|\gamma|}.$$

(iii) Given $\mu_1 < \mu_2$, the mapping $S^{\mu_1} \ni a \mapsto a \in S^{\mu_2}$ is continuous.

(iv) Let $\chi \in C^{\infty}(\mathbb{R}^n)$ for each $\gamma \in \mathbb{Z}^n_+$ satisfy $|\partial^{\gamma} \chi(\xi)| \leq C_{\gamma} \langle \xi \rangle^{-|\gamma|}$, that is, χ is an amplitude of order 0, independent of x and y. For every $\varepsilon > 0$, we set $\chi_{\varepsilon}(\xi) = \chi(\varepsilon\xi)$. Assume also that $a \in S^{\mu}$. Then,

$$\chi_{\varepsilon} a \to_{\varepsilon \to 0} \chi(0) a \quad \text{in} \quad S^{\mu+1}.$$

We prove (iv). In view of (i), it suffices to verify that $\chi_{\varepsilon} \to \chi(0)$ in S^1, which means

$$\sup \left|\partial^{\gamma} \left(\chi_{\varepsilon}(\xi) - \chi(0)\right)\right| \langle \xi \rangle^{|\gamma|-1} \to_{\varepsilon \to 0} 0 \qquad \forall \gamma \in \mathbb{Z}^n_+. \tag{1.1.3}$$

Assume that $\gamma = 0$. Then, (1.1.3) follows from $\left|\chi_{\varepsilon}(\xi) - \chi(0)\right| \leq C|\varepsilon\, \xi| \leq C\varepsilon \langle \xi \rangle$, where $C = \sup |\nabla \chi|$. Suppose now that $\gamma \neq 0$. Since $\varepsilon \to 0$, we can consider $\varepsilon \in (0, 1)$. Then

$$\begin{aligned} \left|\partial^{\gamma} \left(\chi_{\varepsilon}(\xi) - \chi(0)\right)\right| \langle \xi \rangle^{|\gamma|-1} &= |(\partial^{\gamma} \chi)(\varepsilon\, \xi)| \varepsilon^{|\gamma|} \langle \xi \rangle^{|\gamma|-1} \leq \\ &\leq C_{\gamma} \langle \varepsilon\, \xi \rangle^{-|\gamma|} \langle \xi \rangle^{|\gamma|-1} \varepsilon^{|\gamma|} \leq C_{\gamma} \varepsilon; \end{aligned}$$

we have used the relations

$$\begin{aligned} \langle \varepsilon\, \xi \rangle^{-|\gamma|} \langle \xi \rangle^{|\gamma|-1} \leq \langle \varepsilon\, \xi \rangle^{1-|\gamma|} \langle \xi \rangle^{|\gamma|-1} &= \frac{(1 + |\xi|^2)^{(|\gamma|-1)/2}}{(1 + |\varepsilon\, \xi|^2)^{(|\gamma|-1)/2}} = \\ = \varepsilon^{1-|\gamma|} \left(\frac{1 + |\xi|^2}{\varepsilon^{-2} + |\xi|^2}\right)^{(|\gamma|-1)/2} &\leq \varepsilon^{1-|\gamma|}. \end{aligned}$$

Note that property (iv) remains valid when $S^{\mu+1}$ changes for $S^{\mu+\delta}$ with any positive δ. According to the property, every amplitude $a \in S^{\mu}$ can be approximated in the topology of $S^{\mu+\delta}$ by amplitudes whose supports are compact with respect to ξ.

1.1.2 Pseudodifferential Operators

Denote by $C_c^\infty(\Omega)$ the subspace in $C^\infty(\Omega)$ consisting of functions with compact supports. By definition, $u_j \to u$ in $C_c^\infty(\Omega)$ if and only if $\{\operatorname{supp} u_j\} \subset K$ for a certain compact $K \subset \Omega$ and $\partial^\alpha u_j \rightrightarrows \partial^\alpha u$ on K for all α. Recall also that $u_j \to u$ in $C^\infty(\Omega) \Leftrightarrow \partial^\alpha u_j \rightrightarrows \partial^\alpha u$ on each compact K for any multi-index α.

Proposition 1.1.2 *Assume that $a \in S(\Omega, \Omega)$ and $u \in C_c^\infty(\Omega)$. Then, the formula*

$$v(x) = \iint e^{i(x-y)\xi} a(x, y, \xi) u(y)\, dy d\xi \tag{1.1.4}$$

defines a function v in $C^\infty(\Omega)$ (by the integral is meant an iterated integral).

Proof Let $K_1 \subset \Omega$ be an arbitrary compact and let $K_2 = \operatorname{supp} u$. We also set $\langle D_x \rangle^2 = 1 - \Delta = 1 + (D_{x_1}^2 + \ldots + D_{x_n}^2)$. Taking account of the equality

$$e^{i(x-y)\xi} = \langle D_y \rangle^{2N} e^{i(x-y)\xi} \langle \xi \rangle^{-2N}, \quad N \in \mathbb{N},$$

we obtain

$$v(x) = \iint_{\mathbb{R}^n \times K_2} e^{i(x-y)\xi} \langle D_y \rangle^{2N} (a(x, y, \xi) u(y)) \langle \xi \rangle^{-2N}\, dy d\xi. \tag{1.1.5}$$

Since $a \in S(\Omega, \Omega)$, we have $a \in S^\mu(\Omega, \Omega)$ with a certain μ. From the amplitude properties it follows that

$$|\langle D_y \rangle^{2N} (a(x, y, \xi) u(y)) \langle \xi \rangle^{-2N}| \leq C \langle \xi \rangle^{\mu - 2N}, \quad (x, y, \xi) \in K_1 \times K_2 \times \mathbb{R}^n.$$

For large N such that $\mu - 2N < -n$, there holds the equality

$$\iint_{\mathbb{R}^n \times K_2} \langle \xi \rangle^{\mu - 2N}\, dy d\xi = \operatorname{mes} K_2 \int \langle \xi \rangle^{\mu - 2N}\, d\xi < +\infty.$$

Hence, for $x \in K_1$ the integrand in (1.1.5) has a summable majorant independent of x. Therefore, integral (1.1.5) uniformly with respect to $x \in K_1$ converges and the function

$$\begin{aligned} K_1 \ni x \mapsto & \iint_{\mathbb{R}^n \times K_2} e^{i(x-y)\xi} \langle D_y \rangle^{2N} (a(x, y, \xi) u(y)) \langle \xi \rangle^{-2N}\, dy d\xi \\ & \equiv \iint e^{i(x-y)\xi} a(x, y, \xi) u(y)\, dy d\xi \end{aligned}$$

is continuous. The compact K_1 has arbitrarily been chosen, so formula (1.1.4) defines $v \in C(\Omega)$.

Now let α be any multi-index. From (1.1.5), it formally follows

$$\partial^\alpha v(x) = \iint_{\mathbb{R}^n \times K_2} e^{i(x-y)\xi} \langle D_y \rangle^{2N} (b(x, y, \xi)u(y)) \langle \xi \rangle^{-2N} \, dy d\xi, \tag{1.1.6}$$

where

$$b(x, y, \xi) = \sum_{\beta \le \alpha} \binom{\alpha}{\beta} (i\xi)^\beta \partial_x^{\alpha-\beta} a(x, y, \xi)$$

is an amplitude of order $\mu + |\alpha|$. According to the first part of the proof, for $N > (\mu + |\alpha| + n)/2$, integral (1.1.6) converges uniformly with respect to $x \in K_1$, while $K_1 \subset \Omega$ is any compact. This implies formula (1.1.6) and the inclusion $\partial^\alpha v \in C(\Omega)$. □

Integrating ("in the opposite direction") the right-hand side of (1.1.6) by parts, we obtain

$$\partial^\alpha v(x) = \iint \partial_x^\alpha (e^{i(x-y)\xi} a(x, y, \xi)u(y)) \, dy d\xi.$$

Therefore, when calculating the derivatives of function (1.1.4), one can commute the differentiation and integration.

Assume that $a \in S(\Omega, \Omega)$. According to Proposition 1.1.2, the formula

$$Au(x) = (2\pi)^{-n} \iint e^{i(x-y)\xi} a(x, y, \xi)u(y) \, dy d\xi \tag{1.1.7}$$

defines a linear operator $A : C_c^\infty(\Omega) \to C^\infty(\Omega)$, which will sometimes be denoted by $\mathrm{Op}\, a$.

Definition 1.1.3 An operator of the form (1.1.7) is called a pseudodifferential operator (ΨDO) in the domain Ω. In the case of $a \in S^\mu(\Omega, \Omega)$ and $\mu \in [-\infty, +\infty)$, the number μ is called the order of ΨDO A. The collection of all ΨDO of order μ is denoted by $\Psi^\mu(\Omega)$. We also set $\Psi^{-\infty}(\Omega) = \bigcap_\mu \Psi^\mu(\Omega)$ and $\Psi(\Omega) = \bigcup_\mu \Psi^\mu(\Omega)$. (Because of the inclusion $\Psi^{\mu_1}(\Omega) \subset \Psi^{\mu_2}(\Omega)$ for $\mu_1 \le \mu_2$, the ΨDO order is not uniquely determined.)

Proposition 1.1.4 *The bilinear mapping*

$$S^\mu(\Omega, \Omega) \times C_c^\infty(\Omega) \ni (a, u) \mapsto (\mathrm{Op}\, a)u \in C^\infty(\Omega)$$

is continuous for any $\mu \in [-\infty, +\infty)$.

Proof Suppose that $a_j \to a$ in S^μ and $u_j \to u$ in $C_c^\infty(\Omega)$. We have to show that $(\mathrm{Op}a_j)u_j \to (\mathrm{Op}a)u$ in $C^\infty(\Omega)$. Because of

$$(\mathrm{Op}a_j)u_j - (\mathrm{Op}a)u = (\mathrm{Op}(a_j - a))u_j + (\mathrm{Op}a)(u_j - u),$$

we can assume that at least one of the sequences $\{a_j\}$ or $\{u_j\}$ tends to zero. Then, we have to verify $(\mathrm{Op}a_j)u_j \to 0$ in $C^\infty(\Omega)$.

We set $v_j = (\mathrm{Op}a_j)u_j$. Let α be any multi-index and choose a number $N \in \mathbb{N}$ such that $\mu + |\alpha| - 2N < -n$. When proving Proposition 1.1.2, we established that

$$\partial^\alpha v_j(x) = (2\pi)^{-n} \iint e^{i(x-y)\xi} \sigma_j(x, y, \xi)\, dy d\xi,$$

where

$$\sigma_j(x, y, \xi) = \langle D_y \rangle^{2N} (b_j(x, y, \xi) u_j(y)) \langle \xi \rangle^{-2N},$$

$$b_j(x, y, \xi) = \sum_{\beta \le \alpha} \binom{\alpha}{\beta} (i\xi)^\beta \partial_x^{\alpha-\beta} a_j(x, y, \xi).$$

Each function u_j can be taken as a zero order amplitude independent of (x, ξ), while the convergence $u_j \to u$ in $C_c^\infty(\Omega)$ implies that in $S^0(\Omega, \Omega)$. This and assertions (i)–(iii) in 1.1.1 lead to

$$\sigma_j \to_{j\to\infty} 0 \text{ in } S^{\mu+|\alpha|-2N}. \tag{1.1.8}$$

Let $K_1 \subset \Omega$ be any compact. Denote by K_2 $(\subset \Omega)$ a compact containing each of the sets $\operatorname{supp} u_j$. By virtue of (1.1.8),

$$|\sigma_j(x, y, \xi)| \le C_j \langle \xi \rangle^{\mu+|\alpha|-2N}, \quad (x, y, \xi) \in K_1 \times K_2 \times \mathbb{R}^n,$$

where $C_j \to 0$ as $j \to \infty$. Hence

$$|\partial^\alpha v_j(x)| \le C_j \mathrm{mes} K_2 \int \langle \xi \rangle^{\mu+|\alpha|-2N}\, d\xi, \quad x \in K_1.$$

Since $\mu + |\alpha| - 2N < -n$, the integral is finite. Therefore $\partial^\alpha v \rightrightarrows 0$ on K_1. Because α and K_1 have arbitrarily been chosen, we obtain $v_j \to 0$ in $C^\infty(\Omega)$. □

Corollary 1.1.5 *For each fixed $a \in S(\Omega, \Omega)$, there is the continuous mapping*

$$C_c^\infty(\Omega) \ni u \mapsto (\mathrm{Op}\, a)u \in C^\infty(\Omega);$$

for each fixed $u \in C_c^\infty(\Omega)$, *there is a continuous mapping*

$$S^\mu(\Omega, \Omega) \ni a \mapsto (\mathrm{Op}\, a)u \in C^\infty(\Omega).$$

Corollary 1.1.6 *Assume that* $\chi \in C_c^\infty(\mathbb{R}^n)$ *and* $\chi(0) = 1$. *For any* $\varepsilon > 0$ *we set* $\chi_\varepsilon(\xi) = \chi(\varepsilon\xi)$. *Then*

$$\big(\mathrm{Op}\,(a\chi_\varepsilon)\big)u \to_{\varepsilon \to 0} (\mathrm{Op}\, a)u \;\; in \;\; C^\infty(\Omega)$$

for $a \in S(\Omega, \Omega)$ *and* $u \in C_c^\infty(\Omega)$.

Proof Let μ be the order of the amplitude a. By virtue (iv) in 1.1.1, $a\chi_\varepsilon \to a$ in $S^{\mu+1}$. It remains to apply Corollary 1.1.5. □

Now, we show that $S(\Omega, \Omega) \ni a \mapsto \mathrm{Op}\, a \in \Psi(\Omega)$ is not a one-to-one mapping.

Proposition 1.1.7 *For any* ΨDO A *there exist infinitely many amplitudes* a *such that* $A = \mathrm{Op}\, a$.

Proof Let an amplitude a and a polynomial P be such that the function

$$(x, y, \xi) \mapsto a(x, y, \xi)/P(x - y) \tag{1.1.9}$$

is an amplitude as well (for instance, this is the case if $P(x) \neq 0\ \forall x$). By Corollary 1.1.6, setting $A = \mathrm{Op}\, a$, we have

$$\begin{aligned} Au(x) &= \lim_{\varepsilon \to 0} (2\pi)^{-n} \iint \big(P(D_\xi) e^{i(x-y)\xi}\big) \frac{a(x, y, \xi)}{P(x - y)} \chi_\varepsilon(\xi) u(y)\, dy d\xi = \\ &= \lim_{\varepsilon \to 0} (2\pi)^{-n} \iint e^{i(x-y)\xi} P(-D_\xi) \Big(\frac{a(x, y, \xi)}{P(x - y)} \chi_\varepsilon(\xi)\Big) u(y)\, dy d\xi. \end{aligned}$$

Let μ be the order of amplitude (1.1.9). Then,

$$P(-D_\xi)\Big(\frac{a(x, y, \xi)}{P(x - y)} \chi_\varepsilon(\xi)\Big) \to P(-D_\xi) \frac{a(x, y, \xi)}{P(x - y)}$$

in $S^{\mu+1}$, so

$$Au(x) = (2\pi)^{-n} \iint e^{i(x-y)\xi} P(-D_\xi) \frac{a(x, y, \xi)}{P(x - y)} u(y)\, dy d\xi.$$

Therefore, $A = \operatorname{Op} b$ with $b(x, y, \xi) = P(-D_\xi)\big(a(x, y, \xi)/P(x - y)\big)$. Varying P, we arrive at the needed conclusion. □

For a ΨDO $A = \operatorname{Op} a$, we set

$$^t a(x, y, \xi) = a(y, x, -\xi), \quad a^*(x, y, \xi) = \overline{a(x, y, \xi)}$$

and introduce the transposed operator $^tA = \operatorname{Op} {}^t a$ and the "adjoint" operator $A^* = \operatorname{Op} a^*$. The operators ^tA and A^* belong to the same class $\Psi^\mu(\Omega)$ as does A. There hold the equalities

$$\langle Au, v\rangle = \langle u, {}^tAv\rangle, \quad (Au, v) = (u, A^*v), \quad u, v \in C_c^\infty(\Omega),$$

where the pairings $\langle\cdot, \cdot\rangle$ and $(\cdot, \cdot)$ are defined by

$$\langle u, v\rangle = \int u(x)v(x)\, dx, \quad (u, v) = \int u(x)\overline{v(x)}\, dx.$$

Examples

(1) A differential operator

$$P(x, D) = \sum_{|\alpha| \le \mu} p_\alpha(x) D^\alpha (q_\alpha(x)\, \cdot), \quad p_\alpha, q_\alpha \in C^\infty(\Omega),$$

is a ΨDO of order μ:

$$P(x, D)u(x) = (2\pi)^{-n} \iint e^{i(x-y)\xi}\Big(\sum_{|\alpha| \le \mu} p_\alpha(x)\xi^\alpha q_\alpha(y)\Big)u(y)\, dy d\xi.$$

In the case of $q_\alpha = 1$ for all α, the formula takes the form

$$\begin{aligned} P(x, D)u(x) &= (2\pi)^{-n} \iint e^{i(x-y)\xi} P(x, \xi)u(y)\, dy d\xi \\ &= (2\pi)^{-n/2} \int e^{ix\xi} P(x, \xi)\hat{u}(\xi)\, d\xi, \end{aligned}$$

where $\hat{u}$ denotes the Fourier transform of u,

$$\hat{u}(\xi) = (2\pi)^{-n/2} \int e^{-iy\xi} u(y)\, dy.$$

(2) An integral operator with a smooth kernel

$$Au(x) = \int G(x, y)u(y)\,dy, \quad G \in C^\infty(\Omega \times \Omega),$$

is a ΨDO of order $-\infty$. To prove this, we set

$$a(x, y, \xi) = e^{-i(x-y)\xi} G(x, y)\chi(\xi),$$

where $\chi \in \mathcal{S}(\mathbb{R}^n)$ is any function that satisfies $(2\pi)^{-n} \int \chi(\xi)\,d\xi = 1$ (one can even take $\chi \in C_c^\infty(\mathbb{R}^n)$). Then, $a \in S^{-\infty}$ and

$$(\operatorname{Op} a)u(x) = (2\pi)^{-n} \iint e^{i(x-y)\xi} a(x, y, \xi)u(y)\,dyd\xi$$

$$= (2\pi)^{-n} \iint G(x, y)\chi(\xi)u(y)\,dyd\xi = \int G(x, y)u(y)\,dy = Au(x).$$

1.1.3 The Kernel of a Pseudodifferential Operator

Suppose that $a \in S^\mu(\Omega, \Omega)$ and $A = \operatorname{Op} a$. Assuming $\mu < -n - k$ and $k \in \mathbb{Z}_+$, we change the integration order in

$$Au(x) = (2\pi)^{-n} \iint e^{i(x-y)\xi} a(x, y, \xi)u(y)\,dyd\xi, \quad u \in C_c^\infty(\Omega),$$

and obtain

$$Au(x) = \int G_A(x, y)u(y)\,dy, \tag{1.1.10}$$

where

$$G_A(x, y) = (2\pi)^{-n} \int e^{i(x-y)\xi} a(x, y, \xi)\,d\xi. \tag{1.1.11}$$

The function $G_A \in C^k(\Omega \times \Omega)$ is called the kernel of A. It can be identified with an element in $\mathcal{D}'(\Omega \times \Omega)$ acting on the functions $w \in C_c^\infty(\Omega \times \Omega)$ by the formula

$$\langle G_A, w\rangle = \iint G_A(x, y)w(x, y)\,dxdy, \tag{1.1.12}$$

where $\mathcal{D}'(\Omega \times \Omega)$ stands for the general function space adjoint to $C_c^\infty(\Omega \times \Omega)$. In more detail,

$$\langle G_A\,, w\rangle = (2\pi)^{-n} \iiint e^{i(x-y)\xi} a(x, y, \xi) w(x, y)\, dxdyd\xi; \tag{1.1.13}$$

because of $aw \in L^1(\Omega \times \Omega \times \mathbb{R}^n)$, the integration order plays no role. For $u, v \in C_c^\infty(\Omega)$ we set $(v \otimes u)(x, y) = v(x)u(y)$ and in (1.1.12) choose $w = v \otimes u$. Taking into account (1.1.10), we arrive at

$$\langle G_A\,, v \otimes u\rangle = \langle Au, v\rangle. \tag{1.1.14}$$

For arbitrary μ, the kernel of ΨDO A is, by definition, the distribution G_A given by (1.1.13); however, this time the integral means the iterated integral $\int d\xi \iint \ldots dxdy$. To prove that G_A is a distribution, we have to verify the existence of such an integral and its continuous dependence on w. It can be done by means of the formula

$$\langle G_A\,, w\rangle = (2\pi)^{-n} \iiint e^{i(x-y)\xi} \langle D_y\rangle^{2N} (a(x, y, \xi) w(x, y)) \langle \xi\rangle^{-2N}\, dxdyd\xi$$

(compare with the proofs of Propositions 1.1.2 and 1.1.4). In fact, for any $\mu \in \mathbb{R}$, the integral (1.1.13) is a continuous function of $(a, w) \in S^\mu(\Omega, \Omega) \times C_c^\infty(\Omega \times \Omega)$.

Assume that $\chi \in C_c^\infty(\mathbb{R}^n)$, $\chi(0) = 1$, and $\chi_\varepsilon(\xi) := \chi(\varepsilon\xi)$ for $\varepsilon > 0$. As $\varepsilon \to 0$, we have $\chi_\varepsilon a \to a$ in $S^{\mu+1}$, μ being the amplitude order of a. Hence,

$$\begin{aligned}\langle G_A, w\rangle &= \lim_{\varepsilon\to 0} (2\pi)^{-n} \iiint e^{i(x-y)\xi} a(x, y, \xi)\chi_\varepsilon(\xi) w(x, y)\, dxdyd\xi \\ &= \lim_{\varepsilon\to 0} (2\pi)^{-n} \iint w(x, y)\, dxdy \int e^{i(x-y)\xi} a(x, y, \xi)\chi_\varepsilon(\xi)\, d\xi. \end{aligned} \tag{1.1.15}$$

Therefore,

$$G_A(x, y) = \lim_{\varepsilon\to 0} (2\pi)^{-n} \int e^{i(x-y)\xi} a(x, y, \xi)\chi_\varepsilon(\xi)\, d\xi$$

in the sense of the convergence in $\mathcal{D}'(\Omega \times \Omega)$. This formula has usually been written in the form (1.1.11). We substitute $w = v \otimes u$ into (1.1.15) and obtain the formula

$$\begin{aligned}\langle G_A, v \otimes u\rangle &= \lim_{\varepsilon\to 0} (2\pi)^{-n} \int v(x)\, dx \iint e^{i(x-y)\xi} a(x, y, \xi)\chi_\varepsilon(\xi) u(y)\, dyd\xi \\ &= \int v(x)\, Au(x)\, dx = \langle Au\,, v\rangle,\end{aligned}$$

which coincides with (1.1.14). Now, it is easily seen that

(1) Distinct operators have distinct kernels.
(2) For $H \in \mathcal{D}'(\Omega \times \Omega)$ such that $\langle H, v \otimes u\rangle = \langle Au, v\rangle$ with any $u, v \in C_c^\infty(\Omega)$, there holds the equality $H = G_A$. In particular, G_A is independent of the choice of the amplitude a in (1.1.15); it goes without saying that the equality $A = \operatorname{Op} a$ must be fulfilled. (Recall that the linear combinations of functions of the form $v \otimes u$ are dense in $C_c^\infty(\Omega \times \Omega)$).

We also note that the kernel of the transposed operator $\Psi DO\ {}^tA$ is defined by the equality

$$\langle G_{{}^tA}, w\rangle = \langle G_A, {}^t w\rangle, \quad \text{where } {}^t w(x, y) = w(y, x),$$

and the kernel of the adjoint $\Psi DO\ A^*$ satisfies

$$\langle G_{A^*}, w\rangle = \overline{\langle G_A, w^*\rangle}, \quad \text{where } w^*(x, y) = \overline{w(y, x)}.$$

In the case that the kernel belongs to $L^1_{loc}(\Omega \times \Omega)$, equality (1.1.14) can be written in the form

$$\int v(x)\, dx \int G_A(x, y)u(y)\, dy = \int v(x)Au(x)\, dx,$$

which implies (1.1.10).

Let Ω_1 and Ω_2 be domains in $\mathbb{R}^n$ and $G \in \mathcal{D}'(\Omega_1 \times \Omega_2)$. For a fixed $u \in C_c^\infty(\Omega_2)$, the mapping

$$C_c^\infty(\Omega_1) \ni v \mapsto G(v \otimes u) \tag{1.1.16}$$

is a linear continuous functional on $C_c^\infty(\Omega_1)$; in other words, it belongs to $\mathcal{D}'(\Omega_1)$. Therefore, we have defined an operator $C_c^\infty(\Omega_2) \to \mathcal{D}'(\Omega_1)$ that takes $u \in C_c^\infty(\Omega_2)$ to functional (1.1.16). It is easy to verify that the operator is continuous. We denote this operator by A and have

$$\langle Au, v\rangle = \langle G, v \otimes u\rangle. \tag{1.1.17}$$

Conversely, if $A : C_c^\infty(\Omega_2) \to \mathcal{D}'(\Omega_1)$ is a linear continuous operator, there exists a unique distribution $G \in \mathcal{D}'(\Omega_1 \times \Omega_2)$ that satisfies (1.1.17). The distribution is called the Schwartz kernel of the operator A, and the above statement is the Schwartz kernel theorem.

Let A be a ΨDO in a domain $\Omega \subset \mathbb{R}^n$. Then, A implements a continuous mapping $C_c^\infty(\Omega) \to C^\infty(\Omega)$. The canonical embedding $C^\infty(\Omega) \hookrightarrow \mathcal{D}'(\Omega)$ is continuous, so A

can be considered as continuous operator $C_c^\infty(\Omega) \to \mathcal{D}'(\Omega)$. For any $u, v \in C_c^\infty(\Omega)$, the Schwartz kernel of such an operator satisfies (1.1.17) and consequently coincides with its kernel in the sense of the definitions given in the present section.

1.1.4 Smoothing Operators

Definition A pseudodifferential operator A is called smoothing if its kernel belongs to $C^\infty(\Omega \times \Omega)$.

Proposition 1.1.8 *The following assertions are equivalent:*

(1) *A is a smoothing ΨDO.*
(2) $A \in \Psi^{-\infty}(\Omega)$.
(3) $A \in \overline{\Psi}^{-\infty}(\Omega)$.

Proof (1)⇒(2). For $G \in C^\infty(\Omega \times \Omega)$, the operator

$$Au(x) = \int G_A(x, y)u(y)\,dy$$

is integral with a smooth kernel, so $A \in \Psi^{-\infty}(\Omega)$ (see Example 2 in Sect. 1.1.2).

(2)⇒(3). It is evident.

(3)⇒(1). For any $k \in \mathbb{N}$, there exists an amplitude $a_k \in S^{-n-k}$ such that $A = \operatorname{Op} a_k$. Therefore,

$$G_A(x, y) = (2\pi)^{-n} \int e^{i(x-y)\xi} a_k(x, y, \xi)\,d\xi$$

is a function of the class $C^{k-1}(\Omega \times \Omega)$. Because k is arbitrary, we have $G_A \in C^\infty(\Omega \times \Omega)$. □

Proposition 1.1.9 *Let an amplitude a vanish on $V \times \mathbb{R}^n$, where $V \subset \Omega \times \Omega$ is a neighborhood of the diagonal $\Delta := \{(x, y) \in \Omega \times \Omega : x = y\}$. Then, $\operatorname{Op} a \in \Psi^{-\infty}(\Omega)$.*

Proof From the proposition assumption, it follows that, for any $N \in \mathbb{N}$, the function $(x, y, \xi) \mapsto a(x, y, \xi)/|x - y|^{2N}$ is an amplitude (of same order μ as a). As shown in Sect. 1.1.2, $\operatorname{Op} a = \operatorname{Op} b_N$ with $b_N = (-\Delta_\xi)^N a(x, y, \xi)/|x - y|^{2N}$. Moreover, $b_N \in S^{\mu-2N}$ and N has arbitrarily been chosen, so $\operatorname{Op} a \in \Psi^{-\infty}(\Omega)$. □

Proposition 1.1.10 *The kernel G_A of each ΨDO A is smooth outside the diagonal Δ.*

Proof Assume that $a \in S(\Omega, \Omega)$, $\varphi \in C^\infty(\Omega \times \Omega)$, and set $A = \operatorname{Op} a$ and $B = \operatorname{Op}(\varphi a)$; then $G_B = \varphi G_A$. Indeed, for any $w \in C_c^\infty(\Omega \times \Omega)$, we have

$$\langle G_B, w\rangle = (2\pi)^{-n} \iiint e^{i(x-y)\xi} a(x, y, \xi)\varphi(x, y)w(x, y)\, dxdyd\xi =$$

$$= \langle G_A, \varphi w\rangle = \langle \varphi G_A, w\rangle.$$

Let $\theta \in C^\infty(\mathbb{R})$ satisfy $\theta(t) = 0$ for $t \leq 1$ and $\theta(t) = 1$ for $t \geq 2$. We take an arbitrary $\delta > 0$ and set $\varphi_\delta(x, y) = \theta(|x - y|/\delta)$. It is obvious that $\varphi_\delta \in C^\infty(\mathbb{R}^n \times \mathbb{R}^n)$, $\varphi_\delta = 0$ in the δ-neighborhood V_δ of the diagonal Δ, and $\varphi_\delta = 1$ outside $V_{2\delta}$.

Assume that $A = \operatorname{Op} a$ is any ΨDO and $A_\delta = \operatorname{Op}(a\varphi_\delta)$. Since $a\varphi_\delta = 0$ in a neighborhood of the diagonal, we have $A_\delta \in \Psi^{-\infty}(\Omega)$ (Proposition 1.1.9). Therefore, $G_{A_\delta} \in C^\infty(\Omega \times \Omega)$ (Proposition 1.1.8). According to the remark at the beginning of the proof, $G_{A_\delta} = \varphi_\delta G_A$. Hence, $G_{A_\delta} = G_A$ on $(\Omega \times \Omega) \setminus V_{2\delta}$, so G_A is a smooth function on $(\Omega \times \Omega) \setminus V_{2\delta}$. Because $\delta > 0$ has arbitrarily been chosen, which completes the proof. □

1.1.5 Properly Supported Pseudodifferential Operators

A mapping $f : X \to Y$ of topological spaces is called proper if $f^{-1}(K)$ is compact for any compact $K \subset Y$. A distribution $G \in \mathcal{D}'(\Omega \times \Omega)$ is said to have proper support if both projections

$$\pi_1, \pi_2 : \operatorname{supp} G \to \Omega$$

are proper mappings.

Definition 1.1.11 A pseudodifferential operator is called properly supported (or simply proper) if its kernel has proper support.

Any differential operator is proper because the support of its kernel belongs to the diagonal. In the case that A is a proper ΨDO, the operators ${}^tA$ and A^* are proper as well.

For an amplitude a, we denote by $\operatorname{supp}_{x,y} a$ the closure of the projection of $\operatorname{supp} a$ into $\Omega \times \Omega$. An amplitude a is said to have proper support if both projections

$$\pi_1, \pi_2 : \operatorname{supp}_{x,y} a \to \Omega$$

are proper mappings.

Lemma 1.1.12 *For a ΨDO A (that need not be proper), the following assertions are valid:*

(a) *In the case that* $\chi \in C^\infty(\Omega \times \Omega)$ *and* $\chi = 1$ *in a neighborhood of the set* $\operatorname{supp} G_A$, *there holds the equality* $A = \operatorname{Op}(\chi a)$.

(b) $\operatorname{supp} G_A \subset \operatorname{supp}_{x,y} a$.

Proof

(a) Setting $B = \operatorname{Op}(\chi a)$, we have $G_B = \chi G_A = G_A$, so $A = B$.

(b) From

$$\langle G_A, w\rangle = (2\pi)^{-n} \iiint e^{i(x-y)\xi} a(x, y, \xi) w(x, y)\, dx dy d\xi, \quad w \in C_c^\infty(\Omega \times \Omega),$$

it follows that $\langle G_A, w\rangle = 0$ provided $\operatorname{supp} w \cap \operatorname{supp}_{x,y} a = \emptyset$. Therefore, $G_A = 0$ on $(\Omega \times \Omega) \setminus \operatorname{supp}_{x,y} a$, so $\operatorname{supp} G_A \subset \operatorname{supp}_{x,y} a$.

□

Proposition 1.1.13 *The following assertions are equivalent:*

(1) *A is a proper ΨDO.*

(2) *There exists an amplitude a with proper support such that* $A = \operatorname{Op} a$.

Proof (1)⇒(2). Assume that $A = \operatorname{Op} a$ and $\chi \in C_c^\infty(\Omega \times \Omega)$ with proper support equal to 1 in a neighborhood of the set $\operatorname{supp} G_A$. Then, χa is an amplitude with proper support and (by Lemma) $A = \operatorname{Op}(\chi a)$.

(2)⇒ (1). The amplitude a has proper support and (by Lemma) $\operatorname{supp} G_A \subset \operatorname{supp}_{x,y} a$, therefore, G_A has proper support as well. □

Proposition 1.1.14 *The following assertions are equivalent:*

(1) *A is a proper ΨDO.*

(2) *For any compact* $K_2 \subset \Omega$, *there exists a compact* $K_1 \subset \Omega$ *such that*

$$u \in C_c^\infty(K_2) \Rightarrow Au, {}^t\!Au \in C_c^\infty(K_1). \tag{1.1.18}$$

Proof (1)⇒(2). Assume that $A = \operatorname{Op} a$, where a is an amplitude with proper support. We choose any compact $K_2 \subset \Omega$ and set $K_1' = \pi_1(\pi_2^{-1}(K_2))$, where π_1 and π_2 are the projections $\operatorname{supp}_{x,y} a \to \Omega$. In the case of $x \notin K_1'$, the function $(y, \xi) \mapsto a(x, y, \xi)$ vanishes on the set $K_2 \times \mathbb{R}^n$. Consequently, for $u \in C_c^\infty(K_2)$, we have

$$Au(x) = (2\pi)^{-n} \iint e^{i(x-y)\xi} a(x, y, \xi) u(y)\, dy d\xi = 0 \quad \text{for} \quad x \notin K_1',$$

which means that $Au \in C_c^\infty(K_1')$. The inclusion ${}^tAu \in C_c^\infty(K_1'')$ can for a certain compact $K_1'' \subset \Omega$ be established in a similar way. It remains to set $K_1 = K_1' \cup K_2''$.

(2)$\Rightarrow$(1). Denote by G the kernel of ΨDO A, and by π_1 and π_2 denote the projections $\operatorname{supp} G \to \Omega$. We have to verify that the mappings π_1 and π_2 are proper. Let $E \subset \Omega$ be an arbitrary compact and $V \supset E$ a domain with compact closure $K_2 := \overline{V} \subset \Omega$. By assumption, there exists a compact $K_1 \subset \Omega$ such that there holds (1.1.18). Then,

$$\langle G, v \otimes u\rangle = \langle Au, v\rangle = 0 \text{ for all } u \in C_c^\infty(V),\ v \in C_c^\infty(\Omega \setminus K_1).$$

For such u and v, the linear combinations of functions of the form $v \otimes u$ are dense in $C_c^\infty((\Omega \setminus K_1) \times V)$, so

$$\langle G, w\rangle = 0 \text{ for } w \in C_c^\infty((\Omega \setminus K_1) \times V).$$

Hence, $\operatorname{supp} G \cap ((\Omega \setminus K_1) \times V) = \emptyset$ and

$$\pi_2^{-1}(V) \cap ((\Omega \setminus K_1) \times V) = \emptyset \Rightarrow \pi_2^{-1}(E) \cap ((\Omega \setminus K_1) \times E) = \emptyset$$
$$\Rightarrow \pi_2^{-1}(E) \subset K_1 \times E\,.$$

Being a closed subset of the compact $K_1 \times E$, the set $\pi_2^{-1}(E)$ is compact. This proves that mapping π_2 is proper. Such a fact for the mapping $\pi_2' : \operatorname{supp}{}^tG \to \Omega$, where ${}^tG$ is the kernel of ${}^tA$, can be verified in a similar way. Let s stand for the homeomorphism

$$\operatorname{supp} G \to \operatorname{supp}{}^tG : (x, y) \mapsto (y, x).$$

Then, $\pi_1 = \pi_2' \circ s$, and hence, the mapping π_1 is proper. □

Proposition 1.1.15 *Every proper ΨDO A implements a continuous mapping $C_c^\infty(\Omega) \to C_c^\infty(\Omega)$.*

Proof Assume that $\{u_j\}_{j=1}^\infty \subset C_c^\infty(\Omega)$ and $u_j \to 0$ in $C_c^\infty(\Omega)$. There exists a compact $K_2 \subset \Omega$ such that $\operatorname{supp} u_j \subset K_2$ for all j. We employ Proposition 1.1.14 and obtain

$$\operatorname{supp} Au_j \subset K_1 \quad (\forall\, j),$$

with a certain compact $K_1 \subset \Omega$. Moreover,

$$Au_j \to 0 \text{ in } C^\infty(\Omega).$$

It follows that $Au_j \to 0$ in $C_c^\infty(\Omega)$. □

Proposition 1.1.16 *Let A be a proper ΨDO. Then, A extends to the continuous operator $\overline{A} : C^\infty(\Omega) \to C^\infty(\Omega)$.*

Proof Let a be a proper amplitude such that $A = \operatorname{Op} a$. We set

$$v(x) = (2\pi)^{-n} \iint e^{i(x-y)\xi} a(x, y, \xi) u(y)\, dy d\xi.$$

It was shown in Sect. 1.1.2 that from $u \in C_c^\infty(\Omega)$ there follows $v \in C^\infty(\Omega)$; to prove this, we integrated by parts. This was possible because the support of the function $y \mapsto a(x, y, \xi)u(y)$ was compact owing to the compact support of u. Being proper, the amplitude a provides a compact support for the product au and the support of u need not be compact. Therefore, the inclusion $v \in C^\infty(\Omega)$ can be verified in the same way as in Proposition 1.1.2. Thus, the formula

$$\overline{A}u(x) = (2\pi)^{-n} \iint e^{i(x-y)\xi} a(x, y, \xi) u(y)\, dy d\xi, \quad u \in C^\infty(\Omega),$$

defines a linear operator $\overline{A} : C^\infty(\Omega) \to C^\infty(\Omega)$. We show that the operator is continuous. Assume that $K_1 \subset \Omega$ is any compact and $K_2 = \pi_2(\pi_1^{-1}(K_1))$, $\pi_1, \pi_2 : \operatorname{supp}_{x,y} a \to \Omega$ being the projections. Choose a function $\psi \in C_c^\infty(\Omega)$ to satisfy $\psi|K_2 = 1$. We also suppose that $\{u_j\} \subset C^\infty(\Omega)$ and $u_j \to 0$ in $C^\infty(\Omega)$ as $j \to \infty$. From the equality $a(x, y, \xi) = a(x, y, \xi)\psi(y)$ for $x \in K_1$, it follows that

$$\overline{A}u_j = (\operatorname{Op} a)\psi u_j = A(\psi u_j) \text{ on } K_1.$$

Since $\psi u_j \to 0$ in $C_c^\infty(\Omega)$, we have $A(\psi u_j) \to 0$ in $C^\infty(\Omega)$. In particular, $A(\psi u_j) \rightrightarrows 0$ on K_1 together with all derivatives. The compact $K_1 \subset \Omega$ has arbitrarily been chosen, so $\overline{A}u_j \to 0$ in $C^\infty(\Omega)$ □

The space $C_c^\infty(\Omega)$ is dense in $C^\infty(\Omega)$. (Indeed, let $\{K_j\}$ be an exhaustive sequence of compact subsets in Ω and let the sequence $\{\psi_j\} \subset C_c^\infty(\Omega)$ satisfy $\psi_j|K_j = 1$. Then $u\psi_j \to u$ in $C^\infty(\Omega)$ for $u \in C^\infty(\Omega)$.) Therefore, the extension $\overline{A}$ of A given by Proposition 1.1.16 is unique. In what follows we write A instead of $\overline{A}$.

Note that any ΨDO A admits the representation $A = A_1 + A_2$, where A_1 is a proper ΨDO and A_2 is a smoothing ΨDO. To see that, it suffices to take a properly supported function $\chi \in C^\infty(\Omega \times \Omega)$ equal to 1 in a neighborhood of the diagonal and to set $A_1 = \operatorname{Op}(\chi a)$, $A_2 = \operatorname{Op}((1 - \chi)a)$.

1.1.6 Pseudodifferential Operators in Generalized Function Spaces

Denote by $\mathcal{D}'(\Omega)$ and $\mathcal{E}'(\Omega)$ the general function spaces adjoint to $C_c^\infty(\Omega)$ and $C^\infty(\Omega)$, respectively. The space $\mathcal{E}'(\Omega)$ consists of the elements in $\mathcal{D}'(\Omega)$ with compact supports in Ω. We assume that $A \in \Psi(\Omega)$, fix $u \in \mathcal{E}'(\Omega)$ and define a functional f_u on $C_c^\infty(\Omega)$ by the equality

$$\langle f_u, v \rangle = \langle u, {}^t\!Av \rangle. \tag{1.1.19}$$

Clearly, the functional is continuous, so $f_u \in \mathcal{D}'(\Omega)$. We set

$$\overline{A}u = f_u. \tag{1.1.20}$$

Then, (1.1.19) takes the form

$$\langle \overline{A}u, v \rangle = \langle u, {}^t\!Av \rangle. \tag{1.1.21}$$

If the function u fixed above turns out to be in $C_c^\infty(\Omega)$, the equality $\langle Au, v \rangle = \langle u, {}^t\!Av \rangle$ leads to $\overline{A}u = Au$. Therefore, formula (1.1.20) defines an extension of ΨDO $A : C_c^\infty(\Omega) \to C^\infty(\Omega)$ to the operator $\overline{A} : \mathcal{E}'(\Omega) \to \mathcal{D}'(\Omega)$. This operator is continuous. Indeed, if $u_j \to 0$ in $\mathcal{E}'(\Omega)$, we obtain $\langle u_j, {}^t\!Av \rangle \to 0$ for $v \in C_c^\infty(\Omega)$ and then, taking account of (1.1.21), arrive at $\langle \overline{A}u_j, v \rangle \to 0$. In the sense of functional analysis, the operator $\overline{A}$ is adjoint to the operator ${}^t\!A$:

$$\begin{gathered} \mathcal{D}'(\Omega) \xleftarrow{\overline{A}} \mathcal{E}'(\Omega) \\ C_c^\infty(\Omega) \xrightarrow{{}^t\!A} C^\infty(\Omega). \end{gathered}$$

In what follows, we write A instead of $\overline{A}$.

Now suppose that A is a proper ΨDO. Such an operator implements continuous mappings

$$\begin{gathered} A : C_c^\infty(\Omega) \to C_c^\infty(\Omega) \\ A : C^\infty(\Omega) \to C^\infty(\Omega). \end{gathered}$$

Repeating the preceding arguments, we conclude that A extends to the continuous operators

$$\begin{gathered} A : \mathcal{E}'(\Omega) \to \mathcal{E}'(\Omega), \\ A : \mathcal{D}'(\Omega) \to \mathcal{D}'(\Omega). \end{gathered}$$

1.1.7 Symbols

Definition 1.1.17 A function $a \in C^\infty(\Omega \times \mathbb{R}^n)$ is called a symbol of order $\mu \in \mathbb{R}$ if, for any $\alpha, \beta \in \mathbb{Z}_+^n$ and for each compact $K \subset \Omega$, there exists a constant $C = C(\alpha, \beta, K)$ such that

$$|\partial_x^\alpha \partial_\xi^\beta a(x, \xi)| \le C \langle \xi \rangle^{\mu - |\beta|} \text{ for } (x, \xi) \in K \times \mathbb{R}^n.$$

Denote by $S^\mu(\Omega)$ the set of all symbols of order μ. We also introduce $S^{-\infty}(\Omega) = \bigcap_\mu S^\mu(\Omega)$, $S(\Omega) = \bigcup_\mu S^\mu(\Omega)$.

Every symbol can be considered as an amplitude independent of y, so $S^\mu(\Omega) \subset S^\mu(\Omega, \Omega)$. The topology of $S^\mu(\Omega, \Omega)$ induces that of $S^\mu(\Omega)$. For $\mu > -\infty$, such a topology can be given by the seminorm family

$$p_{\alpha\beta K}^{(\mu)}(a) = \sup\{|\partial_x^\alpha \partial_\xi^\beta a(x, \xi)| \langle \xi \rangle^{|\beta| - \mu} : (x, \xi) \in K \times \mathbb{R}^n\}.$$

For $S^\mu(\Omega)$ endowed with the above topology, assertions (i)–(iii) in 1.1.1 are valid.

For any symbol a, we introduce the ΨDO $A = \operatorname{Op} a$:

$$Au(x) = (2\pi)^{-n} \iint e^{i(x-y)\xi} a(x, \xi) u(y) \, dy d\xi = (2\pi)^{-n/2} \int e^{ix\xi} a(x, \xi) \hat{u}(\xi) \, d\xi,$$

where

$$\hat{u}(\xi) = (2\pi)^{-n/2} \int e^{-iy\xi} u(y) \, dy.$$

In what follows, we use the notation $a(x, D)$ along with $\operatorname{Op} a$.

Examples

(1) $P(x, \xi) = \sum_{|\alpha| \le \mu} a_\alpha(x) \xi^\alpha$ with $a_\alpha \in C^\infty(\Omega)$ is a symbol of order μ. The corresponding ΨDO is the differential operator

$$P(x, D) = \sum_{|\alpha| \le \mu} a_\alpha(x) D^\alpha.$$

(2) Let, for $\mu \in \mathbb{C}$, the function $a \in C^\infty(\Omega \times \mathbb{R}^n)$ be degree μ homogeneous for large $|\xi|$, that is, $a(x, t\xi) = t^\mu a(x, \xi)$, where $t \ge 1$ and $|\xi| \ge 1$. Then, $a \in S^{\operatorname{Re} \mu}(\Omega)$.
(3) The function $(x, \xi) \mapsto \langle \xi \rangle^\mu$ is a symbol of order μ.
(4) Assume that $\tilde{a} \in S^\mu(\Omega, \Omega)$ and set $a(x, \xi) = \tilde{a}(x, x, \xi)$. Then, $a \in S^\mu(\Omega)$.

1.1.8 Asymptotic Expansions in the Classes $S^\mu(\Omega)$

Let $a_j \in S^{\mu_j}(\Omega)$, $j = 0, 1, \dots$, $\mu_j \to -\infty$ for $j \to \infty$ and let $a \in C^\infty(\Omega \times \mathbb{R}^n)$. The relation

$$a(x,\xi) \sim \sum_{j=0}^{\infty} a_j(x,\xi) \tag{1.1.22}$$

means that for any $k \in \mathbb{N}$

$$a(x,\xi) - \sum_{j=0}^{k-1} a_j(x,\xi) \in S^{\overline{\mu}_k}(\Omega), \tag{1.1.23}$$

where $\overline{\mu}_k = \max_{j \geq k} \mu_j$. In particular, from (1.1.23) it follows that $a \in S^{\overline{\mu}_0}$.

It is easy to verify the following assertions:

(1) Relation (1.1.22) defines a up to a term in $S^{-\infty}(\Omega)$.
(2) From (1.1.22), it follows that

$$\partial_x^\alpha \partial_\xi^\beta a(x,\xi) \sim \sum_{j=0}^{\infty} \partial_x^\alpha \partial_\xi^\beta a_j(x,\xi) \quad (\forall \alpha, \beta)$$

(the asymptotic expansions can be differentiated).
(3) Any permutation of the series terms does not violate (1.1.22). By permutation, we can always provide the monotonically decreasing sequence μ_j of the symbol orders and obtain $\overline{\mu}_k = \mu_k$.
(4) The relations (1.1.22) and $a_j(x,\xi) \sim \sum_{k=0}^{\infty} a_{jk}(x,\xi)$ with $j = 0, 1, \dots$ imply

$$a(x,\xi) \sim \sum_{j,k} a_{jk}(x,\xi).$$

Theorem 1.1.18 *If $a_j \in S^{\mu_j}(\Omega)$, $j = 0, 1, \dots$, and $\mu_j \to -\infty$ as $j \to \infty$, there exists a function $a \in S^{\overline{\mu}_0}(\Omega)$ that satisfies* (1.1.22).

Proof We can assume that $\mu_j \geq \mu_{j+1}$ for all j. Choose $\chi \in C^\infty(\mathbb{R}^n)$ such that $\chi(\xi) = 0$ for $|\xi| \leq 1$ and $\chi(\xi) = 1$ for $|\xi| \geq 2$. We also assume $\{K_j\}$ to be an exhaustive sequence of compact subsets in Ω. Since $\chi(\varepsilon\xi) \to_{\varepsilon \to +0} 0$ in $S^1(\Omega)$, for every j there is a number $\varepsilon_j > 0$ such that

$$|\partial_x^\alpha \partial_\xi^\beta (\chi(\varepsilon_j \xi) a_j(x,\xi))| \leq 2^{-j} \langle \xi \rangle^{\mu_j - |\beta| + 1} \tag{1.1.24}$$

for $(x,\xi) \in K_j \times \mathbb{R}^n$ with α and β subject to $|\alpha| + |\beta| \le j$. We can consider that $\varepsilon_j \to 0$ as $j \to 0$. Let us define a function a by the equality

$$a(x,\xi) = \sum_{j=0}^{\infty} \chi(\varepsilon_j \xi) a_j(x,\xi). \tag{1.1.25}$$

The sum on the right-hand side is locally finite, that is, for any compact $K \subset \mathbb{R}^n$, there are only finitely many terms different from zero on $\Omega \times K$, so the definition of a is correct. It is evident that $a \in C^\infty(\Omega \times \mathbb{R}^n)$ and

$$\partial_x^\alpha \partial_\xi^\beta a(x,\xi) = \sum_{j=0}^{\infty} \partial_x^\alpha \partial_\xi^\beta \big(\chi(\varepsilon_j \xi) a_j(x,\xi)\big).$$

Any remainder of series (1.1.25) has similar properties. We show that, for every $k \ge 0$, there holds the inclusion

$$\sum_{j=k}^{\infty} \chi(\varepsilon_j \xi) a_j(x,\xi) \in S^{\mu_k}(\Omega). \tag{1.1.26}$$

Let $K \subset \Omega$ be an arbitrary compact and α and β be any multi-indices. We have to verify the inequality

$$\Big|\sum_{j=k}^{\infty} \partial_x^\alpha \partial_\xi^\beta \big(\chi(\varepsilon_j \xi) a_j(x,\xi)\big)\Big| \le C\langle\xi\rangle^{\mu_k - |\beta|}, \quad (x,\xi) \in K \times \mathbb{R}^n.$$

There exists a number l such that $K \subset K_l$, $|\alpha + \beta| \le l$ and $\mu_l + 1 \le \mu_k$. Because

$$\sum_{j=k}^{\infty} \partial_x^\alpha \partial_\xi^\beta \big(\chi(\varepsilon_j \xi) a_j(x,\xi)\big) = \sum_{j=k}^{l-1} \cdots + \sum_{j=l}^{\infty} \cdots,$$

it suffices to check that both of the sums on the right are majorized by $\langle\xi\rangle^{\mu_k - |\beta|}$. For the first sum, it is evident because the sum belongs to $S^{\mu_k - |\beta|}(\Omega)$. For the second one, the needed estimate follows from (1.1.24):

$$\Big|\sum_{j=l}^{\infty} \partial_x^\alpha \partial_\xi^\beta \big(\chi(\varepsilon_j \xi) a_j(x,\xi)\big)\Big| \le \sum_{j=l}^{\infty} 2^{-j} \langle\xi\rangle^{\mu_j - |\beta| + 1} \le \langle\xi\rangle^{\mu_l - |\beta| + 1} \sum_{j=l}^{\infty} 2^{-j}$$
$$\le 2^{-l+1} \langle\xi\rangle^{\mu_k - |\beta|}.$$

The inclusion (1.1.26) has been established. Now, for any $k \geq 1$, we have

$$a(x,\xi) - \sum_{j=0}^{k-1} a_j(x,\xi) = \sum_{j=0}^{k-1} (\chi(\varepsilon_j\xi) - 1)a_j(x,\xi) + \sum_{j=k}^{\infty} \chi(\varepsilon_j\xi)a_j(x,\xi) \in S^{\mu_k}(\Omega);$$

the first sum on the right vanishes for large $|\xi|$, so it belongs to $S^{-\infty}(\Omega)$, while the second sum is in S^{μ_k} by virtue of (1.1.26). □

1.1.9 The Symbol of Proper Pseudodifferential Operator

Let A be a proper ΨDO. We set

$$\sigma_A(x,\xi) = e_{-\xi} A e_\xi(x),$$

where $e_\xi(x) = e^{ix\xi}$. The function σ_A is called the symbol of A.

Examples

(1) Let $P(x, D)$ be a differential operator in Ω. Taking into account that $P(x, D)e_\xi(x) = P(x,\xi)e_\xi(x)$, we obtain $\sigma_P(x,\xi) = P(x,\xi)$.
(2) Let

$$A : u \mapsto \int G(x, y)u(y)\, dy, \quad u \in C^\infty(\Omega),$$

be an integral operator with a properly supported smooth kernel. Then

$$\sigma_A(x,\xi) = e^{-ix\xi} \int G(x, y)e^{iy\xi}\, dy = (2\pi)^{n/2} e^{-ix\xi} \check{G}(x,\xi),$$

where $\check{G}$ is the inverse Fourier transform of G with respect to the second argument. □

In the case that A is a proper ΨDO and $A = \operatorname{Op} a$ with properly supported amplitude a, we have

$$e_{-\xi} A e_\xi(x) = e_{-\xi}(x)\, (2\pi)^{-n} \iint e^{i(x-y)\eta} a(x, y,\eta)e_\xi(y)\, dy d\eta$$

$$= (2\pi)^{-n} \iint e^{i(x-y)\eta} a(x, y,\eta)e^{i(y-x)\xi}\, dy d\eta = (2\pi)^{-n} \iint e^{i(x-y)(\eta-\xi)} a(x, y,\eta)\, dy d\eta.$$

Therefore,

$$\sigma_A(x,\xi) = (2\pi)^{-n} \iint e^{i(x-y)\theta} a(x,y,\xi+\theta)\, dyd\theta.$$

Theorem 1.1.19 *Let $a \in S^\mu(\Omega,\Omega)$ be a proper amplitude and $A = \operatorname{Op} a$. Then,*

(1) $\sigma_A \in S^\mu(\Omega)$.
(2) $\sigma_A(x,\xi) \sim \sum\limits_{\alpha} \frac{1}{\alpha!} D_y^\alpha \partial_\xi^\alpha a(x,y,\xi)\big|_{y=x}$.
(3) $A = \operatorname{Op} \sigma_A$.

Note that $D_y^\alpha \partial_\xi^\alpha a(x,y,\xi)$ is an amplitude of order $\mu - |\alpha|$. Hence, $D_y^\alpha \partial_\xi^\alpha a(x,y,\xi)\big|_{y=x}$ is a symbol of the same order. Therefore, the terms of series (2) are symbols of decreasing orders.

Before proving Theorem 1.1.19, we verify the following lemma.

Lemma 1.1.20 *Assume that $a \in S^\mu(\Omega,\Omega)$ is a proper amplitude, $h \in C[0,1]$, and $\alpha \in \mathbb{Z}_+^n$. We set*

$$\sigma(x,\xi) = \iiint_0^1 e^{i(x-y)\theta} a(x,y,\xi+t\theta)\theta^\alpha h(t)\, dtdyd\theta. \tag{1.1.27}$$

Then, $\sigma \in S^\mu(\Omega)$.

Proof Using the equality $e^{i(x-y)\theta} = \langle D_y\rangle^{2N} e^{i(x-y)\theta} \langle\theta\rangle^{-2N}$, we rewrite (1.1.27) in the form

$$\sigma(x,\xi) = \iiint_0^1 e^{i(x-y)\theta} \langle D_y\rangle^{2N} a(x,y,\xi+t\theta)\theta^\alpha \langle\theta\rangle^{-2N} h(t)\, dtdyd\theta. \tag{1.1.28}$$

Let $K_1 \subset \Omega$ be any compact and $K_2 = \pi_2(\pi_1^{-1}(K_1))$, where $\pi_1, \pi_2 : \operatorname{supp}_{x,y} a \to \Omega$ are the projections. For $x \in K_1$, we have

$$\begin{aligned} &\langle D_y\rangle^{2N} a(x,y,\xi+t\theta) = 0 \ \text{ for } \ y \notin K_2, \\ &|\langle D_y\rangle^{2N} a(x,y,\xi+t\theta)| \le C_1 \langle \xi + t\theta\rangle^\mu \le C_1\, 2^{|\mu|/2} \langle\xi\rangle^\mu \langle t\theta\rangle^{|\mu|} \\ &\qquad \le C_1\, 2^{|\mu|/2} \langle\xi\rangle^\mu \langle\theta\rangle^{|\mu|} \ \text{ for } \ y \in K_2 \ \ (C_1 = C_1(K_1)). \end{aligned}$$

These relations lead to

$$|\langle D_y\rangle^{2N} a(x,y,\xi+t\theta)\theta^\alpha \langle\theta\rangle^{-2N} h(t)| \le C_2 \chi(y) \langle\xi\rangle^\mu \langle\theta\rangle^{|\alpha|+|\mu|-2N}, \quad x \in K_1, \tag{1.1.29}$$

where χ is the characteristic function of the set K_2. For the case that $|\alpha|+|\mu|-2N < -n$, we obtain

$$\iiint_0^1 \chi(y)\langle\theta\rangle^{|\alpha|+|\mu|-2N}\, dt dy d\theta < +\infty. \tag{1.1.30}$$

Introduce the notation $B_R = \{\xi \in \mathbb{R}^n : |\xi| \le R\}$ with $R > 0$. From (1.1.29) and (1.1.30) it follows that for $(x, \xi) \in K_1 \times B_R$ the integrand in (1.1.28) has the summable majorant $C_2\chi(y)\max_{\xi\in B_R}\langle\xi\rangle^{\mu}\langle\theta\rangle^{|\alpha|+|\mu|-2N}$ independent of (x, ξ). Hence, integral (1.1.28) converges uniformly with respect to $(x, \xi) \in K_1 \times B_R$, and formula (1.1.28) defines a continuous function σ on the set $K_1 \times B_R$. Since the compact K_1 and number $R > 0$ have arbitrarily been chosen, there holds the inclusion $\sigma \in C(\Omega \times \mathbb{R}^n)$. Moreover, (1.1.28) and (1.1.29) provide the inequality

$$|\sigma(x, \xi)| \le C(K_1)\langle\xi\rangle^{\mu}, \quad x \in K_1. \tag{1.1.31}$$

Assume now that β and γ are arbitrary multi-indices. Differentiating (1.1.28), we obtain

$$\partial_x^{\beta}\partial_{\xi}^{\gamma}\sigma(x, \xi) = \sum_{\nu\le\beta}\binom{\beta}{\nu} i^{|\beta-\nu|}\sigma_{\nu}(x, \xi) \tag{1.1.32}$$

with

$$\sigma_{\nu}(x, \xi) = \iiint_0^1 e^{i(x-y)\theta}\langle D_y\rangle^N \partial_x^{\nu}\partial_{\xi}^{\gamma} a(x, y, \xi + t\theta)\theta^{\alpha+\beta-\nu}\langle\theta\rangle^{-2N} h(t)\, dt dy d\theta. \tag{1.1.33}$$

Every function (1.1.33) is of the form (1.1.28) (with various a and α). We choose N to satisfy $|\alpha + \beta| + |\mu| - 2N < -n$. Then, from the first part of the proof, it follows that for all ν integral (1.1.33) uniformly converges on any set of the form $K_1 \times B_R$. Hence, the differentiation in (1.1.28) was admissible and $\sigma_{\nu} \in C(\Omega \times \mathbb{R}^n)$. Because $\partial_x^{\nu}\partial_{\xi}^{\gamma} a \in S^{\mu-|\gamma|}(\Omega, \Omega)$, the analogue of (1.1.31) for σ_{ν} is of the form

$$|\sigma_{\nu}(x, \xi)| \le C(\nu, K_1)\langle\xi\rangle^{\mu-|\gamma|}, \quad x \in K_1.$$

Therefore (see (1.1.32)),

$$|\partial_x^{\beta}\partial_{\xi}^{\gamma}\sigma(x, \xi)| \le C\langle\xi\rangle^{\mu-|\gamma|}, \quad x \in K_1.$$

In view of freedom in choosing the compact K_1, we arrive at $\sigma \in S^{\mu}(\Omega)$. □

Proof of Theorem 1.1.19 Taking into account the equality

$$\sigma_A(x,\xi) = (2\pi)^{-n} \iint e^{i(x-y)\theta} a(x,y,\xi+\theta)\, dyd\theta,$$

and the Taylor expansion

$$a(x,y,\xi+\theta) = \sum_{|\alpha|\le N-1} \frac{1}{\alpha!}\partial_\xi^\alpha a(x,y,\xi)\theta^\alpha + \sum_{|\alpha|=N} \frac{N}{\alpha!}\int_0^1 (1-t)^{N-1}\partial_\xi^\alpha a(x,y,\xi+t\theta)\theta^\alpha\, dt,$$

we obtain

$$\sigma_A(x,\xi) = \sum_{|\alpha|\le N-1} \frac{1}{\alpha!}\sigma^{(\alpha)}(x,\xi) + r_N(x,\xi),$$

where

$$\sigma^{(\alpha)}(x,\xi) = (2\pi)^{-n} \iint e^{i(x-y)\theta}\partial_\xi^\alpha a(x,y,\xi)\theta^\alpha\, dyd\theta = D_y^\alpha \partial_\xi^\alpha a(x,y,\xi)\big|_{y=x},$$

$$r_N(x,\xi) = (2\pi)^{-n} \sum_{|\alpha|=N} \frac{N}{\alpha!} \iiint_0^1 e^{i(x-y)\theta}\partial_\xi^\alpha a(x,y,\xi+t\theta)\theta^\alpha (1-t)^{N-1}\, dtdyd\theta.$$

By Lemma 1.1.20, the terms in the last sum, being considered as functions of (x,ξ), belong to $S^{\mu-N}(\Omega)$. Therefore, for any N

$$\sigma_A(x,\xi) = (2\pi)^{-n} \sum_{|\alpha|\le N-1} \frac{1}{\alpha!} D_y^\alpha \partial_\xi^\alpha a(x,y,\xi)\big|_{y=x} + r_N(x,\xi), \quad r_N \in S^{\mu-N}(\Omega).$$

This proves assertion 2 of the theorem, which implies assertion 1. It remains to verify that $A = \operatorname{Op}\sigma_A$. The representation of σ_A can be written in the form

$$\sigma_A(x,\xi) = (2\pi)^{-n} \iint e^{i(x-y)(\eta-\xi)} a(x,y,\eta)\, dyd\eta.$$

Therefore,

$$(2\pi)^{-n}\int e^{ix\xi}\sigma_A(x,\xi)\hat{u}(\xi)\, d\xi = (2\pi)^{-2n}\int e^{ix\xi}\Big(\iint e^{i(x-y)(\eta-\xi)}a(x,y,\eta)\, dyd\eta\Big)\hat{u}(\xi)\, d\xi,$$

that is,

$$(\operatorname{Op}\sigma_A)u(x) = (2\pi)^{-2n} \iiint e^{ix\eta}e^{-iy(\eta-\xi)} a(x,y,\eta)\hat{u}(\xi)\, dyd\eta d\xi. \tag{1.1.34}$$

Now, we show that for fixed x the function

$$(\xi, \eta) \mapsto \int e^{ix\eta} e^{-iy(\eta-\xi)} a(x, y, \eta) \hat{u}(\xi)\, dy = e^{ix\eta} \hat{u}(\xi) \int e^{-iy(\eta-\xi)} a(x, y, \eta)\, dy \tag{1.1.35}$$

decreases rapidly as $|\xi| + |\eta| \to \infty$. Because of $\hat{u} \in \mathcal{S}(\mathbb{R}^n)$, it suffices to obtain the estimate

$$\left| \iint e^{-iy(\eta-\xi)} a(x, y, \eta)\, dy \right| \leq C_N \langle \xi \rangle^{2N} \langle \eta \rangle^{\mu-2N}$$

for any $N \in \mathbb{N}$. In view of $e^{-y(\eta-\xi)} = \langle D_y \rangle^{2N} e^{-iy(\eta-\xi)} \langle \eta - \xi \rangle^{-2N}$,

$$\begin{aligned} \left| \iint e^{-iy(\eta-\xi)} a(x, y, \eta)\, dy \right| &\leq \int |\langle D_y \rangle^{2N} a(x, y, \eta)| dy \, \langle \eta - \xi \rangle^{-2N} \\ &\leq C_N \langle \eta \rangle^{\mu} \langle \eta \rangle^{-2N} \langle \xi \rangle^{2N} = C_N \langle \xi \rangle^{2N} \langle \eta \rangle^{\mu-2N}, \end{aligned}$$

as we wanted to obtain. Due to the rapid decrease of function (1.1.35), we can in (1.1.34) permute the integrals with respect to η and ξ:

$$(\operatorname{Op} \sigma_A) u(x) = (2\pi)^{-2n} \int \Big(\iint e^{i(x-y)\eta} e^{iy\xi} a(x, y, \eta) \hat{u}(\xi)\, dy d\xi \Big) d\eta.$$

By the integral in brackets is meant a double one (for fixed x and η, the function $(y, \xi) \mapsto a(x, y, \eta)\hat{u}(\xi)$ vanishes for large y and rapidly decreases as $\xi \to \infty$). Integrating over ξ, we arrive at

$$(\operatorname{Op} \sigma_A) u(x) = (2\pi)^{-n} \iint e^{i(x-y)\eta} a(x, y, \eta) u(y)\, dy d\eta = Au(x).$$

□

The asymptotic expansion of the symbol of a proper ΨDO A remains valid even in the case that the amplitude a satisfying the condition $A = \operatorname{Op} a$ is not proper. Indeed, choose a properly supported function $\chi \in C^\infty(\Omega \times \Omega)$ equal to 1 in a neighborhood of the set $\Delta \cup \operatorname{supp} G_A$. Then, $A = \operatorname{Op}(\chi a)$ (see Lemma 1.1.12). Since the amplitude χa is proper, from Theorem 1.1.19, it follows that

$$\sigma_A(x, \xi) \sim \sum_\alpha \frac{1}{\alpha!} D_y^\alpha \partial_\xi^\alpha (\chi a)(x, y, \xi)\big|_{y=x} = \sum_\alpha \frac{1}{\alpha!} D_y^\alpha \partial_\xi^\alpha a(x, y, \xi)\big|_{y=x}.$$

The equalities $\sigma_A(x,\xi) = e_\xi(x)Ae_\xi(x)$ and $A = \operatorname{Op}\sigma_A$ establish a bijection between the proper ΨDO and their symbols. If A is an arbitrary ΨDO, it is usual to introduce $\sigma_A = \sigma_{A_1}$, where A_1 is a proper ΨDO and $A - A_1 \in \Psi^{-\infty}$. Such a symbol is not uniquely defined; however, any two symbols differ by a function in $S^{-\infty}$. More precisely, the correspondence $A \leftrightarrow \sigma_A$ generates the isomorphism $\Psi^\mu(\Omega)/\Psi^{-\infty}(\Omega) \cong S^\mu(\Omega)/S^{-\infty}(\Omega)$ for every μ.

1.1.10 Symbolic Calculus of Pseudodifferential Operators

1°. **Symbol expansions of transposed and adjoint operators.**

Theorem 1.1.21 *Let A be a proper ΨDO and σ_A its symbol. Then,*

$$\sigma_{{}^tA}(x,\xi) \sim \sum_\alpha \frac{1}{\alpha!} D_x^\alpha \partial_\xi^\alpha \sigma_A(x,-\xi), \tag{1.1.36}$$

$$\sigma_{A^*}(x,\xi) \sim \sum_\alpha \frac{1}{\alpha!} D_x^\alpha \partial_\xi^\alpha \overline{\sigma_A(x,\xi)}. \tag{1.1.37}$$

Proof Because $A = \operatorname{Op}\sigma_A$, we have

$${}^tAu(x) = (2\pi)^{-n} \iint e^{i(x-y)\xi} \sigma_A(y,-\xi)u(y)\,dyd\xi, \quad u \in C_c^\infty(\Omega).$$

By Theorem 1.1.19,

$$\sigma_{{}^tA}(x,\xi) \sim \sum_\alpha \frac{1}{\alpha!} D_y^\alpha \partial_\xi^\alpha \sigma_A(y,-\xi)\big|_{y=x},$$

which coincides with (1.1.36). The relation (1.1.37) can in a similar way be derived from the equality

$$A^*u(x) = (2\pi)^{-n} \iint e^{i(x-y)\xi}\, \overline{\sigma_A(y,\xi)}u(y)\,dyd\xi.$$

□

The function $\tilde\sigma_A(x,\xi) := \sigma_{{}^tA}(x,-\xi)$ is sometimes called the dual symbol of A. From (1.1.36), it follows that

$$\tilde\sigma_A(x,\xi) \sim \sum_\alpha \frac{1}{\alpha!} D_x^\alpha (-\partial_\xi)^\alpha \sigma_A(x,\xi).$$

2°. **Symbol expansion of operator composition.** Let A and B be arbitrary ΨDO. In the case that $u \in C_c^\infty(\Omega)$, we have $Bu \in C^\infty(\Omega)$ but generally $Bu \notin C_c^\infty(\Omega)$. Therefore, in the general case, the composition AB makes no sense. However, the composition AB is defined if at least one of the operators, A or B, is proper.

Theorem 1.1.22 *Let A and B be proper ΨDO of order μ_1 and μ_2, respectively. Then,*

1. *AB is a proper ΨDO of order $\mu_1 + \mu_2$.*
2. $\sigma_{AB} \sim \sum_\alpha \frac{1}{\alpha!} \partial_\xi^\alpha \sigma_A(x,\xi) D_x^\alpha \sigma_B(x,\xi)$.

Proof Since $B = {}^t({}^tB)$ and ${}^tB = \operatorname{Op} \sigma_{{}^tB}$, we have

$$Bu(x) = (2\pi)^{-n} \iint e^{i(x-y)\xi} \sigma_{{}^tB}(y, -\xi) u(y)\, dy d\xi$$

$$= (2\pi)^{-n} \int e^{ix\xi}\, d\xi \int e^{-iy\xi} \tilde{\sigma}_B(y,\xi) u(y)\, dy.$$

Therefore,

$$(Bu)^\wedge(\xi) = \int e^{-iy\xi} \tilde{\sigma}_B(y,\xi) u(y)\, dy.$$

Setting $v = Bu$ in the equality

$$Av(x) = (2\pi)^{-n} \int e^{ix\xi} \sigma_A(x,\xi) \hat{v}(\xi)\, d\xi,$$

we obtain

$$ABu(x) = (2\pi)^{-n} \iint e^{i(x-y)\xi} \sigma_A(x,\xi) \tilde{\sigma}_B(y,\xi) u(y)\, dy d\xi. \tag{1.1.38}$$

The function $(x, y, \xi) \mapsto \sigma_A(x,\xi)\tilde{\sigma}_B(y,\xi)$ is an amplitude of order $\mu_1 + \mu_2$, so $AB \in \Psi^{\mu_1+\mu_2}(\Omega)$.

Now, we verify that ΨDO AB is proper. According to Proposition 1.1.14, for any compact $K_2 \subset \Omega$, there exists a compact $K' \subset \Omega$ such that

$$u \in C_c^\infty(K_2) \Rightarrow Bu \in C_c^\infty(K').$$

In turn, for K' there exists a compact $K_1 \subset \Omega$ such that

$$v \in C_c^\infty(K') \Rightarrow Av \in C_c^\infty(K_1).$$

Consequently,

$$u \in C_c^\infty(K_2) \Rightarrow ABu \in C_c^\infty(K_1).$$

The operator ${}^t(AB) = {}^tB\,{}^tA$ has a similar property. Again employing Proposition 1.1.14, we conclude that AB is a proper ΨDO.

The asymptotic expansion of σ_{AB} remains to be obtained. Making use of Theorem 1.1.19, we derive from (1.1.38) that

$$\sigma_{AB}(x,\xi) \sim \sum_\alpha \frac{1}{\alpha!} D_y^\alpha \partial_\xi^\alpha [\sigma_A(x,\xi)\tilde{\sigma}_B(y,\xi)]\big|_{y=x}$$

$$= \sum_\alpha \frac{1}{\alpha!}\partial_\xi^\alpha[\sigma_A(x,\xi) D_x^\alpha \tilde{\sigma}_B(x,\xi)] = \sum_\alpha \frac{1}{\alpha!} \sum_{\beta+\gamma=\alpha} \frac{\alpha!}{\beta!\gamma!} \partial_\xi^\beta \sigma_A \cdot \partial_\xi^\gamma D_x^\alpha \tilde{\sigma}_B$$

$$= \sum_{\beta,\gamma} \frac{1}{\beta!\gamma!} \partial_\xi^\beta \sigma_A \cdot \partial_\xi^\gamma D_x^{\beta+\gamma} \tilde{\sigma}_B.$$

Furthermore,

$$\tilde{\sigma}_B \sim \sum_\delta \frac{1}{\delta!} D_x^\delta (-\partial_\xi)^\delta \sigma_B \Rightarrow \partial_\xi^\gamma D_x^{\beta+\gamma} \tilde{\sigma}_B \sim \sum_\delta \frac{(-1)^{|\delta|}}{\delta!} \partial_\xi^{\gamma+\delta} D_x^{\beta+\gamma+\delta} \sigma_B.$$

Hence,

$$\sigma_{AB} \sim \sum_{\beta,\gamma,\delta} \frac{(-1)^{|\delta|}}{\beta!\gamma!\delta!} \partial_\xi^\beta \sigma_A \cdot \partial_\xi^{\gamma+\delta} D_x^{\beta+\gamma+\delta} \sigma_B = \sum_{\beta,\nu} \frac{1}{\beta!} \partial_\xi^\beta \sigma_A \cdot \partial_\xi^\nu D_x^{\beta+\nu} \sigma_B \sum_{\gamma+\delta=\nu} \frac{(-1)^{|\delta|}}{\gamma!\delta!}.$$

In the equality

$$(x+y)^\nu = \sum_{\gamma+\delta=\nu} \frac{\nu!}{\gamma!\delta!} x^\gamma y^\delta$$

we set $x = -y = (1, 1, \dots, 1)$ and obtain

$$\sum_{\gamma+\delta=\nu} \frac{(-1)^{|\delta|}}{\gamma!\delta!} = 0$$

for $\nu \neq 0$. Therefore,

$$\sigma_{AB} \sim \sum_\beta \frac{1}{\beta!} \partial_\xi^\beta \sigma_A D_x^\beta \sigma_B.$$

□

Corollary 1.1.23 *Let* $A \in \Psi^{\mu_1}(\Omega)$ *and* $B \in \Psi^{\mu_2}(\Omega)$ *be proper* Ψ*DO. Then,* $[A, B] \equiv AB - BA \in \Psi^{\mu_1+\mu_2-1}(\Omega)$.

Let $A \in \Psi^{\mu_1}(\Omega)$ and $B \in \Psi^{\mu_2}(\Omega)$, while the operator B is proper. Then, as was mentioned before Theorem 1.1.22, both compositions AB and BA are defined. We show that they belong to $\Psi^{\mu_1+\mu_2}(\Omega)$. To this end, we represent A in the form $A_1 + A_2$, where A_1 is a proper ΨDO, and A_2 is a smoothing ΨDO. We have

$$AB = A_1B + A_2B\,, \qquad BA = BA_1 + BA_2. \tag{1.1.39}$$

By Theorem 1.1.22, the operators A_1B and BA_1 are proper ΨDO of order $\mu_1+\mu_2$. Denote by G the kernel of A_2. It is easy to verify that A_2B and BA_2 have the smooth kernels ${}^tB_yG(x, y)$ and $B_xG(x, y)$, respectively. Therefore, these operators are smoothing, that is, they belong to $\Psi^{-\infty}(\Omega)$. Taking into account (1.1.39), we obtain the needed result.

1.1.11 Change of Variables in Pseudodifferential Operators

Let Ω and Ω_1 be domains in $\mathbb{R}^n$ and let $f : \Omega \to \Omega_1$ be a C^∞-diffeomorphism. Then, the mappings

$$C^\infty(\Omega_1) \ni u \mapsto u \circ f \in C^\infty(\Omega),$$
$$C_c^\infty(\Omega_1) \ni u \mapsto u \circ f \in C_c^\infty(\Omega)$$

are isomorphisms of the linear topological spaces. A pseudodifferential operator A in Ω is in accordance with a linear continuous operator

$$A_1 : C_c^\infty(\Omega_1) \to C^\infty(\Omega_1)$$

defined by

$$A_1u = [A(u \circ f)] \circ f^{-1}. \tag{1.1.40}$$

Theorem 1.1.24

(1) *The mapping*

$$\Psi^\mu(\Omega) \ni A \mapsto A_1 \tag{1.1.41}$$

defined by (1.1.40), for any μ*, is a linear bijection of* $\Psi^\mu(\Omega)$ *onto* $\Psi^\mu(\Omega_1)$ *sending proper operators to proper ones.*

(2) *There holds the asymptotic expansion*

$$\sigma_{A_1}(f(x), \eta) \sim \sum_{\alpha} \sigma_A^{(\alpha)}(x, {}^t f'(x)\eta)\varphi_\alpha(x, \eta),$$

where $\sigma_A^{(\alpha)}(x, \xi) = \partial_\xi^\alpha \sigma_A(x, \xi)$, $f'(x) = (\partial f_i / \partial x_j)$ *is the Jacobi matrix of the mapping* f, *and* $\varphi_\alpha \in C^\infty(\Omega \times \mathbb{R}^n)$ *are polynomials of* η *with degree* $\leq |\alpha|/2$ *being dependent on* f *but independent of* A.

Let us describe the basic stages of the proof.

1. Clearly, the mapping (1.1.41) is linear. It is a bijection because there exists the inverse mapping given by the formula $Av = [A_1(v \circ f^{-1})] \circ f$. In what follows, we write g instead of f^{-1}.
2. Let $A = \mathrm{Op}\, a$ with $a \in S^\mu(\Omega, \Omega)$ and let $\chi \in C^\infty(\Omega \times \Omega)$ be a properly supported function equal to 1 in a neighborhood of the diagonal. We write an operator A in the form $B + C$, where $B = \mathrm{Op}\,(\chi a)$ is a proper operator and $C = \mathrm{Op}\,((1 - \chi)a)$ is a smoothing operator. Under the mapping (1.1.41), the operator C is in accordance with a ΨDO $C_1 \in \Psi^{-\infty}(\Omega_1)$, so to prove $A_1 \in \Psi(\Omega_1)$ it suffices to verify that $B_1 \in \Psi^\mu(\Omega)$.
3. There exists a neighborhood V of the diagonal $\Delta_1 := \{(x, y) \in \Omega_1 \times \Omega_1 : x = y\}$ such that, for all pairs $(x, y) \in V$, the line segment joining x and y belongs to Ω_1. In the neighborhood V, the equality

$$g(x) - g(y) = h(x, y)(x - y)$$

holds, where h is a matrix-valued function defined by the integral

$$h(x, y) = \int_0^1 g'(y + t(x - y))\, dt.$$

For $x = y$, we have $h(x, x) = g'(x)$. Since $|g'(x)| \neq 0$, by continuity $|h(x, y)| \neq 0$ for all (x, y) in the vicinity of Δ_1. Diminishing the neighborhood V, we can obtain $|h(x, y)| \neq 0$ for $(x, y) \in V$.
4. We choose χ (see stage 2) so that the support of the function χ_1 defined by $\chi_1(x, y) := \chi(g(x), g(y))$ belongs to V. We have

$$B_1 u(x) = [B(u \circ f)] \circ g(x)$$

$$= (2\pi)^{-n} \iint e^{i(g(x)-y)\xi} a(g(x), y, \xi)\chi(g(x), y)(u \circ f)(y)\, dy d\xi$$

$$= (2\pi)^{-n} \iint e^{i(g(x)-g(y))\xi} a(g(x), g(y), \xi)\chi_1(x, y)|g'(y)|u(y)\, dy d\xi$$

$$= (2\pi)^{-n} \iint e^{i(x-y)\,{}^t h(x,y)\xi} a(g(x), g(y), \xi)\chi_1(x, y)|g'(y)|u(y)\, dy d\xi.$$

Thus,

$$B_1 u(x) = (2\pi)^{-n} \iint e^{i(x-y)\eta} a(g(x), g(y), {}^t h(x, y)^{-1}\eta) \chi_1(x, y) \times$$
$$\times |g'(y)||h(x, y)|^{-1} u(y)\, dy d\eta. \qquad (1.1.42)$$

It can be verified that the function $(x, y, \eta) \mapsto a(g(x), g(y), {}^t h(x, y)^{-1}\eta) \times \chi_1(x, y)$ is in $S^\mu(\Omega_1, \Omega_1)$. This proves the inclusion $B_1 \in \Psi^\mu(\Omega_1)$ and, consequently, $A_1 \in \Psi^\mu(\Omega_1)$.

If A is a proper ΨDO, the equality $A = B + C$ implies that C is proper. Since $A_1 = B_1 + C_1$ and both operators on right are proper, A is proper as well. Thus, mapping (1.1.41) sends a proper ΨDO to a proper ΨDO.

5. If A is a proper ΨDO, from the very beginning, we can take σ_A for the role of a. Then, (1.1.11) takes the form

$$B_1 u(x) = (2\pi)^{-n} \iint e^{i(x-y)\eta} a_1(x, y, \eta) \chi_1(x, y) u(y)\, dy d\eta,$$

where

$$a_1(x, y, \eta) = \sigma_A(g(x), {}^t h(x, y)^{-1}\eta)\, |g'(y)|\, |h(x, y)|^{-1}. \qquad (1.1.43)$$

We apply Theorem 1.1.19 and obtain (taking into account $\chi_1 = 1$ in a neighborhood of the diagonal Δ_1)

$$\sigma_{A_1}(x, \eta) \sim \sigma_{B_1}(x, \eta) \sim \sum_\beta \frac{1}{\beta!} D_y^\beta \partial_\eta^\beta a_1(x, y, \eta)\big|_{y=x}; \qquad (1.1.44)$$

the first relation meaning that $\sigma_{A_1} - \sigma_{B_1} \in S^{-\infty}(\Omega_1)$ follows from $A_1 = B_1 + C_1$, where $C_1 \in \Psi^{-\infty}(\Omega)$). From (1.1.43), it follows that the terms of (1.1.44) are sums of terms of the form $c(x)\sigma_A^{(\beta+\gamma)}(g(x), {}^t g'(x)^{-1}\eta) \times \eta^\delta$, where $c \in C^\infty(\Omega_1)$ and $|\gamma| = |\delta| \le |\beta|$. We substitute these sums into (1.1.44), collect similar terms with the same $\beta + \gamma$ in one group, and arrive at

$$\sigma_{A_1}(x, \eta) \sim \sum_\alpha \psi_\alpha(x, \eta) \sigma_A^{(\alpha)}(g(x), {}^t g'(x)^{-1}\eta),$$

where $\psi_\alpha(x, \eta)$ are polynomials in η of degree $\le |\alpha|/2$. Changing x for $f(x)$, we finally obtain

$$\sigma_{A_1}(f(x), \eta) \sim \sum_\alpha \varphi_\alpha(x, \eta) \sigma_A^{(\alpha)}(x, {}^t f'(x)\eta),$$

where $\varphi_\alpha(x, \eta) = \psi_\alpha(f(x), \eta)$ (we took into account $g'(f(x))^{-1} = f'(x)$). $\square$

Remark We set $f''_x(y) = f(y) - f(x) - f'(x)(y-x)$. It can be shown that

$$\varphi_\alpha(x,\eta) = \frac{1}{\alpha!} D_y^\alpha e^{if''_x(y)\eta}\big|_{y=x}.$$

1.1.12 Classes $\Psi_b^\mu(\mathbb{R}^n)$

This section is preparatory for proving the continuity of ΨDO in $L_2(\mathbb{R}^n)$.

1°. **Amplitudes, symbols, and ΨDO.** For any $\mu \in \mathbb{R}$, we denote by $S_b^\mu(\mathbb{R}^n, \mathbb{R}^n)$ the set of all amplitudes $a \in S^\mu(\mathbb{R}^n, \mathbb{R}^n)$ satisfying

$$|\partial_x^\alpha \partial_y^\beta \partial_\xi^\gamma a(x,y,\xi)| \le C_{\alpha\beta\gamma}\langle\xi\rangle^{\mu-|\gamma|}, \quad x, y, \xi \in \mathbb{R}^n,$$

for all $\alpha, \beta, \gamma \in \mathbb{Z}_+^n$. We also let $S_b^\mu(\mathbb{R}^n)$ denote the set of all symbols $\sigma \in S^\mu(\mathbb{R}^n)$ such that

$$|\partial_x^\alpha \partial_\xi^\beta \sigma(x,\xi)| \le C_{\alpha\beta}\langle\xi\rangle^{\mu-|\beta|}, \quad x, \xi \in \mathbb{R}^n.$$

The symbols can be considered as amplitudes independent of y so that $S_b^\mu(\mathbb{R}^n) \subset S_b^\mu(\mathbb{R}^n, \mathbb{R}^n)$. We set $S_b^{-\infty} = \bigcap_\mu S_b^\mu$, $S_b = \bigcup_\mu S_b^\mu$ and $\Psi_b(\mathbb{R}^n) = \bigcup_\mu \Psi_b^\mu(\mathbb{R}^n)$, where

$$\Psi_b^\mu(\mathbb{R}^n) = \{\operatorname{Op} a : a \in S_b^\mu(\mathbb{R}^n, \mathbb{R}^n)\}, \quad \mu \in [-\infty, \infty).$$

Since $S_b^\mu(\mathbb{R}^n, \mathbb{R}^n)$ is a linear subspace in $S^\mu(\mathbb{R}^n, \mathbb{R}^n)$, the set $\Psi_b^\mu(\mathbb{R}^n)$ is a linear subspace in $\Psi^\mu(\mathbb{R}^n)$.

2°. **Symbol of ΨDO**

Proposition 1.1.25 *For any ΨDO $A \in \Psi_b^m(\mathbb{R}^n)$ there exists the only symbol $\sigma \in S_b^m(\mathbb{R}^n)$ such that $A = \operatorname{Op}\sigma$.*

Proof The uniqueness is evident because

$$\int e^{ix\xi}\sigma(x,\xi)\hat{u}(\xi)\,d\xi = 0 \text{ for } \forall\, u \in C_c^\infty(\mathbb{R}^n) \;\Rightarrow\; \sigma = 0$$

(here, the conditions $A \in \Psi_b^\mu(\mathbb{R}^n)$ and $\sigma \in S_b^\mu(\mathbb{R}^n)$ are not used; it suffices to assume that $A \in \Psi^\mu(\mathbb{R}^n)$ and $\sigma \in S^\mu(\mathbb{R}^n)$).

Let us prove the existence. For $\forall\, s \geq 0$, there is an amplitude $a \in S_b^{\mu}(\mathbb{R}^n, \mathbb{R}^n)$ such that $A = \operatorname{Op} a$ and

$$|\partial_x^{\alpha}\partial_y^{\beta}\partial_{\xi}^{\gamma} a(x, y, \xi)| \leq C_{\alpha\beta\gamma} \langle x - y\rangle^{-s} \langle\xi\rangle^{\mu - |\gamma|} \quad (\forall\, \alpha, \beta, \gamma). \tag{1.1.45}$$

Indeed, if $A = \operatorname{Op} \tilde{a}$, $\tilde{a} \in S_b^{\mu}(\mathbb{R}^n, \mathbb{R}^n)$, one can take $a(x, y, \xi) = \langle D_{\xi}\rangle^{2N} \tilde{a}(x, y, \xi)\langle x - y\rangle^{-2N}$, $N > s/2$ (see Sect. 1.1.2).

Let $A = \operatorname{Op} a$, where the amplitude $a \in S_b^{\mu}(\mathbb{R}^n, \mathbb{R}^n)$ satisfies (1.1.45) and $s > n$. We set

$$\sigma(x, \xi) = (2\pi)^{-n} \iint e^{i(x-y)\theta} a(x, y, \xi + \theta)\, dy d\theta \tag{1.1.46}$$

(cp. Sect. 1.1.9). Integrating by parts, we obtain

$$\sigma(x, \xi) = (2\pi)^{-n} \iint e^{i(x-y)\theta} \langle D_y\rangle^{2N} a(x, y, \xi + \theta)\, \langle\theta\rangle^{-2N}\, dy d\theta. \tag{1.1.47}$$

If N is sufficiently large, the integrand in (1.1.47) is summable (see estimate (1.1.48) below) and the integral (1.1.46) exists as an iterated one. We have

$$\begin{aligned} \left|\langle D_y\rangle^{2N} a(x, y, \xi + \theta)\, \langle\theta\rangle^{-2N}\right| &\leq C\, \langle x - y\rangle^{-s}\, \langle \xi + \theta\rangle^{\mu}\, \langle\theta\rangle^{-2N} \\ &\leq 2^{|\mu|/2}\, C\, \langle x - y\rangle^{-s}\, \langle\xi\rangle^{\mu}\, \langle\theta\rangle^{|\mu| - 2N}. \end{aligned} \tag{1.1.48}$$

Therefore,

$$|\sigma(x, \xi)| \leq 2^{|\mu|/2}\, C\, \langle\xi\rangle^{\mu} \iint \langle x - y\rangle^{-s}\, \langle\theta\rangle^{|\mu| - 2N}\, dy d\theta \leq C'\, \langle\xi\rangle^{\mu}.$$

The estimates $|\partial_x^{\alpha}\partial_{\xi}^{\beta}\sigma(x, \xi)| \leq C\langle\xi\rangle^{\mu - |\beta|}$ can be established in a similar way. Thus, $\sigma \in S_b^{\mu}(\mathbb{R}^n)$. The equality $\operatorname{Op}\sigma = A$ is verified in the same way as in Theorem 1.1.19. □

We now assume that $A \in \Psi_b^{\mu}(\mathbb{R}^n)$. A function $\sigma \in S_b^{\mu}(\mathbb{R}^n)$ satisfying $A = \operatorname{Op}\sigma$ will be called a symbol of A and denoted by σ_A. According to Proposition 1.1.25, any operator $A \in \Psi_b(\mathbb{R}^n)$ (which does not need to be proper) has the only symbol. If $A \in \Psi_b(\mathbb{R}^n)$ is a proper ΨDO, the symbol σ_A just defined coincides with σ_A in Sect. 1.1.9. Repeating the proof of Theorem 1.1.19 with evident modifications, we obtain the asymptotic expansion

$$\sigma_A(x, \xi) \sim \sum_{\alpha} \frac{1}{\alpha!} D_y^{\alpha}\partial_{\xi}^{\alpha} a(x, y, \xi)\big|_{y=x}; \tag{1.1.49}$$

here, a is any amplitude that defines A, and (1.1.49) means that $\sigma_A - \sum_{|\alpha|\le k-1} \cdots \in S_b^{\mu-k}(\mathbb{R}^n)$ for all $k \in \mathbb{N}$.

3°. **Symbols of transposed ΨDO and adjoint ΨDO**. For $a \in S_b^\mu$ we have ${}^t a$, $a^* \in S_b^\mu$. Therefore, $A \in \Psi_b^\mu$ implies ${}^t A$ and $A^* \in \Psi_b^\mu$. The expansions

$$\sigma_{{}^tA}(x,\xi) \sim \sum_\alpha \frac{1}{\alpha!} D_y^\alpha \partial_\xi^\alpha \sigma_A(x,-\xi),$$

$$\sigma_{A^*}(x,\xi) \sim \sum_\alpha \frac{1}{\alpha!} D_y^\alpha \partial_\xi^\alpha \overline{\sigma_A(x,\xi)} \tag{1.1.50}$$

follow from (1.1.49) (see Sect 1.1.10).

4°. **Composition of ΨDO.** We assume $A \in \Psi_b^\mu(\mathbb{R}^n)$ and note that in the integral

$$Au(x) = (2\pi)^{-n/2} \int e^{ix\xi} \sigma_A(x,\xi)\hat{u}(\xi)\, d\xi \tag{1.1.51}$$

one can change $u \in C_c^\infty(\mathbb{R}^n)$ for any function in the Schwartz class $\mathcal{S}(\mathbb{R}^n)$.

Proposition 1.1.26 *The operator $A : \mathcal{S}(\mathbb{R}^n) \to \mathcal{S}(\mathbb{R}^n)$ is continuous.*

Proof Since $\hat{u} \in \mathcal{S}$, the integral (1.1.51) is convergent. Integrating by parts, we obtain

$$Au(x) = (2\pi)^{-n/2} \int e^{ix\xi} \langle D_\xi \rangle^{2N} (\sigma_A(x,\xi)\hat{u}(\xi))\, d\xi \cdot \langle x \rangle^{-2N}.$$

We now note that $\langle D_\xi \rangle^{2N}(\sigma_A(x,\xi)\hat{u}(\xi))$ is a linear combination of the functions $\partial_\xi^\gamma \sigma_A(x,\xi) \times \partial_\xi^\delta \hat{u}(\xi)$ with $|\gamma + \delta| \le 2N$ and every of these functions is majorized by const$\langle \xi \rangle^s$ for any $s \in \mathbb{R}$. Therefore, taking $s < -n$, we obtain

$$|Au(x)| \le C \int \langle \xi \rangle^s\, d\xi \cdot \langle x \rangle^{-2N} = C' \langle x \rangle^{-2N}.$$

Because the number N can be chosen arbitrarily, the function Au rapidly decays as $|x| \to \infty$. A similar assertion for the derivatives $\partial^\alpha(Au)$ follows from the fact that $\partial^\alpha(Au)$ is a linear combination of integrals of the form (1.1.51):

$$\partial^\alpha(Au)(x) = (2\pi)^{-n} \sum_{\beta \le \alpha} i^{|\beta|} \binom{\alpha}{\beta} \int e^{ix\xi} \partial_x^{\alpha-\beta} \sigma_A(x,\xi) \xi^\beta \hat{u}(\xi)\, d\xi.$$

Verification of the continuity of the mapping $A : \mathcal{S} \to \mathcal{S}$ is left to the reader. (The convergence $u_j \to 0$ in $\mathcal{S}$ means that, for any $\alpha \in \mathbb{Z}_+^n$ and $s \in \mathbb{R}$, there exists a number sequence $C_j \to 0$ such that $|\partial^\alpha u_j(x)| \le C_j \langle x\rangle^s$). □

In what follows, we assume that all operators in Ψ_b^μ are defined on $\mathcal{S}(\mathbb{R}^n)$. It is convenient because every such operator sends $\mathcal{S}(\mathbb{R}^n)$ to $\mathcal{S}(\mathbb{R}^n)$, and the composition makes sense for any A and B in $\Psi_b(\mathbb{R}^n)$, which do not need to be proper. Let us show for $A \in \Psi_b^{\mu_1}(\mathbb{R}^n)$ and $B \in \Psi_b^{\mu_2}(\mathbb{R}^n)$ that $AB \in \Psi_b^{\mu_1+\mu_2}(\mathbb{R}^n)$. We use

$$ABu(x) = (2\pi)^{-n} \iint e^{i(x-y)\xi} \sigma_A(x,\xi)\tilde{\sigma}_B(y,\xi)u(y)\, dyd\xi,$$

where $\tilde{\sigma}_B(y,\xi) = \sigma_{{}^tB}(y,-\xi)$ (see Sect. 1.1.10). Since $\sigma_A \in S_b^{\mu_1}(\mathbb{R}^n)$ and $\tilde{\sigma}_B \in S_b^{\mu_2}(\mathbb{R}^n)$, the function $(x,y,\xi) \mapsto \sigma_A(x,\xi)\tilde{\sigma}_B(y,\xi)$ belongs to $S_b^{\mu_1+\mu_2}(\mathbb{R}^n,\mathbb{R}^n)$. Therefore, $AB \in \Psi_b^{\mu_1+\mu_2}(\mathbb{R}^n)$. The expansion

$$\sigma_{AB}(x,\xi) \sim \sum_\alpha \frac{1}{\alpha!}\partial_\xi^\alpha \sigma_A(x,\xi) D_x^\alpha \sigma_B(x,\xi) \tag{1.1.52}$$

can be verified in the same way as in Sect. 1.1.10.

1.1.13 The Boundedness of ΨDO in $L_2(\mathbb{R}^n)$

Let (u,v) and $\|u\|$ be the inner product and norm in $L_2(\mathbb{R}^n)$ and let $\mathcal{B}L_2(\mathbb{R}^n)$ be the set of all bounded operators in $L_2(\mathbb{R}^n)$. An operator $A \in \Psi(\mathbb{R}^n)$ is said to be bounded and $A \in \mathcal{B}L_2(\mathbb{R}^n)$ if A extends to a bounded operator in $L_2(\mathbb{R}^n)$. Every following condition is equivalent to $A \in \mathcal{B}L_2(\mathbb{R}^n)$:

(i) $\|Au\| \le C\|u\| \quad \forall u \in C_c^\infty(\mathbb{R}^n)$.
(ii) $|(Au,v)| \le C\|u\|\|v\| \quad \forall u,v \in C_c^\infty(\mathbb{R}^n)$.

For $A \in \Psi_b(\mathbb{R}^n)$, we have $(Au,v) = (u,A^*v)$ with any $u,v \in C_c^\infty(\mathbb{R}^n)$. By continuity, the equality extends to $u,v \in \mathcal{S}(\mathbb{R}^n)$. Since $\|Au\|^2 = (Au,Au) = (A^*Au,u) \le \|A^*Au\|\|u\|$, the inclusion $A^*A \in \mathcal{B}L_2(\mathbb{R}^n) \in \mathcal{B}L_2(\mathbb{R}^n)$ implies $A \in \mathcal{B}L_2(\mathbb{R}^n)$. Moreover, $A^*A \in \mathcal{B}L_2(\mathbb{R}^n)$ follows from $(A^*A)^*(A^*A) = (A^*A)^2 \in \mathcal{B}L_2(\mathbb{R}^n)$ and so on. Therefore, if the operator $(A^*A)^{2^k}$ is bounded for a certain $k \in \mathbb{Z}_+$, we obtain $A \in \mathcal{B}L_2(\mathbb{R}^n)$.

Lemma 1.1.27 *Let $G \in C(\mathbb{R}^{2n})$ and let A be an integral operator given for $u \in C_c^\infty(\mathbb{R}^n)$ by*

$$Au(x) = \int G(x, y)u(y)\, dy.$$

Under the conditions

$$C_1 := \sup_y \int |G(x, y)|\, dx < +\infty \quad \text{and} \quad C_2 := \sup_x \int |G(x, y)|\, dy < +\infty,$$

the operator A belongs to $\mathcal{B}L_2(\mathbb{R}^n)$ and $\|A\| \le \sqrt{C_1 C_2}$.

Proof Since $C_c^\infty(\mathbb{R}^n)$ is dense in $L_2(\mathbb{R}^n)$, it suffices to show that $|(Au, v)| \le \sqrt{C_1 C_2}\|u\|\|v\|$ for all $u, v \in C_c^\infty(\mathbb{R}^n)$. We have

$$|(Au, v)|^2 \le \Big(\iint |G(x, y)||u(y)||v(x)|\, dxdy\Big)^2$$
$$\le \iint |G(x, y)||u(y)|^2 dxdy \iint |G(x, y)||v(x)|^2 dxdy$$
$$\le C_1 \int |u(y)|^2 dy \cdot C_2 \int |v(x)|^2 dx = C_1 C_2 \|u\|^2 \|v\|^2.$$

□

Theorem 1.1.28 *Every operator $A \in \Psi_b^0(\mathbb{R}^n)$ is bounded in $L_2(\mathbb{R}^n)$.*

Proof We consider the following three cases.

(1) $A \in \Psi_b^\mu$, $\mu < -n$. Let $a \in S_b^\mu(\mathbb{R}^n, \mathbb{R}^n)$ be an amplitude satisfying $A = \operatorname{Op} a$ and

$$|a(x, y, \xi)| \le C \langle x - y\rangle^{-(n+1)} \langle \xi \rangle^\mu \tag{1.1.53}$$

(such an amplitude exists, see the proof of Proposition 1.1.25). Since $\mu < -n$, the function $\xi \mapsto a(x, y, \xi)$ is summable for fixed x and y. Therefore,

$$Au(x) = \int G(x, y)u(y)\, dy, \quad u \in C_c^\infty(\mathbb{R}^n),$$

where

$$G(x, y) = (2\pi)^{-n} \int e^{i(x-y)\xi} a(x, y, \xi)\, d\xi.$$

From (1.1.53), it follows that $|G(x, y)| \le C' \langle x - y\rangle^{-(n+1)}$. Then,

$$C_1 := \sup_y \int |G(x, y)|\, dx \le C' \int \langle z\rangle^{-(n+1)} dz < +\infty$$

$$C_2 := \sup_x \int |G(x, y)|\, dy < +\infty.$$

By Lemma 1.1.27, the operator A extends to a continuous operator in $L_2(\mathbb{R}^n)$.

(2) $A \in \Psi_b^{\mu}$, $\mu < 0$. For $k \in \mathbb{Z}_+$ the order of ΨDO $(A^*A)^{2^k}$ is equal to $\mu 2^{k+1}$. If k is chosen to satisfy $\mu 2^{k+1} < -n$, the operator $(A^*A)^{2^k}$ is bounded, and consequently, A is bounded as well.

(3) $A \in \Psi_b^0$. We set $b = (M^2 - |\sigma_A|^2)^{1/2}$ with a constant M such that $\sup |\sigma_A| < M$. It is easy to verify $b \in S_b^0$. Let $B = \operatorname{Op} b$. From the expansions (1.1.50) and (1.1.52), it follows that

$$\sigma_{A^*A} = |\sigma_A|^2 + \sigma_1, \quad \sigma_1 \in S_b^{-1},$$
$$\sigma_{B^*B} = M^2 - |\sigma_A|^2 + \sigma_2, \quad \sigma_2 \in S_b^{-1}.$$

Hence, $\sigma_{A^*A} + \sigma_{B^*B} = M^2 + \sigma$, where $\sigma \in S_b^{-1}$. Therefore,

$$A^*A + B^*B = M^2 I + C, \quad C \in \Psi_b^{-1};$$

here, $I : u \mapsto u$ is the identity operator. Since $M^2 I + C \in \mathcal{B}L_2(\mathbb{R}^n)$, we obtain $A^*A + B^*B \in \mathcal{B}L_2(\mathbb{R}^n)$. Then,

$$\|Au\|^2 \le \|Au\|^2 + \|Bu\|^2 = \big((A^*A + B^*B)u, u\big)$$
$$\le \|A^*A + B^*B\| \|u\|^2.$$

Thus, $A \in \mathcal{B}L_2(\mathbb{R}^n)$.

□

1.1.14 ΨDO in Sobolev Spaces

For $s \in \mathbb{R}$, the Sobolev space $H^s(\mathbb{R}^n)$ is the completion of the set $\mathcal{S}$ in the norm

$$\|u\|_s := \Big((2\pi)^{-n} \int |\hat{u}(\xi)|^2 \langle\xi\rangle^{2s} d\xi\Big)^{1/2} < +\infty.$$

For $s \in \mathbb{Z}_+$, the formula

$$\|u\|_s' = \Big(\int \sum_{|\alpha| \leq s} |\partial^\alpha u|^2 dx\Big)^{1/2}$$

defines an equivalent norm in $H^s(\mathbb{R}^n)$. We set $\langle D\rangle^t = F^{-1}\langle\xi\rangle^t F$ for $t \in \mathbb{R}$. It is easy to see that the operator $\langle D\rangle^t : H^s(\mathbb{R}^n) \to H^{s-t}(\mathbb{R}^n)$ is unitary for all s.

Theorem 1.1.29 *Every ΨDO $A \in \Psi_b^\mu(\mathbb{R}^n)$ extends to a bounded operator A_s : $H^s(\mathbb{R}^n) \to H^{s-\mu}(\mathbb{R}^n)$ $(\forall s)$.*

Proof Let us write A in the form $A = \langle D\rangle^{\mu-s}(\langle D\rangle^{s-\mu} A \langle D\rangle^{-s})\langle D\rangle^s$. The operators $\langle D\rangle^s : H^s(\mathbb{R}^n) \to H^0(\mathbb{R}^n)$ and $\langle D\rangle^{\mu-s} : H^0(\mathbb{R}^n) \to H^{s-\mu}(\mathbb{R}^n)$ are unitary, and the operator in parentheses belongs to $\Psi_b^0(\mathbb{R}^n)$ and consequently extends to a bounded operator $H^0(\mathbb{R}^n) \to H^0(\mathbb{R}^n)$. □

1.1.15 Elliptic Pseudodifferential Operators

Definition 1.1.30 A symbol $a \in S^\mu(\Omega)$ is called an elliptic symbol of order μ if, for any compact $K \subset \Omega$, there exist positive constants c and t such that

$$|a(x,\xi)| \geq c\,\langle\xi\rangle^\mu \quad \text{for} \quad x \in K\,,\ |\xi| \geq t. \tag{1.1.54}$$

We denote by $ES^\mu(\Omega)$ the set of all elliptic symbols of order μ. Unlike $S^\mu(\Omega)$, the classes $ES^\mu(\Omega)$ are pairwise non-intersecting. If $a \in ES^\mu(\Omega)$, the function $(x,\xi) \mapsto a(x,\xi)^{-1}$ is an elliptic symbol for large $|\xi|$. More precisely, let $K \subset \Omega$ be an arbitrary compact and let c and t be constants corresponding to K (see (1.1.54)). Then,

$$|\partial_x^\alpha \partial_\xi^\beta \big(a(x,\xi)^{-1}\big)| \leq C\langle\xi\rangle^{-\mu-|\beta|} \quad \text{for} \quad x \in K\,,\ |\xi| \geq t$$

$(\forall \alpha, \beta; C = C(\alpha, \beta, K))$. Indeed, it is easy to verify by induction that $\partial_x^\alpha \partial_\xi^\beta \big(a(x,\xi)^{-1}\big)$ is a linear combination of functions of the form $a^{-k}b$, $1 \leq k \leq |\alpha + \beta| + 1$, $b \in S^{(k-1)\mu-|\beta|}(\Omega)$. From the last inclusion, it follows that

$$|b(x,\xi)| \leq C\,\langle\xi\rangle^{(k-1)\mu-|\beta|}, \quad (x,\xi) \in K \times \mathbb{R}^n;$$

moreover, $|a(x,\xi)^{-1}| \leq c^{-1}\,\langle\xi\rangle^{-\mu}$ for $x \in K$, $|\xi| \geq t$. Therefore,

$$|a(x,\xi)^{-k} b(x,\xi)| \leq Cc^{-k}\,\langle\xi\rangle^{-\mu-|\beta|}$$

$$\Rightarrow\ |\partial_x^\alpha \partial_\xi^\beta \big(a(x,\xi)^{-1}\big)| \leq C'\langle\xi\rangle^{-\mu-|\beta|}, \quad x \in K,\ |\xi| \geq t,$$

what was needed. The ellipticity of $a(x,\xi)^{-1}$ is evident.

We denote by $E\Psi^{\mu}(\Omega)$ the class of all proper ΨDO $A \in \Psi^{\mu}(\Omega)$ satisfying $\sigma_A \in ES^{\mu}(\Omega)$.

Definition 1.1.31 An operator $A \in \Psi^{\mu}(\Omega)$ is called elliptic if

$$A = A_1 + A_2, \quad A_1 \in E\Psi^{\mu}(\Omega), \quad A_2 \in \Psi^{-\infty}(\Omega).$$

Definition 1.1.32 A proper ΨDO B is called a parametrix of a ΨDO A if

$$BA = I + R_1, \quad AB = I + R_2, \quad R_1, R_2 \in \Psi^{-\infty}(\Omega).$$

A parametrix, if it exists, is uniquely defined modulo $\Psi^{-\infty}(\Omega)$. More precisely, let B be a left parametrix and B' a right parametrix of A, that is, $BA = I + R_1$, $AB' = I + R_2$, and $R_1, R_2 \in \Psi^{-\infty}(\Omega)$). Then, $B - B' \in \Psi^{-\infty}(\Omega)$. Indeed,

$$BAB' = (I + R_1)B' = B' + R_1 B', \quad BAB' = B(I + R_2) = B + BR_2.$$

Therefore, $B - B' = R_1 B' - BR_2 \in \Psi^{-\infty}(\Omega)$.

Theorem 1.1.33 *For an elliptic ΨDO $A \in \Psi^{\mu}(\Omega)$, there exists a parametrix $B \in E\Psi^{-\mu}(\Omega)$.*

Lemma 1.1.34

(1) *For any symbol $\sigma \in S^{\mu}(\Omega)$, there exists a proper ΨDO $B \in \Psi^{\mu}(\Omega)$ such that $\sigma_B - \sigma \in S^{-\infty}(\Omega)$.*
(2) *For any asymptotic series $\sum_{j=0}^{\infty} \sigma_j$, $\sigma_j \in S^{\mu_j}(\Omega)$, there exists a proper ΨDO $B \in \Psi^{\overline{\mu}}(\Omega)$, $\overline{\mu} = \max \mu_j$, such that $\sigma_B \sim \sum_{j=0}^{\infty} \sigma_j$.*

Proof of Lemma

(1) $B = \mathrm{Op}\,(\chi a)$, where $\chi \in C^{\infty}(\Omega \times \Omega)$ is a function with proper support equal to 1 in a neighborhood of the diagonal.
(2) There exists a symbol $\sigma \in S^{\overline{\mu}}(\Omega)$ satisfying $\sigma \sim \sum_j \sigma_j$. It remains to use assertion (1).

□

Proof of Theorem We can assume that $A \in E\Psi^{m}(\Omega)$. Let $\sigma_A \in ES^{m}(\Omega)$ be a symbol of A, $\{\psi_j\}_{j=1}^{\infty} \subset C_c^{\infty}(\Omega)$ a partition of unity in Ω, and c_j, t_j constants corresponding to the compact $\operatorname{supp} \psi_j$ (see (1.1.54)). We choose $\chi \in C^{\infty}(\mathbb{R}^n)$ such that $\chi(\xi) = 0$ for $|\xi| \le 1$ and $\chi(\xi) = 1$ for $|\xi| \ge 2$ and set $\tilde{\sigma}(x, \xi) = \sum_j \psi_j(x)\chi(\xi/t_j)\sigma_A(x, \xi)^{-1}$. Then, $\tilde{\sigma} \in S^{-m}(\Omega)$.

Let $B_0 \in \Psi^{-m}(\Omega)$ be a proper ΨDO such that $\sigma_{B_0} - \tilde{\sigma} \in S^{-\infty}(\Omega)$. Since $\tilde{\sigma}\sigma_A - 1 \in S^{-\infty}(\Omega)$, we have $\sigma_{B_0}\sigma_A - 1 \in S^{-\infty}(\Omega)$. Taking into account $\sigma_{B_0 A} - \sigma_{B_0}\sigma_A \in S^{-1}(\Omega)$ and $\sigma_{AB_0} - \sigma_A\sigma_{B_0} \in S^{-1}(\Omega)$, we obtain

$$B_0 A = I + R_0\,, \quad AB_0 = I + R_0', \quad R_0, R_0' \in \Psi^{-1}(\Omega).$$

We denote by σ_j the symbol of R_0^j, $j \geq 0$. Because $R_0^j \in \Psi^{-j}(\Omega)$, we conclude that $\sigma_j \in S^{-j}(\Omega)$ and the series $\sum_j (-1)^j \sigma_j$ is asymptotic. Now, we consider a proper ΨDO C of order zero such that $\sigma_C \sim \sum_j (-1)^j \sigma_j$. Then, $CB_0 A = I + R_1$ and $R_1 \in \Psi^{-\infty}(\Omega)$. Indeed, for any $N \in \mathbb{N}$ there hold the equalities

$$C = \operatorname{Op}\Big(\sum_{j=0}^{N-1} (-1)^j \sigma_j\Big) + \operatorname{Op}\sigma^{(N)} = \sum_{j=0}^{N-1} (-1)^j R_0^j + \operatorname{Op}\sigma^{(N)},$$

where $\sigma^{(N)} \in S^{-N}(\Omega)$. This implies

$$CB_0 A = \Big(\sum_{j=0}^{N-1} (-1)^j R_0^j + \operatorname{Op}\sigma^{(N)}\Big)(I + R_0) = I - (-1)^{N-1} R_0^N + \operatorname{Op}\sigma^{(N)} \cdot (I + R_0).$$

Therefore, $CB_0 A - I \in \Psi^{-N}(\Omega)$ for all N. Thus, the operator $B := CB_0$ is a left parametrix of A. A right parametrix is defined by $B' = B_0 C'$, where $\sigma_{C'} \sim \sum_j (-1)^j \sigma_j'$ and σ_j' is a symbol of $(R_0')^j$. Since $B - B' \in \Psi^{-\infty}(\Omega)$, the operator B is a right parametrix as well. □

1.2 Meromorphic Pseudodifferential Operators

1.2.1 Integral Transforms on a Sphere

Let $\varphi = (\varphi_1, \ldots, \varphi_n)$ and $\omega = (\omega_1, \ldots, \omega_n)$ be unit vectors in $\mathbb{R}^n$, let $\varphi\omega = \varphi_1\omega_1 + \cdots + \varphi_n\omega_n$, and let S^{n-1} be the unit $(n-1)$-dimensional sphere with center at the coordinate origin. For $\operatorname{Re}\mu > -1$ and $u \in C^\infty(S^{n-1})$, we introduce the operators

$$(J_\mu^\pm u)(\varphi) = \int_{S^{n-1}} (\pm\varphi\omega + i0)^\mu u(\omega)\, d\omega, \tag{1.2.1}$$

where $(\varphi\omega + i0)^\mu = (\varphi\omega)_+^\mu + e^{i\mu\pi}(\varphi\omega)_-^\mu$, $(-\varphi\omega + i0)^\mu = e^{i\mu\pi}(\varphi\omega)_+^\mu + (\varphi\omega)_-^\mu$, and, for example, $(\varphi\omega)_+ = \varphi\omega$ for $\varphi\omega \geq 0$ and $(\varphi\omega)_+ = 0$ for $\varphi\omega < 0$.

Proposition 1.2.1 ([21], Prop. 1.1.2) *The maps $J^{\pm}_{\mu} : C^{\infty}(S^{n-1}) \to C^{\infty}(S^{n-1})$ are continuous. The operator-valued functions $\mu \mapsto J^{\pm}_{\mu}$ admit analytic extension to the complex μ-plain.*

For all complex λ, with the exception of $\lambda = i(k + n/2)$, where $k = 0, 1, \dots$, and for $u \in C^{\infty}(S^{n-1})$, we set

$$(E(\lambda)u)(\varphi) = (2\pi)^{-n/2} e^{i\frac{\pi}{2}(i\lambda+n/2)} \Gamma(i\lambda + n/2) \times \int_{S^{n-1}} (-\varphi\omega + i0)^{-i\lambda-n/2} u(\omega)\, d\omega. \tag{1.2.2}$$

Let us recall that the function $\mu \mapsto \Gamma(\mu)$ is meromorphic throughout the μ-plane. The poles of Γ-function are located at the points $\mu = 0, -1, \dots$, are simple, and

$$\operatorname{res}\Gamma(\mu)|_{\mu=-k} = (-1)^k/k!. \tag{1.2.3}$$

According to Proposition 1.2.1, the operator-valued function $\lambda \mapsto E(\lambda) : C^{\infty}(S^{n-1}) \to C^{\infty}(S^{n-1})$ is analytic throughout except for the simple poles indicated above. The residue at $\lambda = i(k + n/2)$ is a finite-dimensional operator; taking into account (1.2.3), we obtain

$$\operatorname{res}E(\lambda)u|_{\lambda=i(k+n/2)} = \frac{(-i)^{k+1}}{(2\pi)^{n/2}} \sum_{|\gamma|=k} \frac{1}{\gamma!} x^{\gamma} \int_{S^{n-1}} y^{\gamma} u(y)\, dy. \tag{1.2.4}$$

We introduce the Mellin transform for the functions in $C_c^{\infty}(\mathbb{R}^n \setminus 0)$ by

$$(M_{r\to\lambda}u)(\lambda, \varphi) \equiv \tilde{u}(\lambda, \varphi) = (2\pi)^{-1/2} \int_0^{+\infty} r^{-i\lambda-1} u(r, \varphi)\, dr, \ \ \lambda \in \mathbb{C}, \tag{1.2.5}$$

where $r = |x|$ and $\varphi = x/|x|$ with $x \in \mathbb{R}^n \setminus 0$. The inversion formula

$$u(r, \varphi) = (M^{-1}_{\lambda\to r}\tilde{u})(r, \varphi) = (2\pi)^{-1/2} \int_{\operatorname{Im}\lambda=\tau} r^{i\lambda} \tilde{u}(\lambda, \varphi)\, d\lambda \tag{1.2.6}$$

and the Parseval equality

$$\int_{\operatorname{Im}\lambda=\tau} |\tilde{u}(\lambda, \varphi)|^2\, d\lambda = \int_0^{+\infty} r^{2\beta} |u(r, \varphi)|^2\, dr, \ \ \tau = \beta + 1/2 \tag{1.2.7}$$

hold. (These formulas can be obtained from the corresponding properties of the one-dimensional Fourier transform by the change of variable $r = e^t$.)

Proposition 1.2.2 ([21], Prop. 1.2.1) *The equality*

$$F_{x\to\xi}u = M^{-1}_{(in/2-\lambda)\to\rho} E_{\varphi\to\psi}(\lambda) M_{r\to(\lambda+in/2)}u$$

holds for $u \in C_c^\infty(\mathbb{R}^n\backslash 0)$, *where* $\operatorname{Im}\lambda = n/2$, $r = |x|$, $\rho = |\xi|$, $\varphi = x/|x|$, *and* $\psi = \xi/|\xi|$; *the Fourier transform is written in the form*

$$(Fu)(\xi) = (2\pi)^{-n/2}\int e^{-i\xi x}u(x)\,dx.$$

For all $\lambda \in \mathbb{C}$, except for $\lambda = -i(k+n/2)$ with $k = 0, 1, \ldots$, we introduce the operator

$$\begin{aligned}(E(\lambda)^{-1}v)(\varphi) = {} & (2\pi)^{-n/2}e^{i\frac{\pi}{2}(n/2-i\lambda)}\Gamma(n/2-i\lambda) \\ & \times \int_{S^{n-1}} (\varphi\omega + i0)^{i\lambda-n/2}v(\omega)\,d\omega,\end{aligned} \tag{1.2.8}$$

assuming $v \in C^\infty(S^{n-1})$. According to Proposition 1.2.1, the function $\lambda \mapsto E(\lambda)^{-1}$: $C^\infty(S^{n-1}) \to C^\infty(S^{n-1})$ is analytic throughout, with the exception of the mentioned points, which are simple poles. The residue at $\lambda = -i(k+n/2)$ is a finite-dimensional operator

$$\operatorname{res} E(\lambda)^{-1}v|_{\lambda=-i(k+n/2)} = \frac{i^{k+1}}{(2\pi)^{n/2}}\sum_{|\gamma|=k}\frac{1}{\gamma!}x^\gamma\int_{S^{n-1}} y^\gamma v(y)\,dy. \tag{1.2.9}$$

We denote by $E(\lambda)^*$ the operator adjoint to $E(\lambda)$ with respect to the inner product in $L_2(S^{n-1})$.

Proposition 1.2.3 ([21], Props. 1.4.1 and 1.4.4) *The following assertions are valid with* $k = 0, 1, \ldots$*:*

(1) $E(\lambda)$ *and* $E(\lambda)^{-1}$ *are inverse to each other for* $\lambda \neq \pm i(k+n/2)$.
(2) $E(\lambda)^* = E(\bar{\lambda})^{-1}$ *for* $\lambda \neq i(k+n/2)$.

We will consider the operators $E(\lambda)^{\pm 1}$ in the spaces $H^s(\lambda, S^{n-1})$ of (generalized) functions on the sphere with norm depending on $\lambda \in \mathbb{C}$. We first introduce the spaces.

Let $\mathfrak{M}$ be a smooth compact manifold without boundary and let $\{U, \chi\}$ be an atlas on $\mathfrak{M}$, that is, $\{U\}$ is a finite open covering of $\mathfrak{M}$, and $\chi : U \to \mathbb{R}^n$ are coordinate maps. We denote by $\{\zeta\}$ a partition of unity subject to the covering. The space $H^s(\lambda, \mathfrak{M})$ is the completion of $C^\infty(\mathfrak{M})$ with respect to the norm

$$\|u; H^s(\lambda, \mathfrak{M})\| = \left\{\sum_U \int |F(\zeta_\chi u)(\xi)|^2(1+|\xi|^2+|\lambda|^2)^s\,d\xi\right\}^{1/2}, \tag{1.2.10}$$

where $\zeta_\chi u = \zeta u \circ \chi^{-1}$ on $\chi(U)$ and $\zeta_\chi u = 0$ outside $\chi(U)$. For any fixed number λ norm (1.2.10) is equivalent to the norm in the Sobolev space $H^s(\mathfrak{M})$. Another partition of unity and another (equivalent) atlas lead to an equivalent norm in $H^s(\lambda, \mathfrak{M})$.

Proposition 1.2.4 ([21], Prop. 1.5.5 and Cor. 1.5.6) *If $\lambda \neq i(k + n/2)$ (or $\lambda \neq -i(k + n/2)$), where $k = 0, 1, \dots,$ the map $E(\lambda) : H^s(\lambda, S^{n-1}) \to H^{s+\mathrm{Im}\,\lambda}(\lambda, S^{n-1})$ (resp. $E(\lambda)^{-1} : H^s(\lambda, S^{n-1}) \to H^{s-\mathrm{Im}\,\lambda}(\lambda, S^{n-1})$) is continuous. On every closed set $\mathcal{F}$ lying in the strip $\{\lambda \in \mathbb{C} : |\mathrm{Im}\,\lambda| < h\}$ and not containing the points $\lambda = i(k + n/2)$ (resp. $\lambda = -i(k + n/2)$) the estimate*

$$\|E(\lambda); H^s(\lambda, S^{n-1}) \to H^{s+\mathrm{Im}\,\lambda}(\lambda, S^{n-1})\| \le c(\mathcal{F}) \tag{1.2.11}$$

(resp.

$$\|E(\lambda)^{-1}; H^s(\lambda, S^{n-1}) \to H^{s-\mathrm{Im}\lambda}(\lambda, S^{n-1})\| \le c(\mathcal{F})) \tag{1.2.12}$$

holds. The operator $E(\lambda) : L_2(S^{n-1}) \to L_2(S^{n-1})$ is unitary for $\mathrm{Im}\,\lambda = 0$.

1.2.2 Canonical Meromorphic Pseudodifferential Operators

To motivate the definition of a meromorphic ΨDO, we consider the convolution operator $A = F^{-1}_{\xi \to x} \Phi(\xi) F_{y \to \xi}$ taking Φ as a homogeneous function of degree a, $\mathrm{Re}\, a > -n/2$, smooth on S^{n-1}. In view of Proposition 1.2.2,

$$\begin{aligned} \rho^a \Phi(\omega)(Fu)(\rho, \omega) = (2\pi)^{-1/2} \int_{\mathrm{Im}\,\lambda = \mathrm{Re}\, a} & \rho^{i(in/2-\lambda)} \Phi(\omega) \\ & \times E(\lambda - ia)\tilde{u}(\lambda - ia + in/2, \cdot)\, d\lambda. \end{aligned} \tag{1.2.13}$$

The integrand is holomorphic in the strip between the lines $\mathrm{Im}\,\lambda = 0$ and $\mathrm{Im}\,\lambda = \mathrm{Re}\, a$. We suppose that $u \in C_c^\infty(\mathbb{R}^n \setminus 0)$, then the function $\lambda \mapsto \tilde{u}(\lambda + in/2, \cdot)$ rapidly decreases as $\lambda \to \infty$ in any strip $|\mathrm{Im}\,\lambda| < h$. Together with (1.2.11), this allows us to change the integration line in (1.2.13) for $\mathrm{Im}\lambda = 0$. Now, applying the inverse Fourier transform F^{-1} and taking into account Propositions 1.2.2 and 1.2.3, we obtain

$$\begin{aligned} Au(r, \varphi) &= (2\pi)^{-1/2} \int_{-\infty}^{+\infty} r^{i(in/2+\lambda)} E_{\omega \to \varphi}(\lambda)^{-1} \Phi(\omega) E_{\psi \to \omega}(\lambda - ia) \\ &\qquad \times \tilde{u}(\lambda - ia + in/2, \psi)\, d\lambda \\ &= (2\pi)^{-1/2} \int_{\mathrm{Im}\,\lambda = -\mathrm{Re}\, a} r^{i(in/2+\lambda+ia)} \mathfrak{A}_{\psi \to \varphi}(\lambda) \tilde{u}(\lambda + in/2, \psi)\, d\lambda, \end{aligned} \tag{1.2.14}$$

where $\mathfrak{A}_{\psi \to \varphi}(\lambda) = E_{\omega \to \varphi}(\lambda + ia)^{-1} \Phi(\omega) E_{\psi \to \omega}(\lambda)$.

Definition 1.2.5 Let $a \in \mathbb{C}$ and $\Phi \in C^\infty(S^{n-1} \times S^{n-1})$. The operator

$$\mathfrak{A}_{\psi\to\varphi}(\lambda) = E_{\omega\to\varphi}(\lambda + ia)^{-1}\Phi(\varphi, \omega)E_{\psi\to\omega}(\lambda), \tag{1.2.15}$$

where $\lambda \neq i(k + n/2)$ and $\lambda \neq -i(k + a + n/2)$, $k = 0, 1, \dots$, is called a canonical meromorphic ΨDO of order a.

Proposition 1.2.6 ([21], Prop. 3.1.1) *Let $\mathfrak{A}$ be an operator of the form (1.2.15). Then, the estimate*

$$\|\mathfrak{A}(\lambda);\, H^s(\lambda, S^{n-1}) \to H^{s-\mathrm{Re}\,a}(\lambda, S^{n-1})\| \le c(s, \mathcal{F})$$

holds on every closed set $\mathcal{F}$ located in the strip $\{\lambda \in \mathbb{C} : |\mathrm{Im}\,\lambda| < h\}$ and not containing poles of the meromorphic operator-function $\lambda \mapsto \mathfrak{A}(\lambda)$.

If the function Φ in (1.2.15) is independent of φ, this assertion follows from Proposition 1.2.4. In the general case, we expand Φ in a series

$$\Phi(\varphi, \omega) = \sum a_{mk}(\varphi)Y_{mk}(\omega),$$

where Y_{mk} is a spherical function of order m. Since the coefficients a_{mk} rapidly decrease as $m \to \infty$, the series

$$\sum a_{mk}E_{\omega\to\varphi}(\lambda + ia)^{-1}Y_{mk}(\omega)E_{\psi\to\omega}(\lambda)$$

converges in the operator norm $\|\cdot\,;\, H^s(\lambda, S^{n-1}) \to H^{s-\mathrm{Re}\,a}(\lambda, S^{n-1})\|$ uniformly on the set $\mathcal{F}$.

Proposition 1.2.7 *Let $\mathfrak{A}$ be an operator of the form* (1.2.15)*, where* $\Phi(\varphi, \omega) = 0$ *in a neighborhood of the manifold* $\{(\varphi, \omega) : \varphi\omega = 0\}$*. Then, for any $q \in \mathbb{R}$ and every fixed $\lambda \in \mathbb{C}$ such that $\lambda \neq i(k + n/2)$, $\lambda \neq -i(k + a + n/2)$ for $k = 0, 1, \dots$ the map*

$$\mathfrak{A}(\lambda) : H^s(S^{n-1}) \to H^{s+q}(S^{n-1})$$

is continuous.

To verify this assertion, recall (1.2.11) and take into account that the kernel of the integral operator $T(\lambda)v(\varphi) := E_{\omega\to\varphi}(\lambda + ia)^{-1}\Phi(\varphi, \omega)v(\omega)$ belongs to $C^\infty(S^{n-1} \times S^{n-1})$.

We now describe the asymptotic behavior of $e^{-i\mu g(\varphi)}\mathfrak{A}_{\psi\to\varphi}(\lambda)e^{i\mu g(\psi)}u(\psi)$ as $|\lambda|^2 + \mu^2 \to \infty$, where the operator $\mathfrak{A}$ is defined by (1.2.15), u and g are in $C^\infty(S^{n-1})$, and $|\mathrm{Im}\,\lambda| \le h$, where $\mu \in \mathbb{R}$ and g is a real function. In particular, the following theorem justifies the name "pseudodifferential" for operators of the form (1.2.15) (see the "invariant" definition of ΨDO on a manifold given by Hörmander in [14]). We assume for the function Φ in (1.2.15) that the maps $\varphi \mapsto \Phi(\varphi, \omega)$ and $\omega \mapsto \Phi(\varphi, \omega)$ are extended to $\mathbb{R}^n \setminus 0$ as homogeneous functions of degree a, and u and g are homogeneous functions of degree zero.

Theorem 1.2.8 ([21], Prop. 3.5.1) *For any number $P \geq 0$, there exist non-negative numbers N and Q such that*

$$\Big| e^{-i\mu g(\varphi)} \mathfrak{A}_{\psi\to\varphi}(\lambda) e^{i\mu g(\psi)} u(\psi) - \sum_{|\alpha|\leq N} \frac{1}{\alpha!} \Phi^{(\alpha)}(\varphi, \mu\nabla g(\varphi) + \sigma\varphi)$$
$$\times \left(D_y^\alpha [u(\psi)\rho^{i(\lambda+in/2)} \exp\{i\mu(g(\psi) - g(\varphi)) - i(y - \varphi, \mu\nabla g(\varphi) + \sigma\varphi)\}]\right)\Big|_{\rho=1,\,\varphi=\psi}\Big|$$
$$\leq c(\mu^2 + \sigma^2)^{-P/2} \|u; C^Q(S^{n-1})\|, \tag{1.2.16}$$

where $\sigma = \operatorname{Re}\lambda$, $\rho = |y|$, $\psi = y/|y|$, *and* $\Phi^{(\alpha)}(\varphi, \omega) = D_\omega^\alpha \Phi(\varphi, \omega)$.

According to the mentioned invariant definition of ΨDO in [14], to prove that $\mathfrak{A}(\lambda)$ is a ΨDO, it suffices to verify that

$$\exp(-i\mu g(\varphi)) \mathfrak{A}_{\psi\to\varphi}(\lambda) \exp(i\mu g(\psi)) u(\psi)$$

expands in an asymptotic series as $\mu \to \infty$ for fixed λ. Formula (1.2.16) provides such an expansion. Let us show, in particular, how to obtain from (1.2.16) the principal symbol for the operator $\mathfrak{A}(\lambda)$ of order zero. The principal term of the asymptotic expansion is of the form $\Phi(\varphi, \mu\nabla g(\varphi) + \sigma\varphi)u(\varphi)$. Here $\sigma = \operatorname{Re}\lambda$, the functions g and Φ are extended as order zero homogeneous to $\mathbb{R}^n \setminus 0$, and $\nabla g(\varphi)$ is calculated in Cartesian coordinates. The vector $\nabla g(\varphi)$ is tangent to the sphere at φ; hence, φ and $\nabla g(\varphi)$ are orthogonal. A value of the sought principal symbol for the covariant vector tangent to the sphere at φ and directed along $\nabla g(\varphi)$ is equal to

$$\lim_{\mu\to+\infty} \Phi(\varphi, \mu\nabla g(\varphi) + \sigma\varphi) = \Phi(\varphi, \nabla g(\varphi)).$$

Choosing various functions g, we calculate the principal symbol for all pairs (φ, ω), where φ is a point of the sphere, and ω is a cotangent vector at φ. Thus,

$$\mathfrak{A}(\lambda) = E_{\omega\to\varphi}(\lambda)^{-1} \Phi(\varphi, \omega) E_{\psi\to\omega}(\lambda)$$

is a ΨDO of order zero on the sphere S^{n-1} and its principal symbol is the function $V \ni (\varphi, \omega) \mapsto \Phi(\varphi, \omega)$ on the set V of pairs (φ, ω) of orthonormal vectors. If $\Phi(\varphi, \omega) \neq 0$ throughout V, the operator $\mathfrak{A}(\lambda)$ is elliptic and, consequently, Fredholm in $L_2(S^{n-1})$.

In the representation theory of C^*-algebras of pseudodifferential operators, operator-valued functions of the form

$$\mathbb{R} \ni \lambda \mapsto \mathfrak{A}(\lambda) = E_{\omega\to\varphi}(\lambda)^{-1}\Phi(\varphi, \omega)E_{\psi\to\omega}(\lambda) : L_2(S^{n-1}) \to L_2(S^{n-1})$$

provide a series of irreducible representations (see Chap. 2). There arise questions about the existence of the inverse operator $\mathfrak{A}(\lambda)^{-1}$ for all $\lambda \in \mathbb{R}$ and about the inequality

$$\|\mathfrak{A}(\lambda)^{-1}; L_2(S^{n-1}) \to L_2(S^{n-1})\| \leq C$$

with constant C independent of λ. In this connection, the condition

$$\Phi(\varphi, \omega) \neq 0 \ \forall\ (\varphi, \omega) \in V$$

is not sufficient, and the requirement

$$\Phi(\varphi, \omega) \neq 0 \ \forall\ (\varphi, \omega) \in S^{n-1} \times S^{n-1},$$

is necessary, which can be interpreted as "ellipticity with regard to parameter."

1.2.3 The Kernel of a Canonical Pseudodifferential Operator

Let us consider operator (1.2.15). We first assume that $-n/2 < \operatorname{Re} a < 0$, set $f = E(ia + in/2)^{-1}\Phi$, and introduce the homogeneous function G of order $-n - a$,

$$G(x) = r^{-n-a} f(\varphi),\ \ r = |x|,\ \ \varphi = x/|x|.$$

For $u \in C_c^\infty(\mathbb{R}^n \backslash)$ we define the operator

$$(G \star u) = \int G(x - y)u(y)\, dy.$$

Since

$$(FG)(\xi) = |\xi|^a E_{\varphi\to\theta}(ia + in/2) f(\varphi) = |\xi|^a \Phi(\theta)$$

with $\theta = \xi/|\xi|$, we have $G \star u = F^{-1}|\xi|^a \Phi(\theta) F u$. In view of (1.2.14),

$$(G \star u)(x) = (2\pi)^{-1/2} \int_{\operatorname{Im}\lambda = -\operatorname{Re} a} r^{i(in/2+\lambda+ia)} \mathfrak{A}_{\psi\to\varphi}(\lambda) \tilde{u}(\lambda + in/2, \psi)\, d\lambda, \tag{1.2.17}$$

where $\mathfrak{A}_{\psi\to\varphi}(\lambda) = E_{\omega\to\varphi}(\lambda + ia)^{-1}\Phi(\omega)E_{\psi\to\omega}(\lambda)$. We extend f as an order zero homogeneous function to $\mathbb{R}^n \setminus 0$ and apply to $G \star u$ the Mellin transform $M_{r\to\lambda+ia+in/2}$ with $\operatorname{Im}\lambda = 0$. Setting $x = r\varphi$, $y = \rho\psi$, we have

$$\begin{aligned}
&(2\pi)^{-1/2} \int_0^{+\infty} r^{-i(\lambda+ia+in/2)-1} (G \star u)(x)\, dr \\
&= (2\pi)^{-1/2} \int_0^{+\infty} r^{-i(\lambda+ia+in/2)-1}\, dr \int_{S^{n-1}} d\psi \int_0^{+\infty} G(r\varphi - \rho\psi) u(\rho\psi)\rho^{n-1}\, d\rho \\
&= \int_{S^{n-1}} d\psi \int_0^{+\infty} t^{-i(\lambda+ia+in/2)-1} G(t\varphi - \psi)\tilde{u}(\lambda + in/2, \psi)\, dt.
\end{aligned} \tag{1.2.18}$$

From (1.2.17) and (1.2.18), it follows that

$$\mathfrak{A}_{\psi\to\varphi}(\lambda)\tilde{u}(\lambda + in/2, \psi) = \int G(\varphi, \psi; \lambda)\tilde{u}(\lambda + in/2, \psi)\, d\psi,$$

where

$$G(\varphi, \psi; \lambda) = \int_0^{+\infty} r^{-i(\lambda+ia+in/2)-1} G(t\varphi - \psi)\, dt. \tag{1.2.19}$$

Thus,

$$\mathfrak{A}_{\psi\to\varphi}(\lambda) v(\psi) = \int G(\varphi, \psi; \lambda) v(\psi)\, d\psi \tag{1.2.20}$$

for $v \in C^\infty(S^{n-1})$.

We obtained representation (1.2.20) for the real λ under the condition $-n/2 < \operatorname{Re} a < 0$. Let us define the representation for all complex a and λ by analytic extension throughout with the exception of the poles. The operator $E(ia + in/2)^{-1}$ is not defined for $a = -n - l$, $l = 0, 1, \ldots$ and is not an isomorphism for $a = 0, 1, \ldots$ (it annihilates a finite-dimensional space, see [21], Prop. 1.4.3).

Proposition 1.2.9 *Let $a \neq 0, 1, \ldots$ and $a \neq -n - l$, where $l = 0, 1, \ldots$. Let also $\lambda \neq i(n/2 + k)$, $\lambda \neq -i(n/2 + k + a)$, $k = 0, 1, \ldots$. Then, formula* (1.2.20) *holds for the operator $\mathfrak{A}_{\psi\to\varphi}(\lambda) = E_{\omega\to\varphi}(\lambda + ia)^{-1}\Phi(\omega)E_{\psi\to\omega}(\lambda)$. The kernel $G(\varphi, \psi; \lambda)$ is* *defined by* (1.2.19)*, where the integral is understood in the sense of analytic extension in* λ*. (Explicit formulas for the analytic extension are presented in [21], § 3.1.)*

We now consider the operator $\mathfrak{A}(\lambda)$ for the values of a excluded in Proposition 1.2.9. Let a be a non-negative integer. The function Φ can be represented as a sum $\Phi_0 + \Phi_1$, where Φ_0 is subject to the conditions

$$\int_{S^{n-1}} \Phi_0(\theta)\theta^\gamma \, d\theta = 0$$

for all multi-indices γ such that $|\gamma| = a$, and Φ_1 is of the form

$$\Phi_1(\theta) = \sum_{j=0}^{[a/2]} h_{a-2j}(\theta),$$

h_j being a harmonic polynomial of degree j. For $\lambda \neq i(n/2 + k)$, $\lambda \neq -i(n/2 + k + a)$, we have

$$\mathfrak{A}(\lambda) = E(\lambda + ia)^{-1}\Phi_0(\theta)E(\lambda) + E(\lambda + ia)^{-1}\Phi_1(\theta)E(\lambda). \tag{1.2.21}$$

For the first term on the right in (1.2.21), a formula of the form (1.2.20) holds with $G(x) = r^{-n-a}E_{\omega\to\varphi}(ia+in/2)^{-1}\Phi(\omega)$ (note that $E(ia+in/2)^{-1}\Phi = E(ia+in/2)^{-1}\Phi_0$) and the second term is a differential operator of order a on the sphere S^{n-1} in which the parameter λ enters as in a polynomial of degree a.

What has been said in this section on the operator $\mathfrak{A}_{\psi\to\varphi}(\lambda) = E_{\omega\to\varphi}(\lambda + ia)^{-1}\Phi(\omega) \times E_{\psi\to\omega}(\lambda)$ can be generalized for operator (1.2.15). In particular, for $a \neq -n - a$, the function G is defined by $G(\varphi, x) = r^{-n-a}E_{\omega\to\varphi}(ia + in/2)^{-1}\Phi(\varphi, \omega)$, and the kernel $G(\varphi, \psi; \lambda)$ is defined by

$$G(\varphi, \psi; \lambda) = \int_0^{+\infty} r^{-i(\lambda+ia+in/2)-1}G(\varphi, t\varphi - \psi)\, dt.$$

Using a representation of the form (1.2.20), one can verify

Proposition 1.2.10 ([21], Prop. 3.1.6) *Let* $\eta, \zeta \in C^\infty(S^{n-1})$ *and* $\operatorname{supp}\eta \cap \operatorname{supp}\zeta = \emptyset$*. Then, for operator* (1.2.15) *the estimate*

$$\|\zeta\mathfrak{A}(\lambda)\eta;\, H^s(\lambda, S^{n-1}) \to H^{s+p}(\lambda, S^{n-1})\| \le c(\mathcal{F}, p, s) \tag{1.2.22}$$

holds, where $\mathcal{F}$ *is an arbitrary closed set located in a strip of the form* $\{\lambda \in \mathbb{C} : |\operatorname{Im}\lambda| < h\}$ *and not containing poles of the function* $\lambda \mapsto \mathfrak{A}(\lambda)$*;* p *is any real number.*

1.2.4 Operations on Canonical Meromorphic Pseudodifferential Operators

Here, we describe composition and the operations of taking the adjoint, differentiation, and shift with respect to the parameter for canonical meromorphic pseudodifferential operators.

In what follows, we consider meromorphic operator-functions that have in every strip $\{\lambda \in \mathbb{C} : |\operatorname{Im}\lambda| < h < \infty\}$ at most finitely many poles. Any closed set located in a strip $\{\lambda \in \mathbb{C} : |\operatorname{Im}\lambda| < h\}$ and not containing poles of the operators considered will be called admissible.

Let $\sigma \in C^\infty(S^{n-1})$ and $\Phi \in C^\infty(S^{n-1} \times S^{n-1})$. We extend σ to $\mathbb{R}^n \setminus 0$ as a homogeneous function of degree ν and extend the function $\omega \mapsto \Phi(\varphi, \omega)$ as a homogeneous function of degree a; the numbers ν and a are taken arbitrary complex.

Proposition 1.2.11 ([21], Prop. 3.2.1) *For an operator $\mathfrak{A}$ of the form* (1.2.15), *the formula*

$$\mathfrak{A}(\lambda)\sigma = \sum_{|\gamma|=0}^{N} \frac{1}{\gamma!}\partial^\gamma\sigma(\varphi)E(\lambda + i\nu + i(a - |\gamma|))^{-1}D_\omega^\gamma\Phi(\varphi, \omega)E(\lambda + i\nu) + \mathfrak{R}_N(\lambda) \tag{1.2.23}$$

holds with any $N \in \mathbb{Z}_+$; the operator $\mathfrak{R}_N$ is subject to the inequality

$$\|\mathfrak{R}_N(\lambda); H^s(\lambda; S^{n-1}) \to H^{s+N+1-\operatorname{Re} a}(\lambda; S^{n-1})\| \le c(\mathcal{F}, N) \tag{1.2.24}$$

on every admissible $\mathcal{F}$.

(Here and in what follows, the letter $\mathfrak{R}$ denotes various operators that are remainders in "asymptotic" formulas.) Setting $\nu = i\mu$, $\sigma(x) = |x|^{i\mu}$, and $\varphi = x/|x|$ in (1.2.23) and then changing $\lambda - \mu$ for λ, we obtain the following assertion.

Proposition 1.2.12 *Let $\mu \in \mathbb{C}$ and let $\mathfrak{A}$ be an operator of the form* (1.2.15). *Then,*

$$\mathfrak{A}(\lambda + \mu) - \mathfrak{A}(\lambda) = \sum_{|\gamma|=1}^{N} \frac{1}{\gamma!}(\partial_x^\gamma |x|^{i\mu})\big|_{|x|=1} E_{\omega\to\varphi}(\lambda + i(a - |\gamma|))^{-1} \times D_\omega^\gamma\Phi(\varphi, \omega)E_{\psi\to\omega}(\lambda) + \mathfrak{R}_N(\lambda, \mu) \tag{1.2.25}$$

and

$$\|\mathfrak{R}_N(\lambda, \mu); H^s(\lambda; S^{n-1}) \to H^{s+N+1-\operatorname{Re} a}(\lambda; S^{n-1})\| \le c(\mathcal{F}, N)|\mu| \tag{1.2.26}$$

on every admissible set $\mathcal{F}$. If $|\mu| < \delta$, with δ an arbitrary small number, then a set $\mathcal{F}$ that is admissible for the operators $E(\lambda + i(a - |\gamma|))^{-1} D_\omega^\gamma \Phi(\varphi, \omega) E(\lambda)$, $0 \leq |\gamma| \leq N$, is admissible also for all operators $\lambda \mapsto \mathfrak{R}_N(\lambda, \mu)$, and the constant $c(\mathcal{F}, N)$ in (1.2.26) *can be chosen to be independent of μ.*

It follows from (1.2.25) that if $\mathfrak{A}$ is an operator of order a, then the difference $\mathfrak{A}(\lambda + \mu) - \mathfrak{A}(\lambda)$ is an operator of order $a - 1$. Taking into account (1.2.25) and (1.2.26), we obtain

Proposition 1.2.13 *For $N = 1, 2, \ldots$ the derivative of operator* (1.2.15) *admits the representation*

$$\partial_\lambda \mathfrak{A}(\lambda) = \sum_{|\gamma|=1}^{N} \frac{1}{\gamma!} \rho_\gamma(x)\big|_{|x|=1} E_{\omega \to \varphi}(\lambda + i(a - |\gamma|))^{-1}$$

$$\times D_\omega^\gamma \Phi(\varphi, \omega) E_{\psi \to \omega}(\lambda) + \mathfrak{R}_N(\lambda), \tag{1.2.27}$$

where $\rho_\gamma(x) = \lim_{\mu \to 0} \mu^{-1} \partial_x^\gamma |x|^{i\mu}$, while $\mathfrak{R}_N$ is subject to estimate (1.2.24).

We now consider the composition of canonical meromorphic pseudodifferential operators. Let

$$\mathfrak{A}(\lambda) = E(\lambda + ia)^{-1} \Phi(\varphi, \omega) E(\lambda),$$

$$\mathfrak{B}(\lambda) = E(\lambda + ib)^{-1} \Psi(\varphi, \omega) E(\lambda).$$

We will assume that the function Φ is extended to $\mathbb{R}^n \setminus 0$ in each argument φ and ω as a homogeneous function of degree a, and Ψ as a homogeneous function of degree b.

Proposition 1.2.14 *The formula*

$$\mathfrak{A}(\lambda)\mathfrak{B}(\lambda) = \sum_{|\gamma|=0}^{N} \frac{1}{\gamma!} E(\lambda + i(a + b - |\gamma|))^{-1} D_\omega^\gamma \Phi(\varphi, \omega) \partial_\varphi^\gamma \Psi(\varphi, \omega) E(\lambda) + \mathfrak{R}_N(\lambda) \tag{1.2.28}$$

holds for any $N \in \mathbb{Z}_+$. The estimate

$$\|\mathfrak{R}_N(\lambda); H^s(\lambda; S^{n-1}) \to H^{s+N+1-\mathrm{Re}\,(a+b)}(\lambda; S^{n-1})\| \leq c(\mathcal{F}, N)$$

is valid on every admissible set $\mathcal{F}$.

Proof We expand Ψ in a series of spherical harmonics, $\Psi(\varphi, \omega) = \sum a_{mk}(\varphi)Y_{mk}(\omega)$. The coefficients a_{mk} are taken to be homogeneous functions of degree b. The series

$$\mathfrak{B}(\lambda) = \sum a_{mk}(\varphi)E(\lambda + ib)^{-1}Y_{mk}(\omega)E(\lambda) \tag{1.2.29}$$

converges in the norm of operators from $H^s(\lambda, S^{n-1})$ to $H^{s-\operatorname{Re} b}(\lambda, S^{n-1})$. According to (1.2.23),

$$\mathfrak{A}(\lambda)a_{mk} = \sum_{|\gamma|=0}^{N} \frac{1}{\gamma!}\partial_\varphi^\gamma a_{mk}(\varphi)E(\lambda + i(a + b - |\gamma|))^{-1}D_\omega^\gamma\Phi(\varphi, \omega)E(\lambda + ib) + \mathfrak{R}_N(\lambda).$$

It remains to apply (1.2.29). □

Proposition 1.2.15 ([21], Prop. 3.2.5) *Let $\mathfrak{A}(\lambda) = E(\lambda + ia)^{-1}\Phi(\varphi, \omega)E(\lambda)$ and let $\mathfrak{A}(\lambda)^*$ be the operator adjoint to $\mathfrak{A}(\lambda)$ with respect to the duality in $L_2(S^{n-1})$. Then,*

$$\mathfrak{A}(\lambda)^* = \sum_{|\gamma|=0}^{N} \frac{1}{\gamma!}E(\bar{\lambda} + i(\bar{a} - |\gamma|))^{-1}\partial_\varphi^\gamma D_\omega^\gamma\overline{\Phi(\varphi, \omega)}E(\bar{\lambda}) + \mathfrak{R}_N(\bar{\lambda}),$$

where $N = 0, 1, \ldots,$ and the operator $\mathfrak{R}_N$ is subject to inequality (1.2.24). The function Φ is taken to be extended to $\mathbb{R}^n \setminus 0$ with respect to φ and with respect to ω as a homogeneous function of degree a.

1.2.5 General Meromorphic Pseudodifferential Operators

The composition of canonical meromorphic pseudodifferential operators is not, in general, a canonical ΨDO. The operations of taking the adjoint and differentiation with respect to parameter also lead out the class of canonical operators. We define a larger class of meromorphic pseudodifferential operators, which is invariant already under the mentioned operations.

Let $a_0, a_1, \ldots$ be a sequence of complex numbers, $\operatorname{Re} a_j \geq \operatorname{Re} a_{j+1}$, and $\operatorname{Re} a_j \to -\infty$. We denote by $\{\Phi\}_{j=0}^\infty$ a sequence of functions in $C^\infty(S^{n-1} \times S^{n-1})$ assuming that the functions $\varphi \mapsto \Phi_j(\varphi, \theta)$ and $\theta \mapsto \Phi_j(\varphi, \theta)$ are extended to $\mathbb{R}^n \setminus 0$ as homogeneous functions of degree a_j.

Definition 1.2.16 An operator-valued function $\lambda \mapsto \mathfrak{A}(\lambda)$ that is meromorphic in the complex plane is called a meromorphic pseudodifferential operator of order a_0 if every strip of the form $\{\lambda \in \mathbb{C} : |\mathrm{Im}\lambda| < h\}$ contains at most finitely many poles of $\mathfrak{A}$ and the inequality

$$\|\mathfrak{A}(\lambda) - \sum_{j=0}^{N} \mathfrak{A}_j(\lambda); H^s(\lambda, S^{n-1}) \to H^{s-\mathrm{Re}\, a_{N+1}}(\lambda, S^{n-1})\| \leq c(\mathcal{F}, N, s) \tag{1.2.30}$$

holds for any $N \in \mathbb{Z}^+$ on every set $\mathcal{F}$ that is admissible for the operators $\mathfrak{A}$ and $\mathfrak{A}_j$, $j = 0, 1, \dots, N$, where $\mathfrak{A}_j(\lambda) = E_{\theta\to\varphi}(\lambda + ia_j)^{-1}\Phi_j(\varphi, \theta)E_{\psi\to\theta}(\lambda)$.

The formal series $\sum \Phi_j$ is called the complete symbol of $\mathfrak{A}$, and the function Φ_0 will be called the principal symbol. A series $\sum_{j=0}^{\infty} \mathfrak{A}_j$ consisting of canonical meromorphic pseudodifferential operators is called an asymptotic series for a meromorphic pseudodifferential operator $\mathfrak{A}$ if inequalities of the form (1.2.30) hold for all N and $\mathcal{F}$. In the sequel, the notation $\mathfrak{A} \sim \sum \mathfrak{A}_j$ means that $\sum \mathfrak{A}_j$ is an asymptotic series for $\mathfrak{A}$.

We note, for example, that by Proposition 1.2.12 the function $\lambda \mapsto \mathfrak{A}(\lambda + \mu)$ with $\mathfrak{A}$ defined by (1.2.15) is a meromorphic ΨDO of order a with symbol

$$\sum_{|\gamma|=0}^{\infty} \frac{1}{\gamma!} |x|^{-i\mu} \partial_x^{\gamma} |x|^{i\mu} D_{\theta}^{\gamma} \Phi(x, \theta)$$

(the function Φ is homogeneous of degree a with respect to x and with respect to θ).

Theorem 1.2.17 ([21], Thm. 3.3.3) *There exists a meromorphic ΨDO with any given symbol $\sum \Phi_j$.*

1.2.6 Change of Variables in Meromorphic Pseudodifferential Operators

Let $g : S^{n-1} \to S^{n-1}$ be a diffeomorphism, $g(\sigma) = (g(\sigma)_1, \dots, g(\sigma)_n)$, $\sigma \in \mathbb{R}^n$, $|\sigma| = 1$, and $|g(\sigma)| = 1$. We assume that the functions g_j are defined on $\mathbb{R}^n \setminus 0$ and are homogeneous of degree one, that $g'(\sigma) = \|\partial g_j/\partial \sigma_k\|_{j,k=1}^{n}$ and that $|g'(\sigma)|$ is a modulus of the determinant $\det g'(\sigma)$.

Theorem 1.2.18 ([21], Thm. 3.8.1) *Let $\mathfrak{A}$ be an operator of the form* (1.2.15). *Then,*

$$\mathfrak{A}_{\psi\to\varphi}(\lambda)u(\psi) = |g'(\sigma)^{-1}|\, \hat{\mathfrak{A}}_{\tau\to\sigma}(\lambda)\, |g'(\tau)|\, \hat{u}(\tau),$$

where $\varphi = g(\sigma)$, $\psi = g(\tau)$, $\hat{u}(\tau) = u(g(\tau))$, and $\hat{\mathfrak{A}}_{\tau\to\sigma}$ is a meromorphic ΨDO with symbol

$$\sum_{|\gamma|=0}^{\infty} \frac{1}{\gamma!} \partial_\omega^\gamma e^{ih(\omega,\sigma,\tau)}\Big|_{\tau=\sigma} D_\omega^\gamma \Phi(g(\sigma), (g'(\sigma)^{-1})^*\omega),$$

where, moreover, $h(\omega, \sigma, \tau) = (\omega, \tau - g'(\sigma)^{-1} g(\tau)) = O(|\sigma - \tau|^2)$.

1.3 C^*-algebras

1.3.1 C^*-algebras and Their Morphisms

Let $\mathcal{A}$ be an algebra over the field $\mathbb{C}$. We say that a map $x \mapsto x^*$ of $\mathcal{A}$ into itself is an *involution* if the following conditions hold: $(x^*)^* = x$, $(x+y)^* = x^* + y^*$, $(\lambda x)^* = \overline{\lambda} x^*$, and $(xy)^* = y^* x^*$. An algebra endowed with involution is called involutive. An involutive normed algebra is a normed algebra $\mathcal{A}$ with involution such that $\|x^*\| = \|x\|$ for all $x \in \mathcal{A}$. A complete involutive normed algebra is called an involutive Banach algebra. If $\|x^*\|^2 = \|x^* x\|$ for every element $x \in \mathcal{A}$, the involutive Banach algebra is called a C^*-algebra.

Let $\mathcal{A}$ be an involutive algebra. A subalgebra of $\mathcal{A}$ that is mapped into itself under involution is, by definition, an involutive subalgebra. Any closed involutive subalgebra of a C^*-algebra is a C^*-algebra.

Let us denote by $\mathcal{B}H$ the algebra of all bounded linear operators on a Hilbert space H. Involution in $\mathcal{B}H$ means transition to the adjoint operator. Every closed involutive subalgebra of $\mathcal{B}H$ is a C^*–algebra.

Let $\mathcal{A}$ and $\mathcal{B}$ be involutive algebras. A morphism $f : \mathcal{A} \to \mathcal{B}$ is a linear map such that $f(xy) = f(x)f(y)$ and $f(x^*) = f(x)^*$ for all $x, y \in \mathcal{A}$. A bijective morphism is called an isomorphism. It turns out that any morphism of C^*-algebras $\mathcal{A}$ and $\mathcal{B}$ (simply regarded as a morphism of involutive algebras) is continuous, and its norm is no greater than one. If $f : \mathcal{A} \to \mathcal{B}$ is an injective morphism, then $\|f(x)\| = \|x\|$ for all $\|x\| \in \mathcal{A}$. Every C^*-algebra is isomorphic to a closed involutive subalgebra of an algebra $\mathcal{B}H$.

Let $\mathcal{A}$ be a C^*-algebra and J a closed two-sided ideal in $\mathcal{A}$. Then, J is self-adjoint (i. e., is preserved as a set under involution) and the quotient algebra $\mathcal{A}/J$, endowed with the quotient norm and corresponding involution, is a C^*-algebra.

Let $\mathcal{A}$ and $\mathcal{B}$ be C^*-algebras, $\varphi : \mathcal{A} \to \mathcal{B}$ a morphism, and I a kernel of φ. Then, I is a closed ideal in $\mathcal{A}$ and the image $\varphi(\mathcal{A})$ is closed in $\mathcal{B}$. Indeed, the map φ is continuous because any morphism of C^*-algebras is continuous. Hence, I is a closed ideal, and $\mathcal{A}/I$ is a C^*-algebra. The morphism $\mathcal{A}/I \to \mathcal{B}$ obtained by factoring through φ is injective; hence, it is an isometry. Thus, the $\varphi(\mathcal{A})$ is closed and complete in $\mathcal{B}$.

1.3.2 Representations of C^*-algebras

In the sequel, an algebra and morphism will always mean C^*-algebra and $*$-morphism. A representation of an algebra $\mathcal{A}$ in a Hilbert space H is a morphism $\pi : \mathcal{A} \to \mathcal{B}H$. The space H is called the representation space of π, and the (Hilbert) dimension of H is called the dimension of the representation.

Representations π and π' of an algebra $\mathcal{A}$ in spaces H and H' are equivalent, by definition, if there is a Hilbert space isomorphism $U : H \to H'$ such that $U\pi(x) = \pi'(x)U$ for all $x \in \mathcal{A}$.

We say that a vector $\xi \in H$ is cyclic (or totalizing) for a representation π of an algebra $\mathcal{A}$ in H if the closure of the set $\pi(\mathcal{A})\xi$ coincides with H.

A representation π of an algebra $\mathcal{A} \in H$ is called irreducible if π satisfies either one of the following (equivalent) conditions:

(1) The only closed subspaces in H that are invariant under $\pi(\mathcal{A})$ are 0 and H.
(2) The commutant of $\pi(\mathcal{A})$ in $\mathcal{B}H$ consists of scalars.
(3) Either every nonzero vector $\xi \in H$ is totalizing for π or π is the null representation of dimension 1.

A subalgebra $\mathcal{A} \subset \mathcal{B}H$ is called irreducible if its identity representation $\mathcal{A} \to \mathcal{B}H : a \mapsto a$ is irreducible.

The following assertion shows that, for every algebra $\mathcal{A}$, there are "sufficiently many" irreducible representations.

Proposition 1.3.1 ([3], 2.7.3) *There exists a family $\{\pi_i\}$ of irreducible representations of an algebra $\mathcal{A}$ such that $\|a\| = \sup_i \|\pi_i(a)\|$ for all $a \in \mathcal{A}$.*

Proposition 1.3.2 *Let π be an irreducible representation of an algebra $\mathcal{A}$ in a Hilbert space H. Then:*

(1) *If I is a two-sided ideal in $\mathcal{A}$ and $\pi(I) \neq 0$, then the restriction $\pi|I$ of π to I is irreducible.*
(2) *If I_1 and I_2 are two-sided ideals in $\mathcal{A}$ and $\pi(I_1) \neq 0$ and $\pi(I_2) \neq 0$, then $\pi(I_1 I_2) \neq 0$.*

Proof

(1) We set $E = \{x \in H : \pi(I)x = 0\}$. Since $\pi(\mathcal{A})E \subset E$ and by requirement $E \neq H$, we have $E = 0$. Hence, $\pi(I)\xi \neq 0$ regardless the nonzero vector $\xi \in H$. The subspace $\pi(I)\xi$ is invariant under $\pi(\mathcal{A})$. Since π is irreducible, we find $\pi(I)\xi = H$. Thus, any nonzero vector $\xi \in H$ is cyclic for $\pi|I$.
(2) As was proven in (1), $\pi(I_2)H = H$ and $\pi(I_1)\pi(I_2)H = H$. Hence, $\pi(I_1 I_2) \neq 0$.

□

The kernel of an irreducible representation of an algebra $\mathcal{A}$ is called a two-sided primitive ideal in $\mathcal{A}$. Every closed two-sided ideal in $\mathcal{A}$ is the intersection of the primitive ideals containing it ([3], 2.9.7). For example, the ideal com$A$ generated by commutators $[a, b] = ab - ba$, where $a, b \in \mathcal{A}$, coincides with the intersection of the kernels of all one-dimensional representations of $\mathcal{A}$.

Proposition 1.3.3 ([3], 2.11.4) *Let I_1 and I_2 be two-sided ideals in an algebra $\mathcal{A}$ and let I be a primitive ideal. If $I_1 I_2 \subset I$* (in, particular, if $I_1 \cap I_2 \subset I$), then either $I_1 \subset I$ or $I_2 \subset I$.

Proof Assume the contrary, i. e., $I_1 \not\subset I$ and $I_2 \not\subset I$. Applying the second assertion of Proposition 1.3.2 to an irreducible representation π of $\mathcal{A}$ with kernel I, we find $\pi(I_1 I_2) \neq 0$, whence $I_1 I_2 \not\subset I$. □

Proposition 1.3.4 ([3], 4.1.5) *Let $\mathcal{A} = \mathcal{K}H$ be the algebra of compact operators in a Hilbert space H. Then, every nonnull irreducible representation of $\mathcal{A}$ is equivalent to the identity representation.*

We say that $\mathcal{A}$ is an algebra of type I if the set $\pi(\mathcal{A})$ contains $\mathcal{K}H$ for every irreducible representation π of $\mathcal{A}$ in H.

Proposition 1.3.5 ([3], 4.1.10) *Let π be an irreducible representation of an algebra $\mathcal{A}$ in H. If $\pi(\mathcal{A}) \cap \mathcal{K}H \neq 0$, then $\mathcal{K}H \subset \pi(\mathcal{A})$ and every irreducible representation of $\mathcal{A}$ with the same kernel as π is equivalent to π.*

Proposition 1.3.6 ([3], 2.10.2) *(extension of a representation) Let $\mathcal{B}$ be a subalgebra of $\mathcal{A}$ and ρ a representation of $\mathcal{B}$ in a Hilbert space G. Then, there exist a Hilbert space H containing G as a subspace and a representation π of $\mathcal{A}$ in H such that $\rho(x) = \pi(x)|G$ for all $x \in \mathcal{B}$. If ρ is irreducible, π can be chosen to be irreducible.*

1.3.3 Spectrum of C^*-algebra

Let Prim$\mathcal{A}$ be the set of two-sided primitive ideals of an algebra $\mathcal{A}$ and $T \subset$ Prim$\mathcal{A}$. We denote by $I(T)$ the intersection of all ideals in T. The set $I(T)$ is a two-sided ideal in $\mathcal{A}$. Moreover, let $\bar{T}$ be the set of primitive ideals containing $I(T)$. It turns out that there exists a unique topology on Prim$\mathcal{A}$ such that $\bar{T}$ is the closure of T for all $T \subset$ Prim$\mathcal{A}$ in this topology, which is called the Jacobson topology on Prim$\mathcal{A}$.

We introduce the set $\widehat{\mathcal{A}}$ of equivalence classes of nonnull irreducible representations of an algebra $\mathcal{A}$. The map $\pi \mapsto \ker\pi$ defines a canonical surjection $\widehat{\mathcal{A}} \to$ Prim$\mathcal{A}$. The spectrum of $\mathcal{A}$ is the set $\widehat{\mathcal{A}}$ endowed with the topology that is a preimage of the Jacobson topology under the canonical map $\widehat{\mathcal{A}} \to$ Prim$\mathcal{A}$.

We also provide another description of the Jacobson topology on the spectrum $\widehat{\mathcal{A}}$. Let $\{x_i\}$ be a family of elements of $\mathcal{A}$ that is everywhere dense in $\mathcal{A}$ and let $Z_i = \{\pi \in \widehat{\mathcal{A}} : \|\pi(x_i)\| > 1\|$. Then, the set of Z_i is a base of the topology on $\widehat{\mathcal{A}}$. Thus, if the algebra $\mathcal{A}$ is separable, then the topology on $\widehat{\mathcal{A}}$ admits a countable base.

We say that a topological space is a T_0-space if, given two points of this space, there is a neighborhood of one point not containing the other point.

The following three conditions are equivalent ([3], 3.1.6):

(1) $\widehat{\mathcal{A}}$ is a T_0-space.
(2) Two irreducible representations of $\mathcal{A}$ with the same kernel are equivalent.
(3) The canonical map $\widehat{\mathcal{A}} \to \mathrm{Prim}\mathcal{A}$ is a homeomorphism.

Proposition 1.3.7 ([3], 3.2.1 and 3.2.2)

(1) *Let I be a closed two-sided ideal of an algebra $\mathcal{A}$ and let $p : \mathcal{A} \to \mathcal{A}/I$ be the canonical morphism. We set $\widehat{\mathcal{A}}_I = \{\pi \in \widehat{\mathcal{A}} : \pi|I = 0\}$. Then, the maps*

$$(\mathcal{A}/I)\widehat{\ } \ni \pi \mapsto \pi \circ p \in \widehat{\mathcal{A}}_I, \quad \widehat{\mathcal{A}} \setminus \widehat{\mathcal{A}}_I \ni \pi \mapsto \pi|I \in \widehat{I}$$

are homeomorphisms.

(2) *The map $I \mapsto \widehat{I}$ is a bijection of the set of the closed two-sided ideals of $\mathcal{A}$ onto the set of open parts of the spectrum $\widehat{\mathcal{A}}$. Moreover, $I_1 \subset I_2 \Leftrightarrow \widehat{I}_1 \subset \widehat{I}_2$.*

As before, let $\mathcal{K}H$ be the algebra of compact operators in a Hilbert space H and X a locally compact space. We consider the algebra $\mathcal{A} = C_0(X) \otimes \mathcal{K}H$ of the continuous functions taking values in $\mathcal{K}H$ and tending, in the operator norm, to zero at infinity. The norm in $\mathcal{A}$ is defined by

$$\|f\| = \sup_{x \in X} \|f(x)\,;\, \mathcal{B}H\|.$$

We set $\pi(x)f = f(x)$ for $x \in X$. It is evident that $\pi(x)$ is an irreducible representation of $\mathcal{A}$ in H. The following proposition is an instance of assertion 10.4.4 in [3].

Proposition 1.3.8 *The map $x \mapsto \pi(x)$ is a homeomorphism of X onto $(C_0(X) \otimes \mathcal{K}H)^\wedge$.*

A subalgebra $\mathcal{B}$ of an algebra $\mathcal{A}$ is called rich if the following conditions hold:

(1) For every irreducible representation π of $\mathcal{A}$, the representation $\pi|\mathcal{B}$ is irreducible.
(2) If π_1 and π_2 are nonequivalent irreducible representations of $\mathcal{A}$, the representations $\pi_1|\mathcal{B}$ and $\pi_2|\mathcal{B}$ are nonequivalent.

Proposition 1.3.9 ([3], 11.1.4) *If $\mathcal{B}$ is a rich subalgebra of the algebra $C_0(X) \otimes \mathcal{K}H$, then $\mathcal{B} = C_0(X) \otimes \mathcal{K}H$.*

1.3.4 Criteria for an Element of an Algebra to Be Invertible or to Be Fredholm

Proposition 1.3.10 *Let $\mathcal{A}$ be an algebra with identity. An element $a \in \mathcal{A}$ is invertible if and only if the operator $\pi(a)$ is invertible for every irreducible representation π of $\mathcal{A}$ (in the space H_π of the representation).*

Proof The necessity of the condition for an element $a \in \mathcal{A}$ to be invertible is evident. Let us verify the sufficiency. Let us assume that a does not have a right (or left) inverse. Then, the element $b = b^* = aa^*$ (respectively, a^*a) cannot be inverted. We introduce the commutative algebra $\mathcal{B}$ generated by b and the identity of $\mathcal{A}$. It is well-known ([3], 1.5) that $\mathcal{B}$ is isomorphic to the algebra of continuous functions on the spectrum Spb. Since b is noninvertible, 0 belongs to the spectrum Spb, and there is a one-dimensional representation ρ of $\mathcal{B}$ annihilating b. By Proposition 1.3.6, there is a representation $\pi \in \widehat{\mathcal{A}}$ extending ρ in the sense that H_ρ is a one-dimensional subspace of H_π and $\pi(x)|H_\rho = \rho(x)$ for all $x \in \mathcal{B}$. Since $\pi(b)|H_\rho = 0$, the operator $\pi(b)$ (hence also $\pi(a)$) is noninvertible. □

Proposition 1.3.11 ([3], 1.3.10) *Let $\mathcal{B}$ be a subalgebra of an algebra $\mathcal{A}$ that has an identity that, moreover, belongs to $\mathcal{B}$. Then, for any $x \in \mathcal{B}$ we have $\mathrm{Sp}_{\mathcal{A}} x = \mathrm{Sp}_{\mathcal{B}} x$, where $\mathrm{Sp}_{\mathcal{C}} y$ denotes the spectrum of an element y in the algebra $\mathcal{C}$.*

Thus, if an element $x \in \mathcal{B}$ is invertible in $\mathcal{A}$, then it is also invertible in $\mathcal{B}$.

A continuous linear operator $A : E_1 \to E_2$, where E_1 and E_2 are Banach spaces, is called Fredholm if the range ImA is closed and the kernel and cokernel are finite-dimensional, $\dim \ker A < \infty$ and $\dim \operatorname{coker} A < \infty$.

Proposition 1.3.12 *Let $\mathcal{A}$ be a subalgebra of $\mathcal{B}H$ and let $\mathcal{A}$ contain both the ideal $\mathcal{K}H$ and the identity operator. Then, an operator $A \in \mathcal{A}$ is Fredholm if and only if every operator $\pi(A)$ is invertible, where $\pi \in (\mathcal{A}/\mathcal{K}H)^\wedge$.*

Proof According to the Atkinson theorem (e.g., see [19], Theorem 1.4.16), an operator A is Fredholm if and only if the class $[A] \in \mathcal{B}H/\mathcal{K}H$ is invertible. Taking into account Proposition 1.3.11, we obtain that A is Fredholm if and only if the class $[A]$ is invertible in $\mathcal{A}/\mathcal{K}H$. It remains to apply Proposition 1.3.10. □

1.3.5 Continuous Field of C^*-algebras

We denote by $\mathcal{B}(E, E')$ the space of linear operators acting from a Banach space E into a Banach space E'; for $E = E'$, we write $\mathcal{B}(E)$. Let $C(T, L)$ be the space of continuous functions on T with values in L, where T and L are topological spaces.

We assume that for every point $t \in T$, there corresponds a Banach space E_t. Every map

$$x : T \to \prod_{t \in T} E_t,$$

subject to the condition $x(t) \in E_t$ for all $t \in T$, is called a vector field on T. Sometimes it is convenient to identify a vector field x with its image $\{x(t)\} \subset \prod E_t$.

A continuous field $\mathcal{E}$ of Banach spaces on T is a family of Banach spaces $\{E_t\}_{t\in T}$ endowed with a set Γ of vector fields such that

(i) Γ is a linear space.
(ii) The set $\{x(t) : x \in \Gamma\}$ is dense in E_t for every $t \in T$.
(iii) The function $t \mapsto \|x(t)\|$ is continuous for every $x \in \Gamma$.
(iv) If $y \in \prod_{t\in T} E_t$ is a vector field, and for any $t_0 \in T$ and $\varepsilon > 0$, there exists a field $x \in \Gamma$ for which $\|y(t) - x(t)\| < \varepsilon$ in a neighborhood of t_0, then $y \in \Gamma$.

The elements $x \in \Gamma$ are called continuous vector fields in $\mathcal{E}$. The space Γ is a module over the ring $C(T)$ ([3], 10.1.9).

An isomorphism of continuous vector fields $\mathcal{E} = (\{E_t\}_{t\in T}, \Gamma)$ and $\mathcal{E}' = (\{E'_t\}_{t\in T}, \Gamma')$ is a family $\varphi = \{\varphi(t)\}_{t\in T}$ of linear operators $\varphi(t) : E_t \to E'_t$ such that

(1) For every t, the operator $\varphi(t)$ is an isometric isomorphism E_t onto E'_t.
(2) $\varphi(\Gamma) \subset \Gamma'$ (this requirement is equivalent to the condition $\varphi(\Gamma) = \Gamma'$).

Example We consider a product $T \times E$, where E is a Banach space, and assume that $E_t = \{t\} \times E$, $q : T \times E \to E$ is a projection, and $q_t = q|E_t$. The map q allows us to endow every fiber E_t with the structure of a Banach space so that the operators $q_t : E_t \to E$ become isometric isomorphisms. Let us choose as a linear space Γ the set of all vector fields of the form $t \mapsto (t, f(t))$, where $f \in C(T, E)$. Then $(\{E_t\}, \Gamma)$ is a continuous field of Banach spaces; it is called a constant field. We denote this field (not quite correctly) by $T \times E$ and will write $\Gamma(T \times E)$ instead of Γ. A field that is isomorphic to a constant field is called trivial.

Let $\mathcal{E} = (\{E_t\}_{t\in T}, \Gamma)$ be a continuous field of Banach spaces on T and $\Omega \subset T$ an open set. We denote by Γ_Ω the set of vector fields on Ω which are the limits of elements in $\Gamma|\Omega$ with respect to local uniform convergence. Then $(\{E_t\}_{t\in\Omega}, \Gamma_\Omega)$ is a continuous field of Banach spaces on Ω, which is denoted by $\mathcal{E}|\Omega$ and is called a field induced by $\mathcal{E}$ on Ω.

A field $\mathcal{E}$ is called locally trivial if, for any point t, there is a neighborhood Ω such that the field $\mathcal{E}|\Omega$ is trivial.

Let $\mathcal{E}$ be a locally trivial field of Banach spaces on T and let $\{\Omega_i\}_{i\in I}$ be an open covering of T such that all fields $\mathcal{E}|\Omega_i$ are trivial and $\varphi_i : \Omega_i \times E_i \to \mathcal{E}|\Omega_i$ is an isomorphism of trivial fields. A family of cards $\{(\varphi_i , \Omega_i)\}_{i\in I}$ is called a trivializing atlas (or simply an atlas) for $\mathcal{E}$.

Let $\mathcal{E} = (\{E_t\}_{t\in T} , \Gamma)$ be a continuous field of Banach spaces. We say that a set $\Lambda \subset \Gamma$ is total if for every $t \in T$ the set $\{x(t) : x \in \Lambda\}$ is total in E_t, i. e., the closure of its linear hull coincides with E_t. A field $\mathcal{E}$ is called separable if Γ contains a countable total subset.

Lemma 1.3.13 ([3], 10.2.7) *Let T be a separable metrizable space and $\mathcal{E} = (\{E_t\}, \Gamma\}$ a locally trivial continuous field of Banach spaces on T. If every E_t is separable, then also $\mathcal{E}$ is separable.*

Lemma 1.3.14 ([3], 10.8.7) *Let T be a finitely dimensional (in the sense of topological dimension) paracompact space and $\mathcal{E} = (\{E_t\}_{t\in T} , \Gamma)$ a separable continuous field of Hilbert spaces; the dimension of every space E_t is $\aleph_0$. Then, the field $\mathcal{E}$ is trivial.*

We now discuss an isomorphism of trivial fields in more detail and assume $\varphi : T\times E \to T \times E'$ to be such an isomorphism. Then $\varphi(t, e) = (t, A_t e)$ for every $t \in T$, where A_t is a linear isometry $E \to E'$ and $A_t = q_t'\varphi(t)q_t^{-1}$.

Lemma 1.3.15 *Let $\varphi : T \times E \to T \times E'$ be an isomorphism of trivial fields and $A_t = q_t'\varphi(t)q_t^{-1}$ for $t \in T$. Then, the function $t \mapsto A_t \in \mathcal{B}(E, E')$ is continuous with respect to strong operator topology.*

Proof We fix $e \in E$. Since the vector field $t \mapsto (t, e)$ belongs to $\Gamma(T \times E)$, its image $t \mapsto (t, A_t e)$ under the isomorphism φ belongs to $\Gamma(T\times E')$. Hence, the function $t \mapsto A_t e$ is continuous. □

A continuous field of C^*-algebras on T is, by definition, a continuous field $(\{\mathcal{A}_t\}, \Theta)$ of Banach spaces, where every space $\mathcal{A}_t$ is endowed with multiplication and involution that turn it into a C^*-algebra, and Θ is invariant under multiplication and involution. An isomorphism of fields $(\{\mathcal{A}_t\}, \Theta)$ and $(\{\mathcal{A}_t'\}, \Theta')$ is defined by a family $\{\varphi(t)\}_{t\in T}$, where $\varphi(t) : \mathcal{A}_t \to \mathcal{A}_t'$ is an isomorphism for every t and $\varphi(\Theta) \subset \Theta'$. The triviality and local triviality of the field of C^*-algebras are introduced with regard to that definition.

Let $(\{\mathcal{A}_t\}, \Theta)$ be a continuous field of C^*-algebras and every $\mathcal{A}_t$ elementary C^*-algebra (i. e., every algebra $\mathcal{A}_t$ is isomorphic to the ideal of compact operators in a certain Hilbert space). Then, $(\{\mathcal{A}_t\}, \Theta)$ is called a field of elementary C^*-algebras.

Let us assume that T is a locally compact space and $\mathcal{L} = (\{\mathcal{A}_t\}, \Theta)$ is a continuous field of C^*-algebras on T. We denote by $\mathcal{A}$ the set of vector fields $x \in \Theta$ such that $\|x(t)\|$ tends to zero as $t \to \infty$. Then, $\mathcal{A}$ is an involutive subalgebra of Θ. For $x \in \mathcal{A}$, we set

$\|x\| = \sup\{\|x(t)\|\,;\, t \in T\}$. Being endowed with such a norm, $\mathcal{A}$ is a C^*-algebra; it is called a C^*-algebra defined by the field $\mathcal{L}$. If $\mathcal{L}$ is a trivial field, then $\mathcal{A} \cong C_0(T, \mathcal{B}) \cong C_0(T) \otimes \mathcal{B}$, where $\mathcal{B}$ is a certain C^*-algebra. In the specific case of a trivial field $\mathcal{L}$ of elementary algebras, we have $\mathcal{A} \cong C_0(T) \otimes \mathcal{K}H$.

Proposition 1.3.16 ([3], 10.4.3) *Let T be a locally compact space, let $\mathcal{L} = (\{\mathcal{A}_t\}, \Theta)$ be a continuous field of algebras on T, and let $\mathcal{A}$ be the algebra defined by the field $\mathcal{L}$. Moreover, let ρ_π denote the irreducible representation $x \mapsto \pi(x(t))$ of $\mathcal{A}$ for $t \in T$ and $\pi \in \widehat{\mathcal{A}}_t$. Then, $\pi \mapsto \rho_\pi$ is a bijection of the union of the sets $\widehat{\mathcal{A}}_t$ onto $\widehat{\mathcal{A}}$.*

Under the conditions of Proposition 1.3.16, the topology on $\widehat{\mathcal{A}}$ was described in [9]. We present the corresponding result for the case of a trivial field $\mathcal{L}$.

Proposition 1.3.17 ([3], 10.10.2) *Let T be a locally compact space, let $\mathcal{B}$ be a C^*-algebra, and let $\mathcal{A}$ be a C^*-algebra of continuous maps T into $\mathcal{B}$ tending to zero at infinity. Then, the space $\widehat{\mathcal{A}}$ is canonically identified with the space $T \times \widehat{\mathcal{B}}$.*

1.3.6 A Sufficient Triviality Condition for the Fields of Elementary Algebras

Let T be a finite-dimensional paracompact space and H a Hilbert space of $\dim H = \aleph_0$. Considering the trivial fields $\Omega \times H$ and $\Omega \times \mathcal{K}H$, where Ω is an open subset of T, we denote "fibers" $\{t\} \times H$ and $\{t\} \times \mathcal{K}H$ by H_t and $(\mathcal{K}H)_t$.

Let $\mathcal{E} = (\{E_t\}_{t\in T}, \Gamma)$ be a separable continuous Hilbert space field, $\dim E_t = \aleph_0$, and $\mathcal{L} = (\{\mathcal{K}E_t\}_{t\in T}, \Theta)$ a continuous field of elementaryC^*-algebras (for notation uniformity in what follows, we write $\Gamma(\mathcal{L})$ instead of Θ). According to Lemma 1.3.14, there exists an isomorphism $\varphi : T \times H \to \mathcal{E}$. If for any $K(\cdot) \in \Gamma(\mathcal{L})$ the vector field $t \mapsto \varphi(t)^{-1}K(t)\varphi(t)$ belongs to $\Gamma(T \times \mathcal{K}H)$, the field $\mathcal{L}$ is trivial. In particular cases, this simple triviality condition of the field $\mathcal{L}$ is not very useful because of the necessity to construct the isomorphism φ. Usually, in applications, there arise local trivializations of the field $\mathcal{E}$. The following proposition suggests a sufficient triviality condition of the field $\mathcal{L}$ in terms of such trivializations.

Proposition 1.3.18 *Let T, H, $\mathcal{E}$, and $\mathcal{L}$ be the same as above and let $\{(\varphi_i, \Omega_i)\}_{i\in I}$ be an atlas for $\mathcal{E}$. Assume that, for all $i \in I$ and $K \in \Gamma(\mathcal{L})$, the vector field $t \mapsto \varphi_i(t)^{-1}K(t)\varphi_i(t)$ on Ω_i belongs to $\Gamma(\Omega_i \times \mathcal{K}H)$. Then the field $\mathcal{L}$ is trivial, hence the C^*-algebra defined by this field is isomorphic to $C_0(T) \otimes \mathcal{K}H$.*

Before proving this proposition, we present the next lemma, where $U(H)$ denotes the unitary group of a Hilbert space H.

Lemma 1.3.19 *Assume that $t \mapsto R(t) \in \mathcal{K}H$ and $t \mapsto u(t) \in U(H)$ are defined on an open set $\Omega \subset T$, R is continuous in the uniform operator topology, and u is continuous in the strong topology. Then, the function $t \mapsto u(t)R(t)u(t)^{-1}$ is continuous in the uniform topology.*

Proof of the Lemma It is evident that the function uR is continuous in the uniform topology. Then, $uRu^{-1} = (u(uR)^*)^*$ is also continuous in the same topology. □

Proof of the Proposition Let $\varphi = \{\varphi(t)\}_{t\in T}$ be an isomorphism $T \times H \to \mathcal{E}$. For every vector field $K(\cdot) \in \Gamma(\mathcal{L})$, we define the vector field

$$T \ni t \mapsto (t, q_t\varphi(t)^{-1}K(t)\varphi(t)q_t^{-1}) \in (\mathcal{K}H)_t,$$

where q_t is the restriction of the projection $q : T \times H \to H$ to the "fiber" H_t. The proposition will be proven if we verify the continuity of the function

$$t \mapsto Q(t) := q_t\varphi(t)^{-1}K(t)\varphi(t)q_t^{-1}$$

in the uniform topology. To this end, we note that for any $i \in I$ the equality

$$(Q|\Omega_i)(t) = u_i(t)R_i(t)u_i(t)^{-1} \tag{1.3.1}$$

holds, where $R_i(t) = q_t\varphi_i(t)^{-1}K(t)\varphi_i(t)q_t^{-1}$ and $u_i(t) = q_t(\varphi^{-1} \circ \varphi_i)q_t^{-1}$. From the assumption of the proposition it follows that the function R_i belongs to the space $C(\Omega_i, \mathcal{K}H)$. Moreover, the function u_i is continuous in the strong operator topology; this follows from Lemma 1.3.15, since $\varphi^{-1} \circ \varphi_i$ is an isomorphism of the fields $\Omega_i \times H$ and $(T \times H)|\Omega_i$. Taking into account Lemma 1.3.19 and equality (1.3.1), we find that the function $Q|\Omega_i$ is continuous in the uniform topology for any $i \in I$; hence, $Q \in C(T, \mathcal{K}H)$. □

1.3.7 Solvable Algebras

Definition 1.3.20 An algebra $\mathcal{A}$ is called solvable if there exists a composition series $\{0\} = \mathcal{J}_{-1} \subset \mathcal{J}_0 \subset \cdots \subset \mathcal{J}_l = \mathcal{A}$ of ideals such that there is an isomorphism

$$\mathcal{J}_k/\mathcal{J}_{k-1} \simeq C_0(X_k) \otimes \mathcal{K}H_k \tag{1.3.2}$$

for $k = 1, \ldots, l$, where H_k is a Hilbert space, X_k is a locally compact Hausdorff space, and $C_0(X_k)$ is the algebra of continuous functions tending to zero at infinity. A composition series with this property is said to be solving, and the number l is called the length of the series. The least one of the solving series lengths is called the length of the algebra $\mathcal{A}$.

Algebras $\mathcal{L}$ of the form $C_0(X) \otimes \mathcal{K}H \simeq C_0(X, \mathcal{K}H)$ play the role of "elementary" objects. Every nonzero irreducible representation of such an algebra $\mathcal{L}$ is the map $\mathcal{L} \ni f \mapsto f(x) \in \mathcal{K}H$ for a certain $x \in X$. The Jacobson topology on the spectrum $\widehat{\mathcal{L}}$ coincides with the topology of X. Therefore, $\widehat{\mathcal{L}}$ can be identified with the space X (Proposition 1.3.8). Thus, a solvable algebra consists of elementary blocks. In view of (1.1.39), $(\mathcal{J}_j/\mathcal{J}_{j-1})^\wedge = X_j$ and hence $\widehat{\mathcal{J}_j} = G_j$, where $G_j = \bigcup_{i=0}^{j} X_i$. Being the spectrum of an ideal, G_j is an open subset of the spectrum $\widehat{\mathcal{A}}$ of the algebra $\mathcal{A}$.

Let us discuss the solvable algebra definition in more detail. Let $\mathcal{A}$ be an arbitrary C^*-algebra. To every element $a \in \mathcal{A}$ there corresponds the function $\hat{a}$ given on the spectrum $\widehat{\mathcal{A}}$ by $\hat{a}(\pi) = \pi(a)$ ($\in \mathcal{B}H_\pi$). The functions $\hat{a}$ form an algebra isomorphic to $\mathcal{A}$ since

$$\|\hat{a}\| = \sup\{\, \|\hat{a}(\pi)\,;\, \mathcal{B}H_\pi\|\,;\ \pi \in \widehat{\mathcal{A}}\} = \|a\|$$

(Proposition 1.3.1). In general, for every set $\Theta \subset \prod_{\pi \in \widehat{\mathcal{A}}} \mathcal{B}H_\pi$ of vector fields on $\widehat{\mathcal{A}}$ containing $\{\hat{a} : a \in \mathcal{A}\}$, the C^*-algebra field $(\{\pi(\mathcal{A})\}_{\pi \in \widehat{\mathcal{A}}}, \Theta)$ is not continuous. The cause is that, as a rule, the space $\widehat{\mathcal{A}}$ endowed with the Jacobson topology is not Hausdorff and therefore the functions $\pi \mapsto \|\hat{a}(\pi); \mathcal{B}H_\pi\|$ are not continuous (one can guarantee semicontinuity from below only [3, 3.3.2]). Let

$$\emptyset = G_{-1} \subset G_0 \subset, \ldots \subset G_l = \widehat{\mathcal{A}}$$

be open subsets of $\widehat{\mathcal{A}}$ such that the space $X_k := G_k \setminus G_{k-1}$ is Hausdorff for every $k \in \{0, \ldots, l\}$ (we suppose that such G_k exist). We connect with each set G_k the ideal

$$\mathcal{J}_k = \{a \in \mathcal{A} : \hat{a}(\pi) = 0 \ \text{for} \ \pi \in \widehat{\mathcal{A}} \setminus G_k\}.$$

The spectrum of this ideal can be identified with G_k and the spectrum of the quotient algebra $\mathcal{J}_k/\mathcal{J}_{k-1}$ is identified with X_k. Any element a in $\mathcal{J}_k$ defines the vector field $\hat{a}$ on X_k depending on the residue class $a + \mathcal{J}_{k-1}$ only. Let Θ_k be the local closure of the set of such fields. Then, $(\{\pi(\mathcal{J}_k)\}_{\pi \in X_k}, \Theta_k)$ is a C^*-algebra continuous field (the continuity follows from the separability of X_k [3, 3.3.9]).

The solvability of an algebra $\mathcal{A}$ means that the sets G_k subject to the above conditions can be chosen to satisfy, for all k, the requirements:

(i) $\pi(\mathcal{J}_k) = \mathcal{K}H_\pi$ for $\pi \in X_k$.
(ii) The algebra field $\mathcal{L}_k := (\{\mathcal{K}H_\pi\}_{\pi \in X_k}, \Theta_k)$ is trivial.

In this case, the algebra $\mathcal{A}_k$ defined by the field $\mathcal{L}_k$ is isomorphic to $C_0(T_k) \otimes \mathcal{K}H$ (see Sect. 1.3.5). Since $\mathcal{A}_k \cong \mathcal{J}_k/\mathcal{J}_{k-1}$, $k = 1, \ldots, l$ [3, 10.5.4], there is isomorphism (1.3.2) for every k.

The solvability of an algebra in the sense of Definition 1.3.20 will be temporarily called the solvability in the narrow sense. To define the solvability in the wide sense, we change condition (ii) for the condition

(ii)′ The fields $\mathcal{L}_k$ are locally trivial.

Being solvable in the narrow sense, an algebra is also solvable in the wide sense; the converse is generally false. The algebras of pseudodifferential operators in Chaps. 2–4 turn out to be solvable in the narrow sense.

A solving series allows us to determine the collection of symbols $\sigma_1(A), \dots, \sigma_N(A)$ for an operator A in an arbitrary solvable algebra $\mathcal{A}$ (see [6]). The symbol $\sigma_j(A)$ is an operator-valued function on the space X_j. In terms of such symbols, the notions of a j-parametrix and a j-symbol are introduced for A. Solving series plays an important role in studying the groups $K_*(\mathcal{A})$ of operator K-theory related to an algebra $\mathcal{A}$ [34]. These questions are not discussed in the book.

1.3.8 Maximal Radical Series

Definition 1.3.21 The maximal radical $m(\mathcal{A})$ of an algebra $\mathcal{A}$ is the intersection of all two-sided maximal ideals of $\mathcal{A}$. The composition series $\cdots \subset m(m(\mathcal{A})) \subset m(\mathcal{A}) \subset \mathcal{A}$ is called the maximal radical series.

In what follows, we describe sufficient conditions providing the coincidence of the shortest solving series with the maximal radical series.

Definition 1.3.22 A composition series $\{0\} = \mathcal{J}_{-1} \subset \mathcal{J}_0 \subset \dots \mathcal{J}_n$ in an algebra $\mathcal{A}$ is said to be stratified if

$$\pi(\mathcal{J}_k) \neq \pi(\mathcal{J}_{k+1}) \neq \dots \pi(\mathcal{J}_{n+1})$$

for every irreducible representation π of $\mathcal{A}$ satisfying $\pi(\mathcal{J}_k) \neq 0$. (In other words, the requirement $\pi(\mathcal{J}_k) \neq \pi(\mathcal{J}_{k+1})$ has to be fulfilled for any irreducible representation π of the ideal $\mathcal{J}_{k+1}$, $0 \leq k \leq n-1$.)

Proposition 1.3.23 *If a solving composition series is stratified, it coincides with the maximal radical series.*

Proof Assume that $\mathcal{J}_k/\mathcal{J}_{k-1} \simeq C_0(X_k) \otimes \mathcal{K}H_k$. Since

$$m(C_0(X_k) \otimes \mathcal{K}H_k) = m(C_0(X_k)) \otimes \mathcal{K}H_k = 0,$$

we have $m(\mathcal{J}_k/\mathcal{J}_{k-1}) = 0$. This means that $m(\mathcal{J}_k) \subseteq \mathcal{J}_{k-1}$. Indeed, let p be the projection $\mathcal{J}_k \to \mathcal{J}_k/\mathcal{J}_{k-1}$. The preimage $p^{-1}(\mu)$ of an arbitrary maximal ideal $\mu \subset \mathcal{J}_k/\mathcal{J}_{k-1}$ is a maximal ideal in $\mathcal{J}_k$. Taking into account that

$$\left(\bigcap_\mu p^{-1}(\mu)\right)/\mathcal{J}_{k-1} = \bigcap_\mu (p^{-1}(\mu)/\mathcal{J}_{k-1}) = \bigcap \mu = m(\mathcal{J}_k/\mathcal{J}_{k-1}) = 0,$$

we obtain $\mathcal{J}_{k-1} = \cap p^{-1}(\mu) \supseteq m(\mathcal{J}_k)$.

Now, we show that $m(\mathcal{J}_k) \supseteq \mathcal{J}_{k-1}$ for $k = 0, \dots, n$. We have to verify that every maximal ideal $M \subset \mathcal{J}_k$ contains $\mathcal{J}_{k-1}$. Since M is primitive, there exists an irreducible representation π of the ideal $\mathcal{J}_k$ such that $\ker \pi = M$. By our assumption, $\pi(\mathcal{J}_{k-1}) \neq \pi(\mathcal{J}_k)$. On the other hand, $\pi(\mathcal{J}_k) \simeq \mathcal{J}_k/M$ and the quotient algebra is simple, i. e., it contains no nontrivial ideals. Therefore, $\pi(\mathcal{J}_{k-1}) = 0$ and $M \supseteq \mathcal{J}_{k-1}$. □

Here is an example of an algebra whose shortest solving series does not coincide with the maximal radical series. Let X be a compact space and let $\mathcal{A} = \mathcal{K}H \oplus C(X)$. Then, the series $0 \subset \mathcal{K}H \subset \mathcal{K}H \oplus C(X)$ is solving, and the length of $\mathcal{A}$ is equal to 1. However, $m(\mathcal{A}) = 0$.

1.3.9 The Localization Principle

Let $\mathcal{A}$ be a C^*-algebra, let $\mathcal{C}$ be a commutative subalgebra of $\mathcal{A}$, and let $\widehat{\mathcal{C}}$ be the maximal ideal space of $\mathcal{C}$. For every $x \in \widehat{\mathcal{C}}$, we set $\mathcal{I}_x = \{c \in \mathcal{C} : \hat{c}(x) = 0\}$, where $\hat{c} \in C_0(\widehat{\mathcal{C}})$ is the Gelfand transformation of $c \in \mathcal{C}$. Let $\mathcal{J}_x$ be the ideal of $\mathcal{A}$ spanned by the elements in $\mathcal{I}_x$ and let $\mathcal{A}_x = \mathcal{A}/\mathcal{J}_x$. The transition from $\mathcal{A}$ to $\mathcal{A}_x$ is called the localization at x; the algebras $\mathcal{A}_x$ are called the local algebras.

Proposition 1.3.24 *Let J be an ideal in $\mathcal{A}$. Assume that*

(i) $\pi|\mathcal{C} \neq 0$ *for all* $\pi \in \widehat{\mathcal{A}}$.
(ii) *For any* $x_1, x_2 \in \widehat{\mathcal{C}}$ *such that* $x_1 \neq x_2$*, there exist* $c_1, c_2 \in \mathcal{C}$ *satisfying* $\hat{c}_j(x_j) \neq 0$*,* $j = 1, 2$*, and* $c_1\mathcal{A}c_2 \subset J$.
(iii) *For any* $\pi \in \widehat{J}$ *and* $x \in \widehat{\mathcal{C}}$ *there exist* $a \in J$ *and* $c \in \mathcal{I}_x$ *satisfying* $\pi(ac) \neq 0$.

Then,

$$\widehat{\mathcal{A}} = \cup_{x \in \widehat{\mathcal{C}}} \widehat{\mathcal{A}}_x \cup \widehat{J}. \tag{1.3.3}$$

Remark 1.3.25 The requirement (i) is automatically fulfilled if the algebra $\mathcal{A}$ contains a unity $e \in \mathcal{C}$.

Proof Since $\widehat{\mathcal{A}} = (\mathcal{A}/J)\hat{} \cup \widehat{J}$, it suffices to verify that

$$(\mathcal{A}/J)\hat{} = \cup_{x\in\widehat{\mathcal{C}}}\widehat{\mathcal{A}}_x. \tag{1.3.4}$$

Let π be an irreducible representation of $\mathcal{A}/J$; then π can be considered as a representation of $\mathcal{A}$ that vanishes on J. Let us consider the representation $\pi|\mathcal{C}$ of the algebra $\mathcal{C}$. There exists a subset $E \subset \widehat{\mathcal{C}}$ such that $\ker(\pi|\mathcal{C}) = \{c \in \mathcal{C} : \hat{c}|E = 0\}$. The set E is not empty, otherwise, the equality $\ker(\pi|\mathcal{C}) = \mathcal{C}$ would be fulfilled, which contradicts (i). We shall prove that E is a singleton. Let us assume, to the contrary, that $x_1, x_2 \in E$ and $x_1 \neq x_2$. We denote by c_1 and c_2 elements in $\mathcal{C}$ satisfying condition (ii) and by $\mathcal{J}(c_j)$, $j = 1, 2$, the ideal in $\mathcal{A}$ generated by c_j. From (ii), it follows that $\mathcal{J}(c_1)\mathcal{J}(c_2) \subset \ker\pi$ and, because $\ker\pi$ is a primitive ideal, either $\mathcal{J}(c_1) \subset \ker\pi$ or $\mathcal{J}(c_2) \subset \ker\pi$. Therefore, in view of $c_1 \in \mathcal{J}(c_1) \cap \mathcal{C}$ and $c_2 \in \mathcal{J}(c_2) \cap \mathcal{C}$, we have either $c_1 \in \ker(\pi|\mathcal{C})$ or $c_2 \in \ker(\pi|\mathcal{C})$. However, neither of the two last inclusions is possible since $\hat{c}_i|E \neq 0$, $j = 1, 2$.

Thus, $E = \{x\}$, where x is a certain point of the set $\widehat{\mathcal{C}}$. Then, $\ker(\pi|\mathcal{C}) = \mathcal{I}_x$, $\ker\pi \supset \mathcal{J}_x$, and $\pi \in (\mathcal{A}/\mathcal{J}_x)\hat{} \equiv \widehat{\mathcal{A}}_x$. As a result, we find

$$(\mathcal{A}/J)\hat{} \subset \cup_{x\in\widehat{\mathcal{C}}}\widehat{\mathcal{A}}_x.$$

Let us verify the inverse inclusion. We assume that $\pi \in \widehat{\mathcal{A}}_x$ and consider π as a representation of $\mathcal{A}$ such that $\ker\pi \supset \mathcal{J}_x$. We have $\pi|J = 0$. Indeed, otherwise, we would obtain $(\pi|J) \in \widehat{J}$. By virtue of (iii), there exist $a \in J$ and $c \in \mathcal{I}_x$ such that $\pi(ac) \neq 0$. However, $\pi(ac) = \pi(a)\pi(c) = 0$ because $\pi|\mathcal{I}_x = 0$. This contradiction shows that $\pi|J = 0$. Therefore, $\pi \in (\mathcal{A}/J)\hat{}$ and we arrive at (1.3.3). □

We also have to use another version of the localization principle in the situation where an algebra being exposed to localization does not contain a commutative subalgebra needed for such a procedure. Let $\mathcal{B}$ and $\mathcal{C}$ be subalgebras of an algebra $\mathcal{L}$. We suppose that $\mathcal{C}$ is commutative, $\mathcal{I}_x$ is defined as at the beginning of this section, and $\mathcal{J}_x$ is the ideal in $\mathcal{L}$ generated by the set $\mathcal{I}_x$. Let $\mathcal{B}_x$ denote the image of $\mathcal{B}$ under the canonical map $p_x : \mathcal{L} \to \mathcal{L}/\mathcal{J}_x$. The algebras $\mathcal{B}_x$ are called the local algebras. The choice of the algebra $\mathcal{L}$ does not play any role; changing $\mathcal{L}$ to a subalgebra of $\mathcal{L}$ containing $\mathcal{B}$ and $\mathcal{C}$, we would obtain the local algebras isomorphic to the old ones.

Let H be a Hilbert space, $\dim H = \infty$, and let $\varkappa$ denote the canonical map $\mathcal{B}H \to \mathcal{B}H/\mathcal{K}H$.

Proposition 1.3.26 *Let subalgebras $\mathcal{B}, \mathcal{C} \subset \mathcal{B}H$ satisfy the following conditions: (1) $\mathcal{K}H \subset \mathcal{B}$; (2) the algebra $\mathcal{C}$ is commutative and contains the identity operator; (3) if $C_1, C_2 \in \mathcal{C}$ and $C_1C_2 \in \mathcal{K}H$, then $C_1BC_2 \in \mathcal{K}H$ for every $B \in \mathcal{B}$. Moreover, let $\mathcal{C}$ contain a nonscalar operator and let the map $\varkappa : \mathcal{C} \to \mathcal{B}H/\mathcal{K}H$ be an isometry. Then,*

$$\widehat{\mathcal{B}} = \bigcup_{x\in\widehat{\mathcal{C}}} \widehat{\mathcal{B}}_x \cup [Id]. \tag{1.3.5}$$

C^*-algebras of Pseudodifferential Operators on Smooth Manifolds with Discontinuities in Symbols Along a Submanifold

2

This chapter deals with C^*-algebras generated by pseudodifferential operators (ΨDOs) of zero order in the space $L_2(\mathcal{M})$ on a closed smooth manifold $\mathcal{M}$. We discuss the three types of algebras: (1) the algebras spanned by the ΨDOs with smooth symbols; (2) the algebras generated by ΨDOs with smooth symbols and the operators of multiplication by functions (coefficients) that may have discontinuities "of the first kind" at finitely many points; (3) the algebras spanned by smooth ΨDOs and the coefficients with discontinuities along a submanifold $\mathcal{N}$ of the positive dimension.

These algebras contain the ideal $\mathcal{K}L_2(\mathcal{M})$ of compact operators. In the first case, the corresponding algebra $\mathcal{A}$ becomes commutative after taking the quotient by the ideal $\mathcal{K}L_2(\mathcal{M})$. In the second case, the quotient algebra $\mathcal{A}/\mathcal{K}L_2(\mathcal{M})$ is not commutative; besides one-dimensional representations, the spectrum $(\mathcal{A}/\mathcal{K}L_2(\mathcal{M}))\widehat{\ }$ includes infinite-dimensional representations. Namely, to every point $x^0 \in \mathcal{M}$ where the coefficient discontinuities are admitted, there corresponds a series of irreducible representations $\pi(x^0;\lambda) \in (\mathcal{A}/\mathcal{K}L_2(\mathcal{M}))\widehat{\ }$, $\lambda \in \mathbb{R}$, in the space $L_2(S(\mathcal{M})_{x^0})$; here $S(\mathcal{M})_{x^0}$ denotes the unit sphere of the tangent space at x^0. Finally, in the third case, the spectrum $(\mathcal{A}/\mathcal{K}L_2(\mathcal{M}))\widehat{\ }$, in addition to one-dimensional representations and those of the type $\pi(x;\lambda)$ for $x \in \mathcal{N}$, contains a series of representations parametrized by the points of the cospheric bundle $S^*(\mathcal{N})$ over the manifold $\mathcal{N}$. The last series consists of two-dimensional representations if $\operatorname{codim}\mathcal{N} = 1$ and infinite-dimensional ones for $\operatorname{codim}\mathcal{N} > 1$.

For the mentioned algebras, we indicate all equivalence classes of irreducible representations, describe the spectral topology, and present solving composition series that turn out to be maximal radical series. The results of this chapter are essentially used in Chaps. 3 and 5; they are not needed for Chap. 4.

© The Author(s), under exclusive license to Springer Nature Switzerland AG 2023

B. Plamenevskii, O. Sarafanov, *Solvable Algebras of Pseudodifferential Operators*, Pseudo-Differential Operators 15, https://doi.org/10.1007/978-3-031-28398-7_2

2.1 Algebras Generated by Pseudodifferential Operators with Smooth Symbols

Let $\mathcal{M}$ be a smooth n-dimensional compact manifold without boundary and let $\{V_j, v_j\}$ be an atlas on $\mathcal{M}$, where $\{V_j\}$ is an open covering of $\mathcal{M}$ and v_j is a coordinate diffeomorphism. The covering is chosen to be fine so that the union $V_i \cup V_j$ for any two indices i and j can serve as a coordinate neighborhood, possibly not connected. We denote by v_{ij} a diffeomorphism of $V_i \cup V_j$ onto an open subset Ω $(\equiv \Omega_{ij})$ in $\mathbb{R}^n$.

Definition 2.1.1 An operator $A : C^\infty(\mathcal{M}) \to C^\infty(\mathcal{M})$ is called a ΨDO of order m on $\mathcal{M}$ if $(v_{ij}^{-1})^* \chi_i A \chi_j v_{ij}^* \in \Psi^m(\Omega)$ for any i, j. The set of all ΨDOs of order m is denoted by $\Psi^m(\mathcal{M})$.

Let ν_j be the measure on V_j induced by the Lebesgue measure on $v_j(V_j)$. Glueing the local measures by a partition of unity, we define the measure ν on $\mathcal{M}$ and set $L_2(\mathcal{M}) := L_2(\mathcal{M}, \nu)$. The operators in $\Psi^0(\mathcal{M})$ are bounded on $L_2(\mathcal{M})$.

Proposition 2.1.2 *Let $\mathcal{P}$ be the algebra generated by the operators in $\Psi^0(\mathcal{M})$ on $L_2(\mathcal{M})$. Then: (1) the algebra $\mathcal{P}$ is irreducible; (2) $\mathcal{K}L_2(\mathcal{M}) \subset \mathcal{P}$; (3) the quotient algebra $\mathcal{P}/\mathcal{K}L_2(\mathcal{M})$ is commutative.*

Proof

(1) The algebra $\mathcal{P}$ contains every operator of multiplication by a smooth function. Therefore, any subspace invariant for $\mathcal{P}$ is contained in a subspace of the form $\chi_E L_2(\mathcal{M})$, where χ_E is the characteristic function of the set E with $0 < |E| < |\mathcal{M}|$; for example, $|E|$ denotes the measure of E. Indeed, let $\mathcal{G}$ be an invariant subspace and $P : L_2(\mathcal{M}) \to \mathcal{G}$ an orthogonal projection. Then, P commutes with all operators in $\mathcal{P}$, in particular, with the operators of multiplication by functions in $C^\infty(\mathcal{M})$. Consequently, $P(f) = fP(1)$, i.e., P acts as the operator of multiplication by the function $P(1)$. Moreover, $P^2(f) = P(f) = fP(1)$; hence, $P(1)^2 = P(1)$. Thus, the function $P(1)$ can take the values 0 and 1 only, so the function is characteristic for a certain set E.

For $\phi, \psi \in C^\infty(\mathcal{M})$, we set

$$B_{\phi,\psi} u(x) = \phi(x) \int \psi(y) u(y)\, d\nu(y).$$

Let us choose functions $u \in \mathcal{G}$ and ψ so that the integral would not be zero. Then, $\operatorname{supp}(B_{\phi,\psi} u) = \operatorname{supp} \phi$. Since $B_{\phi,\psi} \in \boldsymbol{\Psi}^{-\infty}(\mathcal{M})$, we have $B_{\phi,\psi} \in \mathcal{P}$ and

$\mathrm{supp}(B_{\phi,\psi}u) \subset E$. The last inclusion contradicts the possibility of choosing ϕ arbitrarily in $C^\infty(\mathcal{M})$.

(2) The operator $B_{\phi,\psi}$ is compact in $L_2(\mathcal{M})$, so $\mathcal{K}L_2(\mathcal{M}) \cap \mathcal{P} \neq 0$. It remains to apply Proposition 1.3.5.
(3) If A_1 and A_2 in $\Psi^0(\mathcal{M})$, the commutator $[A_1, A_2] = A_1A_2 - A_2A_1$ belongs to $\Psi^{-1}(\mathcal{M})$ (see Corollary 1.1.23). Therefore, the map $[A_1, A_2] : L_2(\mathcal{M}) \to H^1(\mathcal{M})$ is continuous. Since the embedding $H^1(\mathcal{M}) \subset L_2(\mathcal{M})$ is compact, we obtain $[A_1, A_2] \in \mathcal{K}L_2(\mathcal{M})$.

□

The algebra $C(\mathcal{M})$ of continuous functions on M is a subalgebra in $\mathcal{P}$. For $x \in \mathcal{M}$, we set $\mathcal{I}_x := \{f \in C(M) : f(x) = 0\}$ and denote by $\mathcal{J}_x$ the ideal in $\mathcal{P}$ generated by the set $\mathcal{I}_x$.

Proposition 2.1.3 *The equality*

$$\widehat{\mathcal{P}} = \cup_{x\in\mathcal{M}}(\mathcal{P}/\mathcal{J}_x\widehat{)} \cup [Id]$$

holds, where $[Id]$ *is the equivalence class of the identity representation.*

Proof We are going to apply Proposition 1.3.24 for $\mathcal{A} = \mathcal{P}$, $J = \mathcal{K}L_2(\mathcal{M})$, and $\mathcal{C} = C(\mathcal{M})$. Let us verify that the assumptions of the proposition are fulfilled. It suffices to show that, for $a \in C(\mathcal{M})$ and $A \in \mathcal{P}$, the commutator $[a, A] = aA - Aa$ is compact. We can consider that the inclusions $a \in C^\infty(\mathcal{M})$ and $A \in \Psi^0(\mathcal{M})$ are valid. Then, $[a, A] \in \Psi^{-1}(\mathcal{M})$ (Corollary 1.1.23) and $[a, A] \in \mathcal{K}L_2(\mathcal{M})$. It remains to take into account that $C(\mathcal{M}\widehat{)} = \mathcal{M}$. □

Definition 2.1.4 Let $\bar{\Psi}^0(\mathcal{M})$ be the set of operators in $\Psi^0(\mathcal{M})$ subject to the following additional conditions. If $A \in \bar{\Psi}^0(\mathcal{M})$ and σ is the symbol of an operator of the form $(v_{ij}^{-1})^*\chi_i A\chi_j v_{ij}^*$, then:

(i) There exists the limit $\sigma^0(x, \xi) := \lim_{t\to+\infty} \sigma(x, t\xi)$.
(ii) The function $(x, \xi) \mapsto \sigma(x, \xi) - \zeta(\xi)\sigma^0(x, \xi)$ belongs to the class $S^{-1}(\Omega)$, where $\zeta \in C^\infty(\mathbb{R}^n)$, $\zeta(\xi) = 0$ for $|\xi| \leq 1/2$, and $\zeta(\xi) = 1$ for $|\xi| \geq 1$.

We assume that σ_V is a symbol of $A \in \bar{\Psi}^0(\mathcal{M})$ in a local coordinate system V. According to the definition of the class $\bar{\Psi}^0(\mathcal{M})$, there exists the limit $\sigma_V^0(x, \xi) = \lim_{t\to+\infty} \sigma_V(x, t\xi)$. For a fixed x, the function $\xi \to \sigma_V^0(x, \xi)$ is homogeneous of zero degree. Thus, we obtain the function σ^0 defined on the cospheric bundle $S^*(\mathcal{M})$ over $\mathcal{M}$. We call σ^0 the principal symbol of A.

Let $\mathcal{A}$ denote the algebra generated by the operators of the class $\bar{\Psi}^0(\mathcal{M})$ on $L_2(\mathcal{M})$. Propositions 2.1.2 and 2.1.3 remain valid (as well as their proof) with the algebra $\mathcal{P}$ replaced by $\mathcal{A}$. The next assertion describes the local algebras $\mathcal{A}/\mathcal{J}_x$.

Proposition 2.1.5 *Let $A \in \bar{\Psi}^0(\mathcal{M})$ and let σ^0 be the principal symbol of A. For $z \in \mathcal{M}$, the map $A \mapsto \sigma^0$ extends to an isomorphism $\mathcal{A}/\mathcal{J}_z \cong C(S^*(\mathcal{M})_z)$, where as usual $S^*(\mathcal{M})_z$ is the fiber of $S^*(\mathcal{M})$ over z.*

Proof The map $A \rightarrow \sigma^0$ is compatible with algebraic operations, i.e., it is linear, multiplicative, and commutes with involution. Therefore, it suffices to show that

$$\inf\{\|A + J\|; J \in \mathcal{J}_z\} = \|\sigma^0|S_z^*; C(S_z^*)\|$$

for every $A \in \bar{\Psi}^0(\mathcal{M})$.

We first verify the inequality

$$\inf\{\|A + J\|; J \in \mathcal{J}_z\} \geq \|\sigma^0|S_z^*; C(S_z^*)\|. \tag{2.1.1}$$

Let $\chi \in C_c^\infty(V)$, where V is a coordinate neighborhood of z, $|\chi| \leq 1$, and $\chi(z) = 1$. Since the principal symbols of the operators A and $\chi A \chi$ coincide on S_z^* and $\|A + J\| \geq \|\chi A \chi + \chi J \chi\|$, we can consider that $V = \mathbb{R}^n$ and $z = 0$. Now, the role of $\mathcal{A}$ is played by the algebra spanned by the operators $\chi A \chi$, where $A \in \bar{\Psi}^0(\mathbb{R}^n)$, while $\mathcal{J}_z = \mathcal{J}_0$ is the ideal in $\mathcal{A}$ generated by the operators of multiplication by the functions $\zeta \in C_c^\infty(\mathbb{R}^n)$ equal to 0 at the coordinate origin. Let us introduce the operator $U_t : u(\cdot) \mapsto t^{n/2}u(t\cdot)$ with $t > 0$; the operator is unitary on $L_2(\mathbb{R}^n)$. For $A = \mathrm{Op}\sigma$, we have

$$(U_t A U_t^{-1} u)(x) = (2\pi)^{-n/2} \int e^{ix\xi} \sigma(tx, \xi/t) \hat{u}(\xi)\, d\xi, \ \ u \in C_c^\infty(\mathbb{R}^n).$$

We set $A_0 := F^{-1}\sigma^0(0, \xi)F$ and, for any $J \in \mathcal{J}_0$, obtain $U_t(A + J)U_t^{-1} \rightarrow A_0$ in the strong operator topology as $t \rightarrow 0$. This implies the relation

$$\|A + J\| \geq \|A_0\| = \sup\{|\sigma^0(0, \xi)|; \xi \in S^{n-1}\},$$

which leads to (2.1.1).

To verify the converse inequality, we set

$$Bu(x) = F_{\xi \rightarrow x}^{-1} \sigma^0(0, \xi)(1 - \chi(\xi)) F_{y \rightarrow \xi} u(y),$$

where $\chi \in C_c^\infty(\mathbb{R}^n)$, $0 \le \chi \le 1$, and $\chi(0) = 1$. Moreover, let $b_1(x, \xi) = \sigma(x, \xi) - \sigma(0, \xi)$ and let $b_2(x, \xi) = \sigma(0, \xi) - \sigma^0(0, \xi)(1 - \chi(\xi))$. Then

$$\chi(A - B)\chi = \chi \mathrm{Op} b_1 \chi + \chi \mathrm{Op} b_2 \chi. \tag{2.1.2}$$

The operator $\chi \mathrm{Op} b_1 \chi$ belongs to the ideal $\mathcal{J}_0$. To prove the inclusion $\chi \mathrm{Op} b_2 \chi \in \mathcal{J}_0$, we note that $(1 - \chi)K \in \mathcal{J}_0$ for any compact operator K; hence, $\mathcal{K}L_2(\mathbb{R}^n) \cap \mathcal{J}_0 \neq 0$. Being a nonzero ideal of irreducible algebra, $\mathcal{J}_0$ is irreducible (Proposition 1.3.2, (2)); therefore, $\mathcal{K}L_2(\mathbb{R}^n) \subset \mathcal{J}_0$ (Proposition 1.3.5). From the equality $\lim_{\xi \to \infty} b_2(x, \xi) = 0$, it follows that the operator $\chi \mathrm{Op} b_2 \chi$ is compact and, consequently, belongs to $\mathcal{J}_0$. Now, in view of (2.1.2), we obtain $\chi(A - B)\chi \in \mathcal{J}_0$. Finally,

$$\inf \|\chi A \chi + J\| \le \|\chi B \chi\| \le \sup |\sigma^0(0, \xi)(1 - \chi(\xi))| \le \sup |\sigma^0(0, \xi)|.$$

□

We set $\Upsilon = S^*(\mathcal{M}) \cup e$, where e is a point, and introduce a topology on Υ. We assume that e is an open set and that the fundamental system of neighborhoods of a point $\xi \in S^*(\mathcal{M})$ consists of the sets $\mathcal{V} \cup e$, where $\mathcal{V}$ is a neighborhood of ξ in $S^*(\mathcal{M})$.

Theorem 2.1.6 *Let $\mathcal{A}$ be the algebra of operators on $L_2(\mathcal{M})$ generated by the ΨDOs in $\bar{\Psi}^0(\mathcal{M})$ and let σ_A^0 be the principal symbol of $A \in \bar{\Psi}^0(\mathcal{M})$. Then:*

(1) *The map $A \mapsto \sigma_A^0$ extends to an isomorphism $j : \mathcal{A}/\mathcal{K}L_2(\mathcal{M}) \to C(S^*(\mathcal{M}))$.*
(2) *Any irreducible representation of $\mathcal{A}$ is either one-dimensional or equivalent to the identity representation Id. A one-dimensional representation can be implemented as $\pi(\xi) : A \mapsto \sigma_A^0(\xi)$, where $\sigma_A^0 = j([A])$, $[A]$ is the equivalence class of A in $\mathcal{A}/\mathcal{K}L_2(\mathcal{M})$, and $\xi \in S^*(\mathcal{M})$.*
(3) *The relation $\pi(\xi) \mapsto \xi$, $Id \mapsto e$ determines a bijection from the spectrum $\widehat{\mathcal{A}}$ onto Υ. The topology on Υ coincides with the Jacobson topology.*
(4) *An operator $A \in \mathcal{A}$ is Fredholm if and only if the symbol $j([A])$ is distinct from zero throughout $S^*(\mathcal{M})$.*
(5) *The maximal radical $m(\mathcal{A})$ of the algebra $\mathcal{A}$, i.e., the intersection of all two-sided maximal ideals in $\mathcal{A}$, coincides with $\mathcal{K}L_2(\mathcal{M})$. The composition series $0 \subset \mathcal{K}L_2(\mathcal{M}) \subset \mathcal{A}$ is solving and coincides with the maximal radical series. The length of $\mathcal{A}$ is equal to 1.*

To verify this theorem, one should compare Propositions 2.1.3, 2.1.5, and 1.3.12; details are left to the reader.

2.2 Algebras of Pseudodifferential Operators with Isolated Singularities in Symbols

2.2.1 Algebras $\mathcal{A}$ and $\mathfrak{S}$

We fix a point $x^0 \in \mathcal{M}$ and introduce the set $\mathfrak{M}$ of continuous functions on $\mathcal{M} \setminus x^0$ such that, for $f \in \mathfrak{M}$, there exists a finite limit $\lim_{r \to 0} f(r, \varphi) = f^0(\varphi)$ uniform with respect to φ, where (r, φ) are local spheric coordinates with the origin at x^0. In this section, $\mathcal{A}$ denotes the algebra generated on $L_2(\mathcal{M})$ by the operators in $\bar{\Psi}^0(\mathcal{M})$ and the operators $f\cdot$ of multiplication by functions $f \in \mathfrak{M}$. We can use the same localizing subalgebra $C(\mathcal{M})$ as in Sect. 2.1. Just as in the proof of Proposition 2.1.5, it is easy to see that the local algebra $\mathcal{A}_{x^0} := \mathcal{A}/\mathcal{J}_{x^0}$ is generated on $L_2(\mathbb{R}^n)$ by operators of the form

$$Au(x) = (2\pi)^{-n/2} \int e^{ix\xi} \sigma^0(\xi)\hat{u}(\xi)\, d\xi \tag{2.2.1}$$

and

$$f^0 \cdot u(x) = f^0(x/|x|)u(x), \tag{2.2.2}$$

where σ^0 is a zero degree homogeneous function, while f^0 and σ^0 belong to $C^\infty(S^{n-1})$. In what follows, we write Φ and a instead of σ^0 and f^0. Recollecting Sect. 1.2.2, in particular (1.2.14)), we write (2.2.1) in the form

$$Au(r, \varphi) = (2\pi)^{-1/2} \int_{-\infty}^{+\infty} r^{i(n/2+\lambda)} \mathfrak{A}_{\psi \to \varphi}(\lambda)\tilde{u}(\lambda + in/2, \psi)\, d\lambda, \tag{2.2.3}$$

where

$$\mathfrak{A}_{\psi \to \varphi}(\lambda) = E_{\omega \to \varphi}(\lambda)^{-1} \Phi(\omega) E_{\psi \to \omega}(\lambda). \tag{2.2.4}$$

We apply the Parseval equality to the Mellin transform and obtain that the algebra $\mathcal{A}_{x^0}$ is isomorphic to the algebra $\mathfrak{S}$ spanned by functions of the form

$$\mathbb{R} \ni \lambda \mapsto \mathfrak{B}(\lambda) = a(\varphi) E_{\omega \to \varphi}(\lambda)^{-1} \Phi(\omega) E_{\psi \to \omega}(\lambda), \tag{2.2.5}$$

where a and Φ are in $C^\infty(S^{n-1})$, with pointwise operations and the norm

$$\|\mathfrak{B}; \mathfrak{S}\| = \sup \{ \|\mathfrak{B}(\lambda); \mathcal{B}L_2(S^{n-1})\|, \ \lambda \in \mathbb{R} \}.$$

We intend to use Proposition 1.3.24 for the algebra $\mathfrak{S}$ taking the ideal $C_0(\mathbb{R}) \otimes \mathcal{K}L_2(S^{n-1})$ as J. Therefore, we first verify the inclusion $C_0(\mathbb{R}) \otimes \mathcal{K}L_2(S^{n-1}) \subset \mathfrak{S}$. Before describing the local algebras, we deal with some technical preparations; they mainly relate to the behavior of operator (2.2.4) under local straightening of the sphere S^{n-1}. All irreducible representations (to within equivalence) of the algebra $\mathfrak{S}$ are given by Theorem 2.2.11. Then, we investigate the Jacobson topology on the set $\widehat{\mathfrak{S}}$ and describe the spectrum $\widehat{\mathcal{A}}$.

2.2.2 Proof of the Inclusion $C_0(\mathbb{R}) \otimes \mathcal{K}L_2(S^{n-1}) \subset \mathfrak{S}$

Let $\mathfrak{S}(\lambda)$ denote the algebra spanned by operators of the form (2.2.5) on $L_2(S^{n-1})$, where λ is a fixed real number.

Proposition 2.2.1 *The following assertions are valid:*

(1) *The algebra $\mathfrak{S}(\lambda)$ is irreducible for every $\lambda \in \mathbb{R}$.*
(2) *$\mathcal{K}L_2(S^{n-1}) \subset \mathfrak{S}(\lambda)$, and the quotient algebra $\mathfrak{S}(\lambda)/\mathcal{K}L_2(S^{n-1})$ is commutative.*

Proof

(1) The algebra $\mathfrak{S}(\lambda)$ contains every operator of multiplication by a smooth function. Therefore, any invariant subspace of this algebra is contained in a subspace of the form $\chi_\Omega L_2(S^{n-1})$, where χ_Ω is the characteristic function of the set $\Omega \subset S^{n-1}$ with $0 < |\Omega| < |S^{n-1}|$ (see the proof of Proposition 2.1.2). Let us suppose that there exists a nontrivial invariant subspace and let u be an element of such a subspace. We can assume that the support of u is in an open semisphere and choose a point ω_0 so that the set $\{\psi \in S^{n-1} : |\psi\omega_0| < \varepsilon\}$ is disjoint from $\operatorname{supp} u$ for sufficiently small ε. Denote by $\{G_m\}$ a sequence of smooth averaging kernels such that

$$\int_{S^{n-1}} G_m(\theta, \psi)u(\psi)\, d\psi \to u(\theta) \text{ in } L_2(S^{n-1}).$$

Then,

$$\int_{S^{n-1}} G_m(\theta, \psi)(-\omega_0\psi + i0)^{-i\lambda-n/2}u(\psi)\, d\psi \to (-\omega_0\theta + i0)^{-i\lambda-n/2}u(\theta) \tag{2.2.6}$$

for almost all θ. We fix a point $\theta = \theta_0$ such that the limit (2.2.6) exists and is nonzero. Then, for sufficiently large m,

$$\int_{S^{n-1}} G_m(\theta_0, \psi)(-\omega_0\psi + i0)^{-i\lambda-n/2}u(\psi)\, d\psi \neq 0. \tag{2.2.7}$$

Obviously, the function $\psi \mapsto v(\psi) = G_m(\theta_0, \psi)u(\psi)$ belongs to the invariant subspace.

Denote by $\{\Phi_k\}$ a δ-shaped sequence of smooth functions such that the sets $\operatorname{supp}\Phi_k$ shrink to the point ω_0. Further, let a be a smooth function equal to 0 in a small neighborhood of the equator $\{\varphi : \varphi\omega_0 = 0\}$ and to 1 outside another small neighborhood of this equator. Setting $\mathfrak{A}_k(\lambda) = a(\varphi)E(\lambda)^{-1}\Phi_k(\omega)E(\lambda)$, we have

$$\mathfrak{A}_k(\lambda)v \to C(\lambda)a(\varphi)(\varphi\omega_0 + i0)^{i\lambda-n/2}\int_{S^{n-1}}(-\omega_0\psi + i0)^{-i\lambda-n/2}v(\psi)\,d\psi, \tag{2.2.8}$$

where $C(\lambda) = (2\pi)^{-n}\exp(i\pi n/2)\Gamma(-i\lambda+n/2)\Gamma(i\lambda+n/2)$. From (2.2.7) and the equality

$$(-\varphi\omega_0 + i0)^{-i\lambda-n/2} = e^{-i(i\lambda+n/2)\pi}(\varphi\omega_0)_+^{-i\lambda-n/2} + (\varphi\omega_0)_-^{-i\lambda-n/2},$$

it follows that the right-hand side of (2.2.8) cannot vanish almost everywhere in $S^{n-1}\setminus\Omega$. We have obtained a contradiction.

(2) Let $a \in C^\infty(S^{n-1})$ and let $\mathfrak{A}(\lambda)$ be an operator of the form (2.2.4). From Proposition 1.2.11, it follows that the commutator $[a, \mathfrak{A}(\lambda)]$ is compact on $L_2(S^{n-1})$. Thus, $\mathfrak{S}(\lambda) \cap \mathcal{K}L_2(S^{n-1}) \neq 0$. In view of Proposition 1.3.5, $\mathcal{K}L_2(S^{n-1}) \subset \mathfrak{S}(\lambda)$.

□

Proposition 2.2.2 *Let λ_1 and λ_2 be distinct real numbers. Then, the representations $\pi(\lambda_j) : \mathfrak{B}(\cdot) \mapsto \mathfrak{B}(\lambda_j)$, $j = 1, 2$, of the algebra $\mathfrak{S}$ on $L_2(S^{n-1})$ are nonequivalent.*

Proof Let $\Phi \in C^\infty(S^{n-1} \times S^{n-1})$ and $\mathfrak{A}(\lambda) = E_{\omega\to\varphi}(\lambda)^{-1}\Phi(\varphi,\omega)E_{\psi\to\omega}(\lambda)$. It is clear that $\mathfrak{A} \in \mathfrak{S}$. By virtue of Proposition 1.2.11,

$$\mathfrak{A}(\lambda)\psi_j = \varphi_j\mathfrak{A}(\lambda+i) + E^{-1}(\lambda)D_{\omega_j}\Phi(\varphi,\omega)E(\lambda+i), \tag{2.2.9}$$

where $\varphi = (\varphi_1, \dots, \varphi_n) \in \mathbb{R}^n$, $|\varphi| = 1$, and so on. Let us replace $\Phi(\varphi,\omega)$ by $\varphi\omega\Phi(\varphi,\omega)$ in (2.2.9). Taking into account the definition of the operator $E(\lambda)^{-1}$ (see (1.2.8)) and the relation $(\varphi\omega + i0)^{i\mu}\varphi\omega = (\varphi\omega + i0)^{i(\mu-i)}$, we obtain

$$\mathfrak{A}(\lambda+i) = E(\lambda+i)^{-1}\varphi\omega\Phi(\varphi,\omega)E(\lambda+i) = (\lambda+in/2)E(\lambda)^{-1}\Phi(\varphi,\omega)E(\lambda+i).$$

Now, equality (2.2.9) takes the form

$$\begin{aligned} &E_{\omega\to\varphi}(\lambda)^{-1}\varphi\omega\Phi(\varphi,\omega)E_{\psi\to\omega}(\lambda)\psi_j \\ &= \varphi_j(in/2+\lambda)E_{\omega\to\varphi}(\lambda)^{-1}\Phi(\varphi,\omega)E_{\psi\to\omega}(\lambda+i) \\ &+ E_{\omega\to\varphi}(\lambda)^{-1}D_{\omega_j}(\varphi\omega\Phi(\varphi,\omega))E_{\psi\to\omega}(\lambda+i). \end{aligned} \tag{2.2.10}$$

The definition (1.2.2) of $E(\lambda)$ and the equality $(-\omega\psi+i0)^{-i\mu}\omega\psi = -(-\omega\psi+i0)^{-i(\mu+i)}$ imply that

$$E(\lambda)^{-1}\Phi(\varphi,\omega)E(\lambda)\omega\psi = -(i\lambda - 1 + n/2)E(\lambda)^{-1}\Phi(\varphi,\omega)E(\lambda+i). \quad (2.2.11)$$

We multiply (2.2.11) by φ_j and subtract the result from (2.2.10). Then,

$$\begin{aligned} E(\lambda)^{-1}\varphi\omega\Phi(\varphi,\omega)E(\lambda)\psi_j - \varphi_j E(\lambda)^{-1}\Phi(\varphi,\omega)E(\lambda)\omega\psi \quad (2.2.12)\\ = E(\lambda)^{-1}\Psi(\varphi,\omega)E(\lambda+i), \end{aligned}$$

where $\Psi(\varphi,\omega) = i(n-1)\varphi_j\Phi(\varphi,\omega) + D_{\omega_j}(\varphi\omega\Phi(\varphi,\omega))$. It is obvious that the left-hand sides of (2.2.11) and (2.2.12) belong to the algebra $\mathfrak{S}$. Let us replace in (2.2.11) the function Φ by Ψ. The operator-functions

$$\lambda \mapsto \mathfrak{M}(\lambda) := (i\lambda - 1 + n/2)E(\lambda)^{-1}\Psi(\varphi,\omega)E(\lambda+i),$$

$$\lambda \mapsto \mathfrak{N}(\lambda) := E(\lambda)^{-1}\Psi(\varphi,\omega)E(\lambda+i)$$

also belong to $\mathfrak{S}$. Suppose that, for some distinct λ_1 and λ_2, the representations $\pi(\lambda_1)$ and $\pi(\lambda_2)$ are equivalent. Then, the norms of both the operators $\mathfrak{M}(\lambda_1)$ and $\mathfrak{M}(\lambda_2)$ and those of $\mathfrak{N}(\lambda_1)$ and $\mathfrak{N}(\lambda_2)$ must coincide. If $\lambda_1 \neq -\lambda_2$, this is impossible since $|i\lambda_1 - 1 - n/2| \neq |i\lambda_2 - 1 - n/2|$. In case of $\lambda_1 = -\lambda_2$, the values of the function $\mathfrak{M}(\lambda) - (i\lambda_1 - 1 + n/2)\mathfrak{N}(\lambda) = i(\lambda - \lambda_1)\mathfrak{N}(\lambda)$ have distinct norms at λ_1 and λ_2. □

Proposition 2.2.3 *The algebra $\mathfrak{S}$ contains the ideal $C_0(\mathbb{R}) \otimes \mathcal{K}L_2(S^{n-1})$.*

Proof The intersection $\mathfrak{S} \cap (C_0(\mathbb{R}) \otimes \mathcal{K}L_2(S^{n-1}))$ is a closed two-sided ideal in $\mathfrak{S}$. The restriction of the representations $\pi(\lambda_1)$ and $\pi(\lambda_2)$ to this ideal is irreducible and nonequivalent (Proposition 1.3.7). It follows that $\mathfrak{S} \cap (C_0(\mathbb{R}) \otimes \mathcal{K}L_2(S^{n-1}))$ is a rich subalgebra of the algebra $C_0(\mathbb{R}) \otimes \mathcal{K}L_2(S^{n-1})$; therefore, $\mathfrak{S} \cap (C_0(\mathbb{R}) \otimes \mathcal{K}L_2(S^{n-1})) = C_0(\mathbb{R}) \otimes \mathcal{K}L_2(S^{n-1})$ (Proposition 1.3.9). □

2.2.3 Auxiliary Results

This section is mainly devoted to describing the behavior of operator (2.2.4) under local straightening the sphere S^{n-1}. Keeping in mind further applications, we consider here a somewhat more general situation than it is necessary for studying the algebra $\mathfrak{S}$ introduced in (2.2.5).

Let L be a subalgebra of the algebra $L_\infty(S^{n-1})$ and $C(S^{n-1}) \subset L$. We denote by $\widetilde{\mathfrak{S}}$ the algebra generated by operator-functions of the form (2.2.5), where $a \in L$ and $\Phi \in$

$C^\infty(S^{n-1})$. The norm in $\widetilde{\mathfrak{S}}$ is defined by the equality

$$\|\mathfrak{B}; \widetilde{\mathfrak{S}}\| = \sup\{\|\mathfrak{B}(\lambda); \mathcal{B}L_2(S^{n-1})\|, \lambda \in \mathbb{R}\}.$$

Let $\mathcal{N} = (0, \ldots, 0, 1)$ be the northern pole of the sphere S^{n-1} and let $J(\mathcal{N})$ be the ideal of the algebra $\widetilde{\mathfrak{S}}$ that is spanned by the operators ζId, where $\zeta \in C(S^{n-1})$ and $\zeta(\mathcal{N}) = 0$.

Proposition 2.2.4 *Assume that* $\{\chi_q\}$ *is a sequence of functions in* $C(S^{n-1})$ *such that* $\chi_q(\mathcal{N}) = 1$, $|\chi_q| \leq 1$, *and the sets supp* χ_q *shrink to the point* $\mathcal{N}$. *Then the relation* $\|\chi_q\mathfrak{B}; \widetilde{\mathfrak{S}}\| \to 0$ *is equivalent to the inclusion* $\mathfrak{B} \in J(\mathcal{N})$ *for an element* $\mathfrak{B} \in \widetilde{\mathfrak{S}}$.

Proof We suppose that $\|\chi_q\mathfrak{B}; \widetilde{\mathfrak{S}}\| = \|\mathfrak{B} - (1-\chi_q)\mathfrak{B}; \widetilde{\mathfrak{S}}\| \to 0$. Since the operator-function $\lambda \mapsto (1-\chi_q)\mathfrak{B}(\lambda)$ belongs to the ideal $J(\mathcal{N})$, it follows that $\mathfrak{B} \in J(\mathcal{N})$.

Now, we assume that $\mathfrak{B} \in J(\mathcal{N})$ and prove the relation $\|\chi_q\mathfrak{B}; \widetilde{\mathfrak{S}}\| \to 0$. It suffices to consider that $\mathfrak{B} = \sum_j \prod_k \zeta_{jk}\mathfrak{A}_{jk}\eta_{jk}$, where the operators $\mathfrak{A}_{jk}$ are of the form (2.2.5), $\zeta_{jk}, \eta_{jk} \in C(S^{n-1})$ (j and k take finitely many values), and at least one of the factors ζ_{jk} or η_{jk} in every product $\prod_k \zeta_{jk}\mathfrak{A}_{jk}\eta_{jk}$ vanishes at $\mathcal{N}$. Let us choose a sequence $\{\kappa_q\}$ of functions in $C(S^{n-1})$ that satisfy the same conditions as $\{\chi_q\}$ and, in addition, are subject to the requirement $\kappa_q\chi_q = \chi_q$. For example, if $\eta_{jl}(\mathcal{N}) = 0$, then the j-th term in the sum $\chi_q\mathfrak{B}(\lambda)$ can be written in the form

$$\begin{aligned} &\chi_q \textstyle\prod_k \zeta_{jk}\mathfrak{A}_{jk}(\lambda)\eta_{jk} \\ &= \kappa_q K(\lambda) + \zeta_{j1}\mathfrak{A}_{j1}(\lambda)\eta_{j1}\ldots\zeta_{jl}\mathfrak{A}_{jl}(\lambda)\eta_{jl}\chi_q\ldots\zeta_{jM}\mathfrak{A}_{jM}(\lambda)\eta_{jM}, \end{aligned} \tag{2.2.13}$$

where $K(\lambda) \in \mathcal{K}L_2(S^{n-1})$ and $K(\lambda) \to 0$ as $\lambda \to \infty$. Since $\eta_{jl}\chi_q \to 0$ in $C(S^{n-1})$, the second term in (2.2.13) tends to 0 with respect to the norm in $\widetilde{\mathfrak{S}}$.

It remains to be shown that $\|\kappa_q K; \widetilde{\mathfrak{S}}\| \to 0$. We choose a large number T so that $\|\kappa_q K(\lambda); \mathcal{B}L_2(S^{n-1})\| < \varepsilon$ for $|\lambda| > T$ and a given $\varepsilon > 0$. For any $\lambda \in [-T, T]$, there exists a finite ε-net for the set $Q(\lambda) = \{v : v = K(\lambda)u, \|u; L_2(S^{n-1})\| = 1\}$. This and the continuity of the function $\lambda \mapsto K(\lambda)$ imply the existence of a finite ε-net $\{w\}$ for the union $\cup Q(\lambda)$, $|\lambda| \leq T$. For every $u \in L_2(S^{n-1})$ with unit norm and any $\lambda \in [-T, T]$, we have

$$\|\kappa_q K(\lambda)u; L_2(S^{n-1})\| \leq \|\kappa_q(K(\lambda)u - w); L_2(S^{n-1})\| + \|\kappa_q w; L_2(S^{n-1})\|,$$

where w is such an element of the net that the first term on the right is less than ε. The inequality $\|\kappa_q w; L_2(S^{n-1})\| < \varepsilon$ is fulfilled for all elements of the net for sufficiently large q. In summary, we arrive at $\|\kappa_q K; \widetilde{\mathfrak{S}}\| \leq 2\varepsilon$. □

Let $\widetilde{\mathfrak{S}}_+$ denote the subalgebra of the algebra $\widetilde{\mathfrak{S}}$ generated by the operator-functions $\lambda \mapsto \zeta\mathfrak{A}(\lambda)\eta$, where $\mathfrak{A}$ is a function of the form (2.2.5) and $\zeta, \eta \in C_c(S^{n-1}_+)$; from now on, the inclusion $\chi \in C_c(S_+^{n-1})$ means that $\chi \in C(S^{n-1})$ and $\operatorname{supp}\chi$ belongs to the open semisphere $S_+^{n-1} = \{x = (x', x_n) \in \mathbb{R}^n : |x| = 1, x_n > 0\}$ with pole $\mathcal{N} = (0', 1)$. We set $J_+(\mathcal{N}) = \widetilde{\mathfrak{S}}_+ \cap J(\mathcal{N})$.

Proposition 2.2.5 *The embedding $\widetilde{\mathfrak{S}}_+ \to \widetilde{\mathfrak{S}}$ induces an isomorphism*

$$j : \widetilde{\mathfrak{S}}_+/J_+(\mathcal{N}) \simeq \widetilde{\mathfrak{S}}/J(\mathcal{N}).$$

Proof The continuity of the map $j : \widetilde{\mathfrak{S}}_+/J_+(\mathcal{N}) \to \widetilde{\mathfrak{S}}/J(\mathcal{N})$ follows from the norm definition in a quotient algebra. Therefore, it suffices to verify the inequality $\|j([A]); \widetilde{\mathfrak{S}}/J(\mathcal{N})\| \geq \|[A]; \widetilde{\mathfrak{S}}_+/J_+(\mathcal{N})\|$. Let $\{\chi_q\}$ be the sequence in Proposition 2.2.4. For a certain $K_\varepsilon \in J(\mathcal{N})$, we have

$$\begin{aligned}\|j([A]); \widetilde{\mathfrak{S}}/J(\mathcal{N})\| &\geq \|A + K_\varepsilon; \widetilde{\mathfrak{S}}\| - \varepsilon \geq \|\chi_q A; \widetilde{\mathfrak{S}}\| - \|\chi_q K_\varepsilon; \widetilde{\mathfrak{S}}\| - \varepsilon \\ &\geq \|[A]; \widetilde{\mathfrak{S}}_+/J_+(\mathcal{N})\| - \|\chi_q K_\varepsilon; \widetilde{\mathfrak{S}}\| - \varepsilon.\end{aligned}$$

It remains to note that, by virtue of Proposition 2.2.4, $\|\chi_q K_\varepsilon; \widetilde{\mathfrak{S}}\| \to 0$ as $q \to \infty$. □

The next proposition describes the transformation of $\zeta\mathfrak{A}(\lambda)\eta$ under local straightening the sphere in a neighborhood of the point $\mathcal{N}$; here, $\mathfrak{A}$ is an operator of the form (2.2.4), and ζ and η are in $C_c(S_+^{n-1})$. Let $x = (x', x_n) \in \mathbb{R}^n$, $r(x') = (1 + |x'|^2)^1/2$, and $\varkappa(x') = (x'/r(x'), 1/r(x'))$. Then, the volume element $d\psi$ of S^{n-1} is $r(x')^{-2(n-1)}dx'$, where $\psi = \varkappa(x')$. For $v \in L_2(S_+^{n-1})$, we set

$$(U(\lambda)v)(x') = v(\varkappa(x'))r(x')^{i\lambda-(n-1)}.$$

The values of the function $\mathbb{R} \mapsto U(\lambda) : L_2(S_+^{n-1}) \to L_2(\mathbb{R}^{n-1})$ are unitary operators.

Let $G(x) = f(x/|x|)|x|^{-n}$ with $f \in C^\infty(S^{n-1})$ and let the function f be orthogonal to 1 in $L_2(S^{n-1})$. Then, the operator (2.2.4) with $\Phi(\xi) = F_{x\to\xi}G(x)$ admits the representation

$$(\mathfrak{A}(\lambda)v)(\varphi) = \int_{S^{n-1}} v(\psi)\,d\psi \int_0^\infty t^{-i(\lambda+in/2)-1}G(t\varphi - \psi)\,dt \tag{2.2.14}$$

(see Sect. 1.2.3).

Proposition 2.2.6 *Let $u \in L_2(\mathbb{R}^{n-1})$ and let η and ζ be in $C_c(S_+^{n-1})$. Then,*

$$\begin{aligned}&(U(\lambda)\zeta\mathfrak{A}(\lambda)\eta U(\lambda)^{-1}u)(x') = \zeta(\varkappa(x'))r(x')^{1-n/2} \\ &\times \int_{\mathbb{R}^{n-1}} u(y')\eta(\varkappa(y'))r(y')^{1-n/2}\,dy' \int_0^\infty t^{-i(\lambda+in/2)-1}G(tx' - y', t-1)\,dt. \quad (2.2.15)\end{aligned}$$

Proof According to (2.2.14),

$$(U(\lambda)\zeta\mathfrak{A}(\lambda)\eta U(\lambda)^{-1}u)(x') = \zeta(\varkappa(x'))r(x')^{i\lambda-n+1}\int_{S^{n-1}} u(\varkappa^{-1}(\psi))$$

$$\times\eta(\psi)r(\varkappa^{-1}(\psi))^{-i\lambda+n-1}\,d\psi\int_0^\infty t^{-i(\lambda+in/2)-1}G(t\varkappa(x')-\psi)\,dt$$

$$= \zeta(\varkappa(x'))r(x')^{i\lambda-n+1}\int_{\mathbb{R}^{n-1}} u(y')\eta(\varkappa(y'))r(y')^{-i\lambda-n+1}\,dy'$$

$$\times\int_0^\infty t^{-i(\lambda+in/2)-1}G(t\varkappa(x')-\varkappa(y'))\,dt. \quad (2.2.16)$$

Taking into account the definition of $\varkappa$ and the homogeneity of the function G, we write the inner integral in the form

$$\int_0^{+\infty} t^{-i(\lambda+in/2)-1}G(t\varkappa(x')-\varkappa(y'))\,dt$$

$$= r(y')^n\int_0^{+\infty} t^{-i(\lambda+in/2)-1}G(tr(y')x'/r(x')-y', tr(y')/r(x')-1)\,dt$$

$$= r(y')^n(r(x')/r(y'))^{-i(\lambda+in/2)}\int_0^{+\infty} t^{-i(\lambda+in/2)-1}G(tx'-y', t-1)\,dt.$$

It remains to substitute this expression into (2.2.16). □

We set

$$(\mathfrak{B}(\lambda)u)(x') = r(x')^{1-n/2}\int_{\mathbb{R}^{n-1}} u(y')r(y')^{1-n/2}\,dy'\int_0^{+\infty} t^{-i(\lambda+in/2)-1}G(tx'-y', t-1)\,dt,$$

$$(\mathfrak{D}(\lambda)u)(x') = \int_{\mathbb{R}^{n-1}} u(y')\,dy'\int_{-\infty}^{+\infty} e^{-i\lambda s}G(x'-y', s)\,ds.$$

Proposition 2.2.7 *Let $\{\chi_q\}$ be a sequence of functions in $C_c(\mathbb{R}^{n-1})$ such that $\chi_q(0) = 1$ and $|\chi_q| \le 1$. Moreover, let the supports of χ_q shrink to the coordinate origin 0. Then, the relation*

$$\sup\{\|\chi_q(\mathfrak{B}(\lambda)-\mathfrak{D}(\lambda))\sigma; \mathcal{B}L_2(\mathbb{R}^{n-1})\|; \lambda\in\mathbb{R}\} \to 0$$

holds for any function $\sigma \in C_c(\mathbb{R}^{n-1})$ as $q \to \infty$.

Proof We represent $K_1(\lambda) := \chi_q(\mathfrak{B}(\lambda)-\mathfrak{D}(\lambda))\sigma$ in the form

$$K_1(\lambda) = \chi_q(K_2(\lambda)+K_3(\lambda)+K_4(\lambda))\sigma, \quad (2.2.17)$$

where

$$(K_2(\lambda)u)(x') = \int_{\mathbb{R}^{n-1}} u(y')\,dy' \int_0^\infty (t^{-i(\lambda+in/2)-1} - e^{-i\lambda(t-1)})G(tx'-y',t-1)\,dt, \tag{2.2.18}$$

$$(K_3(\lambda)u)(x') = \int_{\mathbb{R}^{n-1}} u(y')\,dy' \int_{-\infty}^\infty e^{-i\lambda s}\big(G((1+s)x'-y',s) - G(x'-y',s)\big)\,ds, \tag{2.2.19}$$

$$(K_4(\lambda)u)(x') = -\int_{\mathbb{R}^{n-1}} u(y')\,dy' \int_{-\infty}^{-1} e^{-i\lambda s} G((1+s)x'-y',s)\,ds. \tag{2.2.20}$$

We will prove that

$$\sup\{\|\chi_q K_j(\lambda)\sigma;\mathcal{B}L_2(\mathbb{R}^{n-1})\|;\lambda\in\mathbb{R}\}\to 0 \tag{2.2.21}$$

for $q\to\infty$, $j=2,3,4$:

(A) We consider operator (2.2.18). Assume that $\zeta\in C^\infty(\mathbb{R})$, $\zeta(t)=1$ for $|t-1|<1/2$, and $\zeta(t)=0$ for $|t-1|>3/4$. Integrating by parts, we write the inner integral in (2.2.18) in the form

$$\begin{aligned}&\int_0^\infty (1-\zeta(t))(t^{-i(\lambda+in/2)-1} - e^{-i\lambda(t-1)})G(tx'-y',t-1)\,dt\\&+\int_0^\infty \zeta'(t)h(t,\lambda)G(tx'-y',t-1)\,dt + \int_0^\infty \zeta(t)h(t,\lambda)\frac{\partial}{\partial t}G(tx'-y',t-1)\,dt,\end{aligned} \tag{2.2.22}$$

where

$$h(t,\lambda) = (t^{-i(\lambda+in/2)}-1)/(i\lambda-n/2) - (e^{-i\lambda(t-1)}-1)/i\lambda.$$

According to (2.2.22), $K_2(\lambda)$ can be written as the sum $K_2(\lambda)=K_{21}(\lambda)+K_{22}(\lambda)+K_{23}(\lambda)$. In the two first integrals in (2.2.22), the factor at $G(tx'-y',t-1)$ vanishes if $|t-1|<1/2$. Therefore,

$$\sigma_1(K_{21}(\lambda)+K_{22}(\lambda))\sigma_2 \in \mathcal{K}L_2(\mathbb{R}^{n-1})$$

for all $\lambda \in \mathbb{R}$ and $\sigma_j \in C_c(\mathbb{R}^{n-1})$. Moreover, $\lim \|\sigma_1 K_{2j}(\lambda)\sigma_2\| = 0$ as $\lambda \to \infty$, $j = 1, 2$. (It is obvious for K_{22}; for K_{21}, it follows from the fact that the first term in (2.2.22) reduces to the form

$$\int_0^\infty \left(\frac{t^{-i\lambda+n/2}}{i\lambda - n/2} - \frac{e^{-i\lambda(t-1)} - e^{i\lambda}}{i\lambda} \right) ((1 - \zeta(t))G(tx' - y', t - 1))'_t \, dt$$

by means of integration by parts.) We show that K_{23} has the same properties. Let us first verify the inequality $|(t - 1)^{-2}\zeta(t)h(t, \lambda)| \le c|\zeta(t)|$. To this end, we consider the function $g : (\alpha, \beta) \mapsto (e^{-i\alpha\beta} - 1)/(i\alpha)$ for $\alpha \in \mathbb{C}$, $\beta \in \mathbb{R}$, and $\mathrm{Im}\,(\alpha\beta) \le M$. For any pair of points (α_j, β_j), $j = 1, 2$,

$$|g(\alpha_1, \beta_1) - g(\alpha_2, \beta_2)| \le C(M)(|\beta_1 - \beta_2| + \beta_2^2|\alpha_1 - \alpha_2|). \tag{2.2.23}$$

Indeed, $|\partial g/\partial \beta| = e^{\mathrm{Im}\,(\alpha\beta)}$; hence,

$$|g(\alpha_1, \beta_1) - g(\alpha_1, \beta_2)| \le C(M)|\beta_1 - \beta_2|.$$

Moreover,

$$|g(\alpha_1, \beta_2) - g(\alpha_2, \beta_2)| = |\beta_2||(e^{-i\alpha_1\beta_2} - 1)/(i\alpha_1\beta_2) - (e^{-i\alpha_2\beta_2} - 1)/(i\alpha_2\beta_2)|.$$

Since the derivative of the function $z \mapsto (e^{-iz} - 1)/(iz)$ is $O(e^{\mathrm{Im} z})$, we have

$$|g(\alpha_1, \beta_2) - g(\alpha_2, \beta_2)| \le C(M)\beta_2^2|\alpha_1 - \alpha_2|,$$

which leads to (2.2.23). We set in (2.2.23) $\alpha_1 = \lambda + in/2$, $\alpha_2 = \lambda$, $\beta_1 = \ln t$, and $\beta_2 = t - 1$ and obtain

$$|\zeta(t)h(t, \lambda)| \le C(|\ln t - (t - 1)| + (t - 1)^2 n/2)|\zeta(t)| \le C_1(t - 1)^2|\zeta(t)|.$$

Thus, $|(t - 1)^{-2}\zeta(t)h(t, \lambda)| \le \mathrm{const}|\zeta(t)|$.

Now, using this inequality, we prove the estimate

$$k(\lambda, x', y') := \left| \int_0^\infty \zeta(t)h(t, \lambda)\frac{\partial}{\partial t}G(tx' - y', t - 1)\, dt \right| \le C|x' - y'|^{2-n} \tag{2.2.24}$$

under the condition $n \geq 3$ and $|x'| + |y'| \leq$ const. For $n = 2$, the estimate is valid with right-hand side $C \ln |x' - y'|^{-1} + C_1$. For $x = (x', 1)$ and $y = (y', 1)$,

$$k(\lambda, x', y') \leq C \int_0^\infty (t-1)^2 |\zeta(t) \frac{\partial}{\partial t} G(tx - y)|\, dt$$

$$\leq C \int_0^\infty (t-1)^2 |\zeta(t)| |tx - y|^{-1-n}\, dt \leq C \int_0^\infty |\zeta(t)| |tx - y|^{1-n}\, dt$$

because $|tx - y|^2 \geq (t-1)^2$. Furthermore,

$$|tx - y| = |x|((t-a)^2 + b^2)^{1/2},$$

where $a = xy/|x|^2$ and $b^2 = (|x|^2|y|^2 - (xy)^2)/|x|^4$. Therefore,

$$k(\lambda, x', y') \leq C|x|^{1-n} \int_0^\infty |\zeta(t)|((t-a)^2 + b^2)^{(1-n)/2}\, dt.$$

Changing the integration variable, we have

$$k(\lambda, x', y') \leq C|x|^{1-n} b^{2-n} \int_{-\infty}^{+\infty} |\zeta(bt + a)|(t^2 + 1)^{(1-n)/2}\, dt. \tag{2.2.25}$$

If $n \geq 3$, the integral on the right in (2.2.25) is bounded uniformly with respect to a and b. Since

$$b^2 = ((1 + |x'|^2)(1 + |y'|^2) - (1 + x'y')^2)/|x|^4$$
$$\geq (|x'|^2 - 2x'y' + |y'|^2)/|x|^4 = |x' - y'|^2/|x|^4,$$

from (2.2.25), we obtain

$$k(\lambda, x', y') \leq C|x|^{n-3}|x' - y'|^{2-n}$$

and estimate (2.2.24).

Suppose now that $n = 2$. Then, (2.2.25) takes the form

$$k(\lambda, x', y') \leq C|x|^{-1} \int_{-\infty}^{+\infty} |\zeta(bt + a)|(t^2 + 1)^{-1/2}\, dt \leq C \int_{(\alpha - a)/b}^{(\beta - a)/b} (t^2 + 1)^{-1/2}\, dt,$$

where $[\alpha, \beta]$ is a finite interval containing $\operatorname{supp} \zeta$. The points x' and y' do not quit a bounded set, so

$$M = \max\{\sup |\alpha - a|, \sup |\beta - a|\} < \infty.$$

Therefore,

$$k(\lambda, x', y') \le C \int_0^{M/b} (t^2+1)^{-1/2}\, dt \le C(1+\ln_+ M/b) \le C(1+\ln_+(M|x|^2|x'-y'|^{-1}).$$

Increasing M, if needed, and noticing that $|x| \ge 1$, we obtain inequality (2.2.24) with right-hand side replaced by $C \ln |x' - y'|^{-1} + C_1$.

From (2.2.24), it follows that $\sigma_1 K_{23}(\lambda)\sigma_2 \in \mathcal{K}L_2(\mathbb{R}^{n-1})$ for all $\lambda \in \mathbb{R}$ and $\sigma_j \in C_c(\mathbb{R}^{n-1})$. We show that $\lim \|\sigma_1 K_{23}(\lambda)\sigma_2\| = 0$ as $\lambda \to \infty$. Let $\kappa_j \in C(\mathbb{R})$ with $j = 1, 2$, $0 \le \kappa_j \le 1$, and $\kappa_1 + \kappa_2 = 1$, while $\kappa_1(t) = 1$ for $|t| < \varepsilon$ and $\kappa_1(t) = 0$ for $|t| > 2\varepsilon$. We represent $K_{23}(\lambda)$ as the sum $Q_1(\lambda) + Q_2(\lambda)$, where $Q_j(\lambda)$ is the operator with kernel

$$\kappa_j(|x' - y'|) \int_0^{\infty} h(t, \lambda)\zeta(t)\frac{\partial}{\partial t} G(tx' - y', t - 1)\, dt.$$

The kernel of $\sigma_1 Q_2(\lambda)\sigma_2$ has compact support on $\mathbb{R}^{n-1} \times \mathbb{R}^{n-1}$ and does not exceed $C|\lambda|^{-1}$, so $\|\sigma_1 Q_2(\lambda)\sigma_2\| \to 0$ as $\lambda \to \infty$. Moreover, it can be deduced from (2.2.24) that $\|\sigma_1 Q_1(\lambda)\sigma_2\|$ is $O(\varepsilon)$ in the case $n \ge 3$ and $O(\varepsilon \ln \varepsilon^{-1})$ in the case $n = 2$; hence, $\lim \|\sigma_1 K_{23}(\lambda)\sigma_2\| = 0$.

Thus, $\sigma_1 K_{2l}(\cdot)\sigma_2 \in C_0(\mathbb{R}) \otimes \mathcal{K}L_2(\mathbb{R}^{n-1})$ for $l = 1, 2, 3$; therefore, $\sigma_1 K_2(\cdot)\sigma_2 \in C_0(\mathbb{R}) \otimes \mathcal{K}L_2(\mathbb{R}^{n-1})$. The last inclusion implies (2.2.21) for $j = 2$; this can be established as in the proof of Proposition 2.2.4 (with $\mathcal{K}L_2(S^{n-1})$ changed for $\mathcal{K}L_2(\mathbb{R}^{n-1})$).

(B) Let us turn to operator (2.2.19). We have

$$\begin{aligned}(K_3(\lambda)u)(x') &= \int_{\mathbb{R}^{n-1}} \hat{u}(\xi')\, d\xi' \int_{-\infty}^{+\infty} \hat{G}(\xi', s)(e^{i(1+s)\xi' x'} - e^{i\xi' x'})e^{-i\lambda s}\, ds \\ &= \sqrt{2\pi} \int_{\mathbb{R}^{n-1}} \hat{u}(\xi')(\Phi(\xi', \lambda - \xi' x') - \Phi(\xi', \lambda))e^{ix'\xi'}\, d\xi',\end{aligned}$$

where

$$\Phi(\xi) = F_{x\to\xi} G(x), \quad \hat{u}(x') = F_{x'\to\xi'} u(x'), \quad \hat{G}(\xi', s) = F_{x'\to\xi'} G(x', s).$$

(The same letter F denotes the Fourier transform in $\mathbb{R}^n$ and that in $\mathbb{R}^{n-1}$.) Let $\chi \in C_c^\infty(\mathbb{R}^{n-1})$, $|\chi| \le 1$, $\chi(0) = 1$, and $\chi_q(x') := \chi(qx')$ for $q = 1, 2, \ldots$. (By virtue of Proposition 2.2.4, we can consider that the same sequence participates in the statement of Proposition 2.2.7.) Let

$$\zeta(x', \xi', \lambda) = \sqrt{2\pi}(\Phi(\xi', \lambda - \xi' x') - \Phi(\xi', \lambda))$$

and

$$\eta_q(z', \xi', \lambda) = F_{x' \to z'}(\chi_q(x')\zeta(x', \xi', \lambda)).$$

Then,

$$(\chi_q K_3(\lambda)u)^\wedge(z') = \int_{\mathbb{R}^{n-1}} \eta_q(z' - \xi', \xi', \lambda)\hat{u}(\xi')\, d\xi'.$$

To prove (2.2.21), it suffices to verify that, for any $u,\ v \in L_2(\mathbb{R}^{n-1})$, the estimate

$$|(\chi_q K_3(\lambda)u)^\wedge, v)| \le c(q)\|u\|\|v\| \tag{2.2.26}$$

holds, where $(\cdot, \cdot)$ is the inner product in $L_2(\mathbb{R}^{n-1})$, $c(q)$ is independent of u and v, and $c(q) \to 0$ as $q \to \infty$. We have

$$\begin{aligned} |(\chi_q K_3(\lambda)u)^\wedge, v)| \le & \int |\hat{u}(\xi')|^2 |\eta_q(z' - \xi', \xi', \lambda)|\, dz'\, d\xi' \qquad (2.2.27)\\ & \times \int |v(z')|^2 |\eta_q(z' - \xi', \xi', \lambda)|\, dz'\, d\xi'. \end{aligned}$$

It turns out that for any $s \ge 0$

$$|\eta_q(z', \xi', \lambda)| \le Cq^{2s-n}(1 + |z'|^2)^{-s} \tag{2.2.28}$$

(we will check the inequality below). Employing (2.2.28) for $s \in](n-1)/2, n/2[$, we find

$$\int |\eta_q(z' - \xi', \xi', \lambda)|\, dz' \le Cq^{2s-n} \int (1 + |z' - \xi'|^2)^{-s}\, dz',$$

$$\int |\eta_q(z' - \xi', \xi', \lambda)|\, d\xi' \le Cq^{2s-n} \int (1 + |z' - \xi'|^2)^{-s}\, d\xi'.$$

The right-hand sides of the two last inequalities tend to 0 as $q \to \infty$; therefore, (2.2.27) implies (2.2.26).

We now prove estimate (2.2.28). First, assume that s is a non-negative integer. Since $(1 - \Delta_{x'})^s e^{-iz'x'} = (1 + |z'|^2)^s e^{-iz'x'}$, we have

$$\eta_q(z', \xi', \lambda) = (2\pi)^{(1-n)/2}(1 + |z'|^2)^{-s} \int (1 - \Delta_{x'})^s(\chi_q(x')\zeta(x', \xi', \lambda))e^{-iz'x'}\, dx'.$$

It suffices to verify

$$\left| \int \partial_{x'}^{\beta} \chi_q(x') \partial_{x'}^{\gamma} \zeta(x', \xi', \lambda)) e^{-iz'x'} \, dx' \right| \leq C q^{2s-n} \tag{2.2.29}$$

for all multi-indices β and γ such that $|\beta + \gamma| \leq 2s$. We have $|\partial_{x'}^{\gamma} \zeta(x', \xi', \lambda)| \leq C|x'|^{\tau(\gamma)}$, where $\tau(\gamma) = 0$ for $|\gamma| \geq 1$ and $\tau(\gamma) = 1$ for $\gamma = 0$. Therefore,

$$|\partial_{x'}^{\beta} \chi_q(x') \partial_{x'}^{\gamma} \zeta(x', \xi', \lambda))| \leq C|x'|^{\tau(\gamma)} q^{|\beta|} |(\partial^{\beta} \chi)(qx')|.$$

It follows that the left-hand side of (2.2.29) is no greater than

$$C q^{|\beta|} \int |x'|^{\tau(\gamma)} |(\partial^{\beta} \chi)(qx')| \, dx' \leq C q^{|\beta| - n + 1 - \tau(\gamma)} \leq C q^{2s-n}.$$

Thus, inequality (2.2.28) has been established for integers $s \geq 0$. Let s be any non-negative number. We choose integers s_1, s_2 and non-negative μ_1, μ_2 so that $0 \leq s_1 \leq s \leq s_2$, $\mu_1 + \mu_2 = 1$, and $\mu_1 s_1 + \mu_2 s_2 = s$. We write (2.2.28) for $s = s_j$, raise it to the power μ_j with $j = 1, 2$, and multiply the obtained inequalities. The result is (2.2.28) with given $s \geq 0$.

(C) Finally, consider operator (2.2.20). Integrating by parts in the inner integral, we obtain $\sigma_1 K_4(\cdot) \sigma_2 \in C_0(\mathbb{R}) \otimes \mathcal{K}L_2(\mathbb{R}^{n-1})$ for $\sigma_j \in C_c(\mathbb{R}^{n-1})$ with $j = 1, 2$. This, as in part B) of the proof, leads to (2.2.21) (for $j = 4$).

□

Now we list for later use properties of the operators

$$F_{\eta \to x'}^{-1} \Phi(\eta, \lambda) F_{y' \to \eta} : L_2(\mathbb{R}^{n-1}) \to L_2(\mathbb{R}^{n-1}), \tag{2.2.30}$$

where F is the Fourier transform in $\mathbb{R}^{n-1}$, Φ is a zero degree homogeneous function, $\Phi | S^{n-1} \in C^{\infty}(S^{n-1})$, and $\lambda \in \mathbb{R}$.

Proposition 2.2.8 *Let $\mathfrak{D}(\lambda)$ be an operator of the form (2.2.30). Then, the following assertions are valid:*

(1) *The function $\mathbb{R} \setminus 0 \ni \lambda \mapsto \mathfrak{D}(\lambda)$ is continuous with respect to the operator norm.*
(2) *$\mathfrak{D}(\lambda) \to \mathfrak{D}(0)$ in the strong operator topology as $\lambda \to \infty$; $\|\mathfrak{D}(\lambda)\| \leq \max \|\mathfrak{D}(\pm 1)\|$ for all $\lambda \in \mathbb{R}$.*
(3) *If $\chi \in C_c^{\infty}(\mathbb{R}^{n-1})$ and $\lambda \in \mathbb{R}$, the commutator $[\chi, \mathfrak{D}(\lambda)]$ belongs to $\mathcal{K}L_2(\mathbb{R}^{n-1})$.*
(4) *$[\chi, \mathfrak{D}(\lambda)] \to 0$ with respect to the operator norm as $\lambda \to \infty$.*

Proof

(1) The equality $\Phi(\eta, \lambda) - \Phi(\eta, \mu) = \Phi_n(\eta, \lambda, \mu)(\lambda - \mu)$ holds, where

$$\Phi_n(\eta, \lambda, \mu) = \int_0^1 \partial_n \Phi(\eta, \mu + t(\lambda - \mu))\, dt,\ \partial_n \Phi(\eta_1, \ldots, \eta_n) = (\partial \Phi / \partial \eta_n)(\eta_1, \ldots, \eta_n).$$

The function $\mathbb{R}^{n-1} \ni \eta \mapsto \Phi_n(\eta, \lambda, \mu)$ is bounded uniformly with respect to μ in a neighborhood of the point $\lambda \neq 0$.

(2) For $t > 0$,

$$\begin{aligned}\mathfrak{D}(t\lambda)u(x') &= (2\pi)^{-(n-1)/2} \int e^{ix'\eta} \Phi(\eta, t\lambda)\hat{u}(\eta)\, d\eta \\ &= (2\pi)^{-(n-1)/2} \int e^{itx'\eta} \Phi(\eta, t\lambda) t^{n-1} \hat{u}(t\eta)\, d\eta = U_t \mathfrak{D}(\lambda) U_t^{-1} u(x'),\end{aligned}$$

where $U_t u(x') = t^{(n-1)/2} u(tx')$. The operator U_t is unitary on $L_2(\mathbb{R}^{n-1})$; hence, $\|\mathfrak{D}(\lambda)\| = \|\mathfrak{D}(t\lambda)\|$. If $t \to 0$, then $\mathfrak{D}(t\lambda) \to \mathfrak{D}(0)$ in the strong operator topology; hence, $\|\mathfrak{D}(0)\| \leq \max \|\mathfrak{D}(\pm 1)\|$.

(3) Let $\zeta \in C_c^\infty(\mathbb{R}^{n-1})$ and $\zeta(\eta) = 1$ in a neighborhood of the coordinate origin. We represent $\mathfrak{D}(\lambda)$ in the form

$$\mathfrak{D}(\lambda) = F^{-1} \zeta(\eta) \Phi(\eta, \lambda) F + F^{-1}(1 - \zeta(\eta)) \Phi(\eta, \lambda) F =: \mathfrak{D}_1(\lambda) + \mathfrak{D}_2(\lambda).$$

The commutator $[\chi, \mathfrak{D}_2(\lambda)]$ is a ΨDO of order -1. For a ball B with finite radius in $\mathbb{R}^{n-1}$, the embedding $H^1(B) \subset L_2(B)$ is compact; therefore, $[\chi, \mathfrak{D}_2(\lambda)]$ belongs to $\mathcal{K}L_2(\mathbb{R}^{n-1})$. The operator $\mathfrak{D}_1(\lambda)$ is a convolution with a smooth function. Therefore, the inclusion $[\chi, \mathfrak{D}_1(\lambda)] \in \mathcal{K}L_2(\mathbb{R}^{n-1})$ follows from the just mentioned embedding theorem and the compactness of $\operatorname{supp} \chi$.

(4) The kernel of the integral operator $[\chi, \mathfrak{D}(\lambda)]$ can be written in the form

$$\begin{aligned}&\int_{-\infty}^{+\infty} e^{i\lambda s} (\chi(x') - \chi(y')) G(x' - y', s)\, ds \\ &= \frac{i}{\lambda} \int_{-\infty}^{+\infty} e^{i\lambda s} \sum_{i=1}^{n} \chi_i(x', y')(x_i - y_i) \partial_s G(x' - y', s)\, ds =: k(x', y', \lambda)/\lambda,\end{aligned}$$

where $G(x) = F^{-1}_{\xi \to x} \Phi(\xi)$ and

$$\chi_i(x', y') = -\int_0^1 \partial_i \chi(x' + t(y' - x'))\, dt.$$

The operator $\mathfrak{D}'(\lambda)$ with kernel $k(x', y', \lambda)$ is bounded uniformly with respect to λ.

□

2.2.4 The Spectrum of Algebra $\mathfrak{S}$

The study of the irreducible representations of the algebra $\mathcal{A}$ in essence reduces to that of the algebra $\mathfrak{S}$ introduced in Sect. 2.2.1. We apply to $\mathfrak{S}$ the localization principle in Proposition 1.3.24. The role of the ideal J is played by $C_0(\mathbb{R})\otimes\mathcal{K}(L_2(S^{n-1}))$. This allows to find all equivalence classes of irreducible representations of the algebra $\mathfrak{S}$. Then, we describe the spectral topology for $\mathfrak{S}$ and prove that $\mathfrak{S}$ is a solvable algebra of length 1.

The next assertion justifies the use of the localization principle for the algebra $\mathfrak{S}$.

Proposition 2.2.9 *Let J denote the ideal $C_0(\mathbb{R}) \otimes \mathcal{K}L_2(S^{n-1})$ in $\mathfrak{S}$ and $\mathcal{C} = C(S^{n-1})$. Then, $\mathcal{C}$ is a commutative subalgebra in $\mathfrak{S}$ and the assumptions of Proposition 1.3.24 are fulfilled.*

Proof It is obvious that the operator of multiplication by a function $\alpha \in \mathcal{C}$ belongs to $\mathfrak{S}$, i.e., $\mathcal{C} \subset \mathfrak{S}$. The algebra $\mathcal{C}$ contains unity, so assumption (i) of Proposition 1.3.24 is fulfilled. The sphere S^{n-1} coincides with the maximal ideal space $\hat{\mathcal{C}}$ of $\mathcal{C}$. Let $\varphi_1, \varphi_2 \in S^{n-1}$ and $\varphi_1 \neq \varphi_2$. We choose elements c_1 and c_2 in $\mathcal{C}$ so that $c_j(\varphi_j) = 1$ for $j = 1, 2$ and $\operatorname{supp} c_1 \cap \operatorname{supp} c_2 = \emptyset$. Let us show that $c_1\mathfrak{S}c_2 \subset J$. It suffices to verify that, for any generator $\mathfrak{A}$ of the algebra $\mathfrak{S}$ and function $\alpha \in \mathcal{C}$, the commutator $[\mathfrak{A}, \alpha]$ belongs to J. Therefore, as $\mathfrak{A}$, we can take an operator-function of the form (2.2.5). According to Proposition 1.2.11, the commutator $[\mathfrak{A}, \alpha]$ belongs to $C_0(\mathbb{R}) \otimes \mathcal{K}L_2(S^{n-1})$; hence, assumption (ii) of Proposition 1.3.24 is also fulfilled. Finally, assumption (iii) follows from the fact that every irreducible representation of the ideal J is equivalent to a representation $\pi(\lambda) : \mathfrak{A} \mapsto \mathfrak{A}(\lambda)$. □

Let $L^{\mathcal{N}}$ be a subalgebra of the algebra $L_\infty(S^{n-1})$, let $C(S^{n-1}) \subset L^{\mathcal{N}}$, and let every element $a \in L^{\mathcal{N}}$ satisfy the following condition. Denote by $S^{n-2}_{\mathcal{N}}$ the unit sphere in the tangent space for S^{n-1} at the point $\mathcal{N}$ (the northern pole $(0', 1)$ of S^{n-1}). Almost everywhere on $S^{n-2}_{\mathcal{N}}$, there exists a limit

$$\lim_{\varphi\to\mathcal{N}} a(\varphi) = a(\mathcal{N}; \theta),$$

which is uniform with respect to the approach direction θ of φ to the pole $\mathcal{N}$. We introduce the algebra $\mathfrak{S}_{\mathcal{N}}$ generated by the operator-functions of the form (2.2.5), where $a \in L^{\mathcal{N}}$ and $\Phi \in C^\infty(S^{n-1})$. The algebra $\mathfrak{S}_{\mathcal{N}}$ is endowed with the norm

$$\|\mathfrak{A}; \mathfrak{S}_{\mathcal{N}}\| = \sup\{\|\mathfrak{A}(\lambda); \mathcal{B}L_2(S^{n-1})\|; \lambda \in \mathbb{R}\}.$$

Let $J(\mathcal{N})$ be the ideal in $\mathfrak{S}_{\mathcal{N}}$ spanned by the functions ζ Id, where $\zeta \in C(S^{n-1})$ and $\zeta(\mathcal{N}) = 0$. We denote by $\mathcal{L}$ the algebra generated by the functions of the form

$$\mathbb{R} \ni \lambda \mapsto a(\mathcal{N}; \cdot) F^{-1}_{\eta\to x'} \Phi(\eta, \lambda) F_{y'\to\eta} : L_2(\mathbb{R}^{n-1}) \to L_2(\mathbb{R}^{n-1}),$$

where $a(\mathcal{N}; \cdot)$ is the function $\mathbb{R}^{n-1} \setminus 0 \ni x' \mapsto a(\mathcal{N}; x'/|x'|)$ and

$$\|\mathfrak{D}; \mathcal{L}\| = \sup\{\|\mathfrak{D}(\lambda); \mathcal{B}L_2(\mathbb{R}^{n-1})\|; \lambda \in \mathbb{R}\}$$

for $\mathfrak{D} \in \mathcal{L}$.

When describing local algebras, we will use the following proposition many times in later sections; for studying the algebra $\mathfrak{S}$, we need only the special case $L^{\mathcal{N}} = C(S^{n-1})$.

Proposition 2.2.10 *Let $\mathfrak{A}$ be an operator-function of the form (2.2.5). The map*

$$\mathfrak{A}(\lambda) \mapsto a(\mathcal{N}; \cdot) F^{-1}_{\eta\to x'} \Phi(\eta, \lambda) F_{y'\to\eta}$$

extends to an isomorphism $\mathfrak{S}_{\mathcal{N}}/J(\mathcal{N}) \simeq \mathcal{L}$.

Proof We set

$$\mathfrak{A}_{jk}(\lambda) = E_{\omega\to\varphi}(\lambda)^{-1} \Phi_{jk}(\omega) E_{\psi\to\omega}(\lambda),$$
$$\mathfrak{D}_{jk}(\lambda) = F^{-1}_{\eta\to x'} \Phi_{jk}(\eta, \lambda) F_{y'\to\eta}$$

with $\Phi_{jk} \in C^\infty(S^{n-1})$ and $a_{jk} \in L^{\mathcal{N}}$. It suffices to show that

$$\Big\| \sum_j \prod_k a_{jk} \mathfrak{A}_{jk}; \mathfrak{S}_{\mathcal{N}}/J(\mathcal{N}) \Big\| = \Big\| \sum_j \prod_k a_{jk}(\mathcal{N}; \cdot) \mathfrak{D}_{jk}; \mathcal{L} \Big\|, \tag{2.2.31}$$

where j and k run over finitely many values.

We first verify that

$$\Big\| \sum_j \prod_k a_{jk} \mathfrak{A}_{jk}; \mathfrak{S}_{\mathcal{N}}/J(\mathcal{N}) \Big\| \le \Big\| \sum_j \prod_k a_{jk}(\mathcal{N}; \cdot) \mathfrak{D}_{jk}; \mathcal{L} \Big\|. \tag{2.2.32}$$

In view of Proposition 2.2.5, this inequality is equivalent to

$$\Big\| \sum_j \prod_k \zeta_{jk} a_{jk} \mathfrak{A}_{jk} \eta_{jk}; \mathfrak{S}_+/J_+(\mathcal{N}) \Big\| \le \Big\| \sum_j \prod_k a_{jk}(\mathcal{N}; \cdot) \mathfrak{D}_{jk}; \mathcal{L} \Big\| \tag{2.2.33}$$

with functions ζ_{jk} and η_{jk} that belong to $C_c^\infty(S_+^{n-1})$ and equal 1 in a neighborhood of the pole $\mathcal{N}$. Without loss of generality, we suppose that the functions Φ_{jk} are orthogonal to 1 in $L_2(S^{n-1})$ and introduce $G_{jk}(x) = F^{-1}_{\xi\to x}\Phi_{jk}(\xi)$. Let

$$(\mathfrak{B}_{jk}(\lambda)u)(x') = r(x')^{1-n/2}\int_{\mathbb{R}^{n-1}} r(y')^{1-n/2}u(y')\,dy' \times \int_0^{+\infty} t^{-i(\lambda+in/2)-1}G_{jk}(tx'-y',t-1)\,dt. \tag{2.2.34}$$

Setting $\mathfrak{P}(\lambda) = \sum_j \prod_k \chi_q\zeta_{jk}a_{jk}\mathfrak{A}_{jk}\eta_{jk}$, we have

$$\Big\|\sum_j\prod_k \zeta_{jk}a_{jk}\mathfrak{A}_{jk}\eta_{jk};\ \mathfrak{S}_+/J_+(\mathcal{N})\Big\| \le \|\mathfrak{P};\ \mathfrak{S}_{\mathcal{N}}\|. \tag{2.2.35}$$

According to Proposition 2.2.6,

$$\|\mathfrak{P}(\lambda);\ \mathfrak{B}L_2(S^{n-1})\| = \Big\|\sum_j\prod_k(\chi_q\zeta_{jk}a_{jk})\circ\kappa\mathfrak{B}_{jk}\eta_{jk}\circ\kappa;\ \mathcal{B}L_2(\mathbb{R}^{n-1})\Big\|.$$

To simplify notation, in what follows, we write η_{jk} instead of $\eta_{jk}\circ\kappa$ and so on. By virtue of Propositions 2.2.7 and 2.2.8,

$$\|\mathfrak{P}(\lambda);\ \mathfrak{B}L_2(S^{n-1})\| = \Big\|\sum_j\prod_k(\chi_q\zeta_{jk}a_{jk}\mathfrak{D}_{jk}(\lambda)\eta_{jk} + \chi_q\mathfrak{Q}_{jk}(\lambda));\ \mathcal{B}L_2(\mathbb{R}^{n-1})\Big\|,$$

where $\|\chi_q\mathfrak{Q}_{jk};\ \mathfrak{S}_{\mathcal{N}}\| \to 0$ as $q\to\infty$. For sufficiently large q, we obtain the equality $\chi_q\zeta_{jk} = \chi_q$ and the estimate

$$\|\mathfrak{P}(\lambda);\ \mathfrak{B}L_2(S^{n-1})\| \le \Big\|\sum_j\prod_k \chi_q a_{jk}(\mathcal{N};\cdot)\mathfrak{D}_{jk}(\lambda)\eta_{jk};\ \mathcal{B}L_2(\mathbb{R}^{n-1})\Big\| + \varepsilon \tag{2.2.36}$$

for a given $\varepsilon > 0$ and all $\lambda\in\mathbb{R}$. We can assume in these formulas that k runs over the same set of values for every j (otherwise, we could add several factors ζ_{jk} and η_{jk} that coincide with the unity in the quotient algebra $\mathfrak{S}_{\mathcal{N}}/J(\mathcal{N})$). Thus, by Proposition 2.2.8,

$$\sum_j\prod_k \chi_q a_{jk}(\mathcal{N};\cdot)\mathfrak{D}_{jk}(\lambda)\eta_{jk} = \Big(\prod_k\chi_q\Big)\sum_j\prod_k a_{jk}(\mathcal{N};\cdot)\mathfrak{D}_{jk}(\lambda) + \chi_q\mathfrak{Q}(\lambda), \tag{2.2.37}$$

where $\mathfrak{Q}(\lambda)\in\mathcal{K}L_2(\mathbb{R}^{n-1})$, the function $\mathbb{R}\setminus 0\ni\lambda\mapsto\mathfrak{Q}(\lambda)$ is continuous in the operator norm, and $\|\mathfrak{Q}(\lambda);\ \mathcal{B}L_2(\mathbb{R}^{n-1})\|\to 0$ as $\lambda\to\infty$.

We would like to make the function $\chi_q\mathfrak{Q}(\lambda)$ small for all λ by choosing χ_q. However, there is an obstacle caused by the singularity of $\mathfrak{D}_{jk}$ at $\lambda = 0$. To overcome this difficulty, we consider the restrictions of the functions in the algebra $\mathfrak{S}_{\mathcal{N}}$ (and in the other algebras which we deal with) to the set $\{\lambda \in \mathbb{R} : |\lambda| > \delta\}$ with a positive δ and define the norms $\|\mathfrak{A}; \mathfrak{S}_{\mathcal{N}}\|_\delta = \sup\{\|\mathfrak{A}(\lambda); \mathcal{B}L_2(S^{n-1})\|; |\lambda| > \delta\}$ and so on. The norms $\|\cdot; \mathfrak{S}_{\mathcal{N}}/J(\mathcal{N})\|$ and $\|\cdot; \mathfrak{S}_{\mathcal{N}}\|$ in (2.2.35) can be replaced by the norms $\|\cdot; \mathfrak{S}_{\mathcal{N}}/J(\mathcal{N})\|_\delta$ and $\|\cdot; \mathfrak{S}_{\mathcal{N}}\|_\delta$. We show that

$$\sup\{\|\chi_q\mathfrak{Q}(\lambda); \mathcal{B}L_2(\mathbb{R}^{n-1})\|; |\lambda| > \delta\} < \varepsilon \tag{2.2.38}$$

for sufficiently large q and any fixed positive δ and ε. We choose a large T so that $\|\chi_q\mathfrak{Q}(\lambda); \mathcal{B}L_2(\mathbb{R}^{n-1})\| < \varepsilon$ for $|\lambda| > T$. For every $\lambda \in [-T, -\delta] \cup [\delta, T]$ and any $\varepsilon > 0$, there exists a finite ε-net for the set $H(\lambda) = \{v : v = \mathfrak{Q}(\lambda)u; \|u; L_2(\mathbb{R}^{n-1})\| = 1\}$. In view of the continuity of the function $\lambda \mapsto \mathfrak{Q}(\lambda)$, this implies the existence of a finite $\varepsilon/2$-net $\{w\}$ for the union $\cup H(\lambda)$, $\delta \le |\lambda| \le T$. For every λ and an arbitrary $u \in L_2(\mathbb{R}^{n-1})$ with unit norm

$$\|\chi_q\mathfrak{Q}(\lambda)u; L_2(\mathbb{R}^{n-1})\| \le \|\chi_q(\mathfrak{Q}(\lambda)u - w); L_2(\mathbb{R}^{n-1})\| + \|\chi_q w; L_2(\mathbb{R}^{n-1})\|,$$

where the element w of the net is chosen so that the first term on the right is less than $\varepsilon/2$. The inequality $\|\chi_q w; L_2(\mathbb{R}^{n-1})\| < \varepsilon/2$ is fulfilled for all elements of the net for sufficiently large q. Thus, we arrive at estimate (2.2.38).

Coming back to inequality (2.2.35) (with the mentioned change of the norms) and taking into account (2.2.36)–(2.2.38), we obtain

$$\begin{aligned}\Big\|\sum_j\prod_k \zeta_{jk}a_{jk}\mathfrak{A}_{jk}\eta_{jk}; \mathfrak{S}_+/J_+(\mathcal{N})\Big\|_\delta &\le \Big\|\Big(\prod_k\chi_q\Big)\sum_j\prod_k a_{jk}(\mathcal{N};\cdot)\mathfrak{D}_{jk} + \chi_q\mathfrak{Q}; \mathcal{L}\Big\|_\delta\\ &\le \Big\|\Big(\prod_k\chi_q\Big)\sum_j\prod_k a_{jk}(\mathcal{N};\cdot)\mathfrak{D}_{jk}; \mathcal{L}\Big\|_\delta + \|\chi_q\mathfrak{Q}; \mathcal{L}\|_\delta\\ &\le \Big\|\sum_j\prod_k a_{jk}(\mathcal{N};\cdot)\mathfrak{D}_{jk}; \mathcal{L}\Big\| + \varepsilon.\end{aligned}$$

This leads to inequalities (2.2.33) and (2.2.32) because the numbers δ and ε are arbitrary and the functions $\mathfrak{A}_{jk}$ are continuous.

Now, we prove the inequality that is converse for (2.2.32) or, equivalently, for (2.2.33). Setting $\mathfrak{A}(\lambda) = \sum_j\prod_k \zeta_{jk}a_{jk}\mathfrak{A}_{jk}\eta_{jk}$ and again employing Propositions 2.2.6 and 2.2.7, we obtain

$$\|\mathfrak{A}(\lambda); \mathcal{B}L_2(S^{n-1})\| = \Big\|\sum_j\prod_k \zeta_{jk}a_{jk}\mathfrak{D}_{jk}(\lambda)\eta_{jk} + \mathfrak{Q}(\lambda); \mathcal{B}L_2(\mathbb{R}^{n-1})\Big\|, \tag{2.2.39}$$

where $\sup\{\|\chi_q \mathfrak{Q}(\lambda); \mathcal{B}L_2(\mathbb{R}^{n-1})\|, \lambda \in \mathbb{R}\} \to 0$ as $q \to 0$. The operator U_t defined for $t > 0$ by the equality $(U_t u)(x) = t^{(n-1)/2} u(tx)$ is unitary on $L_2(\mathbb{R}^{n-1})$. The equalities $\mathfrak{D}(\lambda)U_t = U_t \mathfrak{D}(\lambda/t)$ and $a(\cdot)U_t = U_t a(\cdot/t)$ hold for an operator $\mathfrak{D}(\lambda)$ of the form (2.2.30) and a function a on $\mathbb{R}^{n-1}$. Therefore, the right-hand side of (2.2.39) coincides with

$$\begin{aligned}&\Big\|U_t^{-1}\Big(\sum_j \prod_k \zeta_{jk} a_{jk} \mathfrak{D}_{jk}(\lambda)\eta_{jk} + \mathfrak{Q}(\lambda)\Big)U_t\Big\| \\ &= \Big\|\sum_j \prod_k (\zeta_{jk} a_{jk})(\cdot/t)\mathfrak{D}_{jk}(\lambda/t)\eta_{jk}(\cdot/t) + U_t^{-1}\mathfrak{Q}(\lambda)U_t\Big\|. \end{aligned} \tag{2.2.40}$$

Given $\varepsilon > 0$, we choose a large q so that $\|\chi_q \mathfrak{Q}(\lambda); \mathcal{B}L_2(\mathbb{R}^{n-1})\| < \varepsilon$ for all $\lambda \in \mathbb{R}$. From (2.2.39) and (2.2.40), it follows that

$$\begin{aligned}\|\mathfrak{A}(\lambda); \mathcal{B}L_2(S^{n-1})\| \geq \Big\|&\sum_j \prod_k (\zeta_{jk} a_{jk})(\cdot/t)\mathfrak{D}_{jk}(\lambda/t)\eta_{jk}(\cdot/t) \\ &+(1-\chi_q(\cdot/t))U_t^{-1}\mathfrak{Q}(\lambda)U_t; \mathcal{B}L_2(\mathbb{R}^{n-1})\Big\| - \varepsilon.\end{aligned} \tag{2.2.41}$$

Let us change λ for λt in (2.2.41). The operators $(1 - \chi_q(\cdot/t))U_t^{-1}\mathfrak{Q}(\lambda t)U_t$ weakly converge to the zero operator as $t \to +\infty$ for an arbitrary $\lambda \in \mathbb{R}$. Moreover, the operators $\sum_j \prod_k (\zeta_{jk} a_{jk})(\cdot/t)\mathfrak{D}_{jk}(\lambda)\eta_{jk}(\cdot/t)$ converge to $\sum_j \prod_k a_{jk}(\mathcal{N}; \cdot)\mathfrak{D}_{jk}(\lambda)$ in the strong topology. Therefore,

$$\underline{\lim}_{\,t\to+\infty} \|\mathfrak{A}(\lambda t); \mathcal{B}L_2(S^{n-1})\| \geq \Big\|\sum_j \prod_k a_{jk}(\mathcal{N}; \cdot)\mathfrak{D}_{jk}(\lambda); \mathcal{B}L_2(\mathbb{R}^{n-1})\Big\| \tag{2.2.42}$$

for $\lambda \in \mathbb{R}$. If an element $\mathfrak{B} \in \mathfrak{S}_+$ belongs to $J_+(\mathcal{N})$, then $\|\chi_q \mathfrak{B}; \mathfrak{S}_{\mathcal{N}}\| \to 0$ as $q \to \infty$ (Proposition 2.2.4); hence, replacing $\mathfrak{A}$ by $\mathfrak{A} + \mathfrak{B}$ with some $\mathfrak{B} \in J_+(\mathcal{N})$, we do not change the right-hand side of (2.2.42). Thus,

$$\begin{aligned}\sup\{\|\mathfrak{A}(\lambda) + \mathfrak{B}(\lambda); \mathcal{B}L(S^{n-1})\|; \lambda \in \mathbb{R}\} &\geq \underline{\lim}_{\,t\to+\infty} \|\mathfrak{A}(\lambda t) + \mathfrak{B}(\lambda t); \mathcal{B}L_2(S^{n-1})\| \\ &\geq \Big\|\sum_j \prod_k a_{jk}(\mathcal{N}; \cdot)\mathfrak{D}_{jk}(\lambda); \mathcal{B}L_2(\mathbb{R}^{n-1})\Big\|,\end{aligned}$$

which leads to the inequality

$$\Big\|\sum_j \prod_k \zeta_{jk} a_{jk} \mathfrak{A}_{jk} \eta_{jk}; \mathfrak{S}_+/J_+(\mathcal{N})\Big\| \geq \Big\|\sum_j \prod_k a_{jk}(\mathcal{N}; \cdot)\mathfrak{D}_{jk}; \mathcal{L}\Big\|.$$

Taking into account estimate (2.2.33), we complete the proof. □

Theorem 2.2.11 *Let* $\mathfrak{A} \in \mathfrak{S}$ *and* $\mathfrak{A}_{\psi\to\varphi}(\lambda) = E_{\omega\to\varphi}(\lambda)^{-1}\Phi(\varphi,\omega)E_{\psi\to\omega}(\lambda)$*, where* $\Phi \in C^\infty(S^{n-1} \times S^{n-1})$ *and* $\lambda \in \mathbb{R}$*. Then, the maps*

$$\pi(\lambda) : \mathfrak{A} \mapsto \mathfrak{A}(\lambda) : L_2(S^{n-1}) \to L_2(S^{n-1}), \quad \lambda \in \mathbb{R}, \tag{2.2.43}$$

$$\pi(\varphi,\omega) : \mathfrak{A} \mapsto \Phi(\varphi,\omega), \qquad (\varphi,\omega) \in S^{n-1} \times S^{n-1} \tag{2.2.44}$$

extend to irreducible representations of the algebra $\mathfrak{S}$*. These representations are pairwise nonequivalent. Every irreducible representation of* $\mathfrak{S}$ *is equivalent to one of the representations (2.2.43) or (2.2.44).*

Proof According to Proposition 2.2.9, we can apply the localization principle (Proposition 1.3.24) to the algebra $\mathfrak{S}$ setting $J = C_0(\mathbb{R}) \otimes \mathcal{K}L_2(S^{n-1})$ and $\mathcal{C} = C(S^{n-1})$. Then,

$$\widehat{\mathfrak{S}} = \cup_{\varphi \in S^{n-1}} \widehat{\mathfrak{S}}_\varphi \cup \widehat{J}. \tag{2.2.45}$$

Every irreducible representation of the ideal J is a map $\pi(\lambda) : K \mapsto K(\lambda)$ with $K \in J$ (Proposition 1.3.7). The map uniquely extends to a representation of $\mathfrak{S}$ in $L_2(S^{n-1})$, and we obtain $\pi(\lambda)$ in (2.2.43).

To find the spectrum $\widehat{\mathfrak{S}}_\varphi$ of a local algebra, we employ Proposition 2.2.10. The role of $L^{\mathcal{N}}$ is played by the algebra $C(S^{n-1})$. At a fixed point $\varphi \in S^{n-1}$, the algebra $\mathfrak{S}_\varphi$ is isomorphic to the commutative algebra $\mathcal{L}$ of the operator-functions

$$\mathbb{R} \ni \lambda \mapsto \mathfrak{D}(\lambda) = F^{-1}_{\eta\to x}\Phi(\varphi,\eta,\lambda)F_{y\to\eta} : L_2(\mathbb{R}^{n-1}) \to L_2(\mathbb{R}^{n-1})$$

with $\|\mathfrak{D};\mathcal{L}\| = \sup\{\|\mathfrak{D}(\lambda); \mathcal{B}L_2(\mathbb{R}^{n-1})\|, \lambda \in \mathbb{R}\}$. The function $\xi \mapsto \Phi(\varphi,\xi)$ is homogeneous of zero degree and belongs to $C^\infty(\mathbb{R}^n \setminus 0)$. Therefore, the algebra $\mathfrak{S}_\varphi$ is isomorphic to $C(S^{n-1})$, and the map (2.2.44) defines all (to within equivalence) irreducible representations of the algebra $\mathfrak{S}_\varphi$. □

According to Proposition 2.2.1, $\mathcal{K}L_2(S^{n-1}) \subset \mathfrak{S}(\lambda)$, and the algebra $\mathfrak{S}(\lambda)/\mathcal{K}L_2(S^{n-1})$ is commutative. Clearly, every representation of $\mathfrak{S}(\lambda)$ is also a representation of $\mathfrak{S}$. In particular, the following theorem shows that the algebras $\mathfrak{S}(\lambda)/\mathcal{K}L_2(S^{n-1})$ are isomorphic for all $\lambda \in \mathbb{R}$ and differ from $\mathfrak{S}/(C_0(\mathbb{R}) \otimes \mathcal{K}L_2(S^{n-1}))$.

Theorem 2.2.12 *For* $\lambda \in \mathbb{R}$*, the algebra* $\mathfrak{S}(\lambda)/\mathcal{K}L_2(S^{n-1})$ *is isomorphic to the algebra* $C(V(n,2))$ *of the continuous functions on the manifold* $V(n,2)$ *of the pairs* (φ,ω) *of orthogonal unit vectors in* $\mathbb{R}^n$*. The maps*

$$\pi(\varphi,\omega) : \mathfrak{A}(\lambda) \mapsto \Phi(\varphi,\omega), \ (\varphi,\omega) \in V(n,2), \tag{2.2.46}$$

extend to irreducible representations of the algebra $\mathfrak{S}(\lambda)/\mathcal{K}L_2(S^{n-1})$ *(the notations are the same as in Theorem 2.2.11). The representations are pairwise nonequivalent. Any irreducible representation of* $\mathfrak{S}(\lambda)$ *is equivalent either to one of the representations (2.2.46) or to the identity representation.*

Proof can be found in [21], 5.3.

We are going to describe the Jacobson topology on the spectrum $\widehat{\mathfrak{S}}$ of the algebra $\mathfrak{S}$. Denote by Σ the union of the product $S^{n-1} \times S^{n-1}$ and the line $l = \mathbb{R}$. From Theorem 2.2.11, it follows that the correspondence $\pi(\varphi, \omega) \mapsto (\varphi, \omega)$, $\pi(\lambda) \mapsto \lambda$ defines a bijection of the spectrum $\widehat{\mathfrak{S}}$ onto Σ. We introduce a topology on Σ. If $(\varphi, \omega) \in S^{n-1} \times S^{n-1}$ and the scalar product $\varphi\omega$ of unit vectors φ and ω in $\mathbb{R}^n$ satisfies the inequality $\varphi\omega > 0$ (< 0), then a fundamental neighborhood system of the point (φ, ω) comprises the sets $\mathcal{V}(\varphi, \omega) \cup \{\lambda \in l : \lambda > N \ (< N)\}$, where $\mathcal{V}(\varphi, \omega)$ is a neighborhood of (φ, ω) in the product $S^{n-1} \times S^{n-1}$ and $N \in \mathbb{R}$. In the case $\varphi\omega = 0$, the fundamental system is formed by the sets $\mathcal{V}(\varphi, \omega) \cup \mathbb{R}$. A neighborhood of a point $\lambda \in l \subset \Sigma$ is an interval in l containing λ.

Theorem 2.2.13 *The topology of* $\widehat{\mathfrak{S}}$ *induced by that of* Σ *coincides with the Jacobson topology.*

Proof Recall that the base of the Jacobson topology on $\widehat{\mathfrak{S}}$ is formed by the sets of the classes of equivalent irreducible representations π such that $\|\pi\mathfrak{A}\| > 1$, where $\mathfrak{A}$ runs over a dense set in $\mathfrak{S}$ (see Sect. 1.3.3)).

Let λ_0 be an arbitrary point in the line l and let $\mathfrak{A}$ be an element in $\mathfrak{S}$ such that $\|\mathfrak{A}(\lambda_0)\| > 1$. The set $\{\pi : \|\pi\mathfrak{A}\| > 1\}$ is a neighborhood of λ_0 in the Jacobson topology. Since the function $\lambda \mapsto \mathfrak{A}(\lambda)$ is continuous in the norm, the set is a neighborhood of λ_0 in the usual topology on the real axis. On the other hand, if f and g are elements in $L_2(S^{n-1})$ and c is an arbitrary continuous function vanishing at infinity, the operator-function $\lambda \mapsto c(\lambda)(\cdot, f)g$ belongs to $\mathfrak{S}$. Therefore, any open interval on the real axis l is open in the Jacobson topology.

Now, we consider neighborhoods of a point $(\varphi_0, \omega_0) \in S^{n-1} \times S^{n-1}$ assuming that $\varphi_0\omega_0 > 0$. Let $\Phi(\varphi_0, \omega_0) > 1$ and $\mathfrak{A}(\lambda) = E(\lambda)^{-1}\Phi(\varphi, \omega)E(\lambda)$. Since $\pi(\varphi, \omega)\mathfrak{A} = \Phi(\varphi, \omega)$, the inequality $\|\pi(\varphi, \omega)\mathfrak{A}\| > 1$ holds for all (φ, ω) in a neighborhood of (φ_0, ω_0) on $S^{n-1} \times S^{n-1}$. Moreover, $\|\pi(\lambda)\mathfrak{A}\| > 1$ for all $\lambda \in (N, +\infty)$, where N is a sufficiently large positive number. To show that, we choose a function v in $C^\infty(S^{n-1})$ with support in a small neighborhood of φ_0 such that $\|v; L_2(S^{n-1})\| = 1$ and denote by g a smooth function on the sphere extended to $\mathbb{R}^n \setminus 0$ as a zero degree homogeneous function. In addition, assume that the gradient ∇g does not vanish on the support of v. From Theorem 1.2.8, it follows that

$$e^{-i\mu g(\varphi)}\mathfrak{A}_{\psi\to\varphi}(\lambda)e^{i\mu g(\psi)}v(\psi) - \Phi(\varphi, \mu\nabla g(\varphi) + \sigma\varphi)v(\varphi) = O((|\mu| + |\sigma|)^{-1}) \tag{2.2.47}$$

as $|\mu| + |\sigma| \to \infty$, where $\mu \in \mathbb{R}$ and $\sigma = \Re\lambda$. Note that $\varphi\nabla g(\varphi) = 0$ and $\varphi_0\omega_0 = 0$. Therefore, given large $\sigma > 0$, we can choose μ and g so that the vectors $\mu\nabla g(\varphi_0) + \sigma\varphi_0$ and ω_0 are not parallel. Then, in view of (2.2.47), $\|\pi(\lambda)\mathfrak{A}\| > 1$ for all $\lambda > N$. Thus, a neighborhood of (φ_0, ω_0) in the Jacobson topology is also a neighborhood in the topology on Σ.

The following fact (Lemma 5.4.8 in [21]) will be used below. Assume that $\mathfrak{A}(\lambda) = E^{-1}(\lambda)\Phi(\varphi, \omega)E(\lambda)$, where $\Phi(\varphi, \omega) = 0$ for $\varphi\omega \le 0$ ($\varphi\omega \ge 0$). Then, $\|\mathfrak{A}(\lambda)\| = O(e^{-\pi|\lambda|})$, as $\lambda \to -\infty$ (resp., $\lambda \to +\infty$).

Let $\mathcal{V}(\varphi_0, \omega_0) \cup \{\lambda \in l : \lambda > N\}$ be an arbitrary neighborhood of (φ_0, ω_0) in the space Σ. We verify that it is also a neighborhood of the point in the Jacobson topology. Let us choose a function $\Phi \in C^\infty(S^{n-1} \times S^{n-1})$ so that $\Phi(\varphi_0, \omega_0) > 1$ and $\Phi = 0$ outside the set $\mathcal{V}(\varphi_0, \omega_0) \cap \{(\varphi, \omega) : \varphi\omega > \delta\}$ with small positive δ. Then, for $\mathfrak{A}(\lambda) = E^{-1}(\lambda)\Phi E(\lambda)$, we obtain $\{(\varphi, \omega) : \|\pi(\varphi, \omega)\mathfrak{A}\| > 1\} \subset \mathcal{V}(\varphi_0, \omega_0)$. From the fact formulated above, it follows that $\pi(\lambda)\mathfrak{A} \to 0$ for $\lambda \to -\infty$. Moreover, according to Proposition 1.2.7, the operator $\mathfrak{A}(\lambda)$ is compact for every λ. In view of Proposition 2.2.3, the function $\lambda \mapsto c(\lambda)\mathfrak{A}(\lambda)$, where $c \in C_0^\infty(\mathbb{R})$, belongs to $\mathfrak{S}$. We choose the function c so that $0 \le c \le 1$ and $c(\lambda) = 1$ for $\lambda \in [-M, M]$, where M is a large number. Changing, if needed, $\mathfrak{A}$ for $\mathfrak{A} - c\mathfrak{A}$, we obtain an operator-function subject to the relation $\{\pi : \|\pi\mathfrak{A}\| > 1\} \subset \mathcal{V}(\varphi_0, \omega_0) \cup \{\lambda \in l : \lambda > N\}$.

We have shown that the neighborhood systems of a point (φ_0, ω_0) with $\varphi_0\omega_0 > 0$ in the topology Σ and in the Jacobson topology coincide. Such a coincidence for the points (φ_0, ω_0) with $\varphi_0\omega_0 < 0$ can be verified in a similar way.

Assume now that $\varphi_0\omega_0 = 0$, $\Phi(\varphi_0, \omega_0) > 1$, and $\mathfrak{A}(\lambda) = E(\lambda)^{-1}\Phi E(\lambda)$. We choose the function g in (2.2.47) so that $\omega_0 = \nabla g(\varphi_0)$. Letting μ to infinity, we obtain the inequality $\|\pi(\lambda)\mathfrak{A}\| > 1$ for all $\lambda \in l$. Therefore, a neighborhood of (φ_0, ω_0) in the Jacobson topology is a neighborhood in the space Σ as well. The converse is evident. □

Thus, the topology on the spectrum of the algebra $\widehat{\mathfrak{S}}$ is not separable; $\widehat{\mathfrak{S}}$ is a T_0-space (one of the two any points of the space has a neighborhood not containing the other point). The reason why the topology is not Hausdorff is already evident for $n = 1$. In this case, in the operators $E(\lambda)^{\pm}$ (see (1.2.2) and (1.2.8)), the "integral" over a 0-dimensional sphere is the sum of the integrand values at the points ± 1. Therefore,

$$\begin{pmatrix} (E(\lambda)u(1)) \\ (E(\lambda)u)(-1) \end{pmatrix} = \frac{\Gamma(1/2 + i\lambda)}{\sqrt{2\pi}} \begin{pmatrix} e^{-i(\pi/2)(1/2+i\lambda)} & e^{i(\pi/2)(1/2+i\lambda)} \\ e^{i(\pi/2)(1/2+i\lambda)} & e^{-i(\pi/2)(1/2+i\lambda)} \end{pmatrix} \times \begin{pmatrix} u(1) \\ u(-1) \end{pmatrix}, \tag{2.2.48}$$

$$\begin{pmatrix} (E(\lambda)^{-1}v(1)) \\ (E(\lambda)^{-1}v)(-1) \end{pmatrix} = \frac{\Gamma(1/2 - i\lambda)}{\sqrt{2\pi}} \begin{pmatrix} e^{i(\pi/2)(1/2-i\lambda)} & e^{-i(\pi/2)(1/2-i\lambda)} \\ e^{-i(\pi/2)(1/2-i\lambda)} & e^{i(\pi/2)(1/2-i\lambda)} \end{pmatrix} \times \begin{pmatrix} v(1) \\ v(-1) \end{pmatrix}. \tag{2.2.49}$$

Setting $c(\varphi) = (\Phi(\varphi, 1) + \Phi(\varphi, -1))/2$ and $d(\varphi) = (\Phi(\varphi, 1) - \Phi(\varphi, -1))/2$ for $\varphi = \pm 1$, we obtain

$$\mathfrak{A}(\lambda) = E(\lambda)^{-1}\Phi(\varphi, \omega)E(\lambda) = \begin{pmatrix} c(1) + d(1)\tanh \pi\lambda & id(1)/\cosh \pi\lambda \\ -id(-1)/\cosh \pi\lambda & c(-1) - d(-1)\tanh \pi\lambda \end{pmatrix}. \tag{2.2.50}$$

The maps (2.2.43) and (2.2.44) take the form

$$\pi(\lambda) : \mathfrak{A} \mapsto \mathfrak{A}(\lambda) : \mathbb{C}^2 \to \mathbb{C}^2,$$

$$\pi(\pm 1, \omega) : \mathfrak{A} \mapsto c(\pm 1) + d(\pm 1)\operatorname{sgn} \omega, \quad \omega = \pm 1.$$

It is obvious that the representations $\pi(\lambda)$ are irreducible for all real λ. Comparing the traces, we find that the representations are pairwise nonequivalent. Moreover,

$$\lim_{\lambda \to +\infty} \mathfrak{A}(\lambda) = \operatorname{diag}(c(1) + d(1), c(-1) - d(-1)),$$

$$\lim_{\lambda \to -\infty} \mathfrak{A}(\lambda) = \operatorname{diag}(c(1) - d(1), c(-1) + d(-1)).$$

In other words, at the points $\pm\infty$, the irreducible representation $\pi(\lambda)$ has, as a "limit," the sum of two irreducible one-dimensional representations. This explains the non-separability at the points $\pi(\pm, \omega)$.

To prove the solvability of the algebra $\mathfrak{S}$, we need:

Proposition 2.2.14 *Let* $\mathfrak{A}(\lambda) = E^{-1}(\lambda)\Phi(\varphi, \omega)E(\lambda)$, *where* $\Phi \in C^\infty(S^{n-1} \times S^{n-1})$. *The map* $p : \mathfrak{A} \mapsto \Phi$ *extends to an isomorphism*

$$p : \mathfrak{S}/(C_0(\mathbb{R}) \otimes \mathcal{K}L_2(S^{n-1})) \simeq C(S^{n-1} \times S^{n-1}). \tag{2.2.51}$$

Proof Let $J = C_0(\mathbb{R}) \otimes \mathcal{K}L_2(S^{n-1})$, $\mathfrak{A}_{jk}(\lambda) = E^{-1}(\lambda)\Phi_{jk}(\varphi, \omega)E(\lambda)$, and $\Phi_{jk} \in C^\infty(S^{n-1} \times S^{n-1})$, where j and k run over finite sets. According to Proposition 1.2.14,

$$\sum_j \prod_k \mathfrak{A}_{jk}(\lambda) = E(\lambda)^{-1} \sum_j \prod_k \Phi_{jk}(\varphi\omega)E(\lambda) + T(\lambda),$$

where $T \in J$. Representations (2.2.44) annihilate the ideal J; they can be considered as representations of the quotient algebra $\mathfrak{S}/J$. Every representation $\mathfrak{S}/J$ is equivalent to one of representations (2.2.44) (Theorem 2.2.11). By virtue of Proposition 1.3.1,

$$\|\sum_j \prod_k \mathfrak{A}_{jk}; \mathfrak{S}/J\| = \sup\{\|\pi(\varphi,\omega) \sum_j \prod_k \mathfrak{A}_{jk}\|; (\varphi,\omega) \in S^{n-1} \times S^{n-1}\}$$
$$= \|\sum_j \prod_k \Phi_{jk}; C(S^{n-1} \times S^{n-1})\|.$$

Therefore, the map p extends to an isomorphism (2.2.51). □

The next result immediately follows from Propositions 2.2.14 and 1.3.23.

Theorem 2.2.15 *The composition series* $0 \subset C_0(\mathbb{R}) \otimes \mathcal{K}L_2(S^{n-1}) \subset \mathfrak{S}$ *is solving and* $\mathfrak{S}/(C_0(\mathbb{R}) \otimes \mathcal{K}L_2(S^{n-1})) \simeq C(S^{n-1} \times S^{n-1})$. *This series coincides with the maximal radical series. The length of the algebra* $\mathfrak{S}$ *is equal to 1.*

Remark 2.2.16 Denote by com $\mathfrak{S}$ the ideal of $\mathfrak{S}$ generated by the commutators $[\mathfrak{A}, \mathfrak{B}] := \mathfrak{A}\mathfrak{B} - \mathfrak{B}\mathfrak{A}$, where $\mathfrak{A}, \mathfrak{B} \in \mathfrak{S}$. Then, com $\mathfrak{S} = C_0(\mathbb{R}) \otimes \mathcal{K}L_2(S^{n-1})$. Therefore, the composition series in Theorem 2.2.15 can be written in the form $0 \subset \text{com}\,\mathfrak{S} \subset \mathfrak{S}$.

Indeed, com $\mathfrak{S} \subset J := C_0(\mathbb{R}) \otimes \mathcal{K}L_2(S^{n-1})$ because the algebra $\mathfrak{S}/J$ is commutative. For any $\lambda \in \mathbb{R}$, the algebra $\mathfrak{S}(\lambda)$ is noncommutative, and the restriction $\pi(\lambda)|\text{com}\mathfrak{S}$ is nonzero. According to Proposition 1.3.7, the maps $\pi(\lambda_1)|\text{com}\mathfrak{S}$ and $\pi(\lambda_2)|\text{com}\mathfrak{S}$ are irreducible and nonequivalent representations of the ideal com $\mathfrak{S}$ for $\lambda_1 \neq \lambda_2$, i.e., com $\mathfrak{S}$ is a rich subalgebra in J. Thus, com$\mathfrak{S} = J$ (Proposition 1.3.9).

2.2.5 The Spectrum of Algebra $\mathcal{A}$

Recall that the algebra $\mathcal{A}$ is generated on $L_2(\mathcal{M})$ by the operators of the class $\bar{\Psi}^0(\mathcal{M})$ and the operators of multiplication $f\cdot$ by the functions f in $\mathfrak{M}$ (see the beginning of Sect. 2.2.1). The principal symbol σ^0 of an operator $A \in \bar{\Psi}^0(\mathcal{M})$ is a smooth function on the cospheric bundle $S^*(\mathcal{M})$. For $f \in \mathfrak{M}$, we set $f^0(\varphi) := \lim_{r\to 0} f(r,\varphi)$, where (r,φ) are local spheric coordinates centered at the point x^0. For such generators f and A of the algebra $\mathcal{A}$, we introduce the maps

$$\pi(x,\omega) : A \mapsto \sigma^0(\omega), \quad \pi(x,\omega) : f \mapsto f(x), \tag{2.2.52}$$

where $\omega \in S^*(\mathcal{M})_x$ and $x \in \mathcal{M} \setminus x^0$;

$$\pi(\varphi,\omega) : A \mapsto \sigma^0(\omega), \quad \pi(\varphi,\omega) : f \mapsto f^0(\varphi), \tag{2.2.53}$$

where $(\varphi, \omega) \in S(\mathcal{M})_{x^0} \times S^*(\mathcal{M})_{x^0}$;

$$\pi(\lambda) : A \mapsto E_{\omega \to \varphi}(\lambda)^{-1} \sigma^0(\omega) E_{\psi \to \omega}(\lambda) : L_2(S(\mathcal{M})_{x^0}) \to L_2(S(\mathcal{M})_{x^0}), \quad (2.2.54)$$

$$\pi(\lambda) : f \mapsto f^0 \cdot : L_2(S(\mathcal{M})_{x^0}) \to L_2(S(\mathcal{M})_{x^0}),$$

where $f^0\cdot$ denotes the operator of multiplication by a function f^0 on the space $L_2(S(\mathcal{M})_{x^0})$.

Taking into account Sect. 2.2.1, Propositions 1.3.26 and 2.1.5, and Theorem 2.2.11, we obtain the following result.

Theorem 2.2.17 *Maps (2.2.52)–(2.2.54) extend to representations of the quotient algebra $\mathcal{A}/\mathcal{K}L_2(\mathcal{M})$. These representations are irreducible and pairwise nonequivalent. Every irreducible representation of the algebra $\mathcal{A}$ is equivalent to one of the representations (2.2.52)–(2.2.54) or to the identity representation. The algebra $\mathcal{A}$ is an algebra of type I.*

Let us discuss the spectral topology. We denote by Ξ the disjoint union of the sets $S^*(\mathcal{M})|(\mathcal{M} \setminus x^0)$, $S(\mathcal{M})_{x^0} \times S^*(\mathcal{M})_{x^0}$, and $\mathbb{R}$. By Theorem 2.2.17, the correspondence $\pi(x, \omega) \mapsto \omega$, $\pi(\varphi, \omega) \mapsto (\varphi, \omega)$, and $\pi(\lambda) \mapsto \lambda$ defines a bijection of $(\mathcal{A}/\mathcal{K}L_2(\mathcal{M}))^\wedge$ onto Ξ. Therefore, it suffices to introduce a topology on Ξ.

A basis of neighborhoods for a point $\lambda \in \mathbb{R}$ is formed by the open intervals in $\mathbb{R}$ containing λ. Neighborhoods of a point in $S^*(\mathcal{M})|(\mathcal{M} \setminus x^0)$ are also defined in a usual way. To describe neighborhoods of a point $(\varphi, \omega) \in S(\mathcal{M})_{x^0} \times S^*(\mathcal{M})_{x^0}$, we consider the union $S^*(\mathcal{M})|(\mathcal{M} \setminus x^0) \cup (S(\mathcal{M})_{x^0} \times S^*(\mathcal{M})_{x^0})$ as a compact manifold X with boundary. If $\varphi\omega > 0$, a neighborhood of a point (φ, ω) is introduced as the union of a neighborhood $\mathcal{W}(\varphi, \omega)$ of this point in X and a set of the form $\{\lambda \in \mathbb{R} : \lambda > N\}$ with some N. In the case $\varphi\omega < 0$, a neighborhood of (φ, ω) is the union $\mathcal{W}(\varphi, \omega) \cup \{\lambda \in \mathbb{R} : \lambda < N\}$. Finally, for $\varphi\omega = 0$, a basis of neighborhoods is formed by the sets $\mathcal{W}(\varphi, \omega) \cup \mathbb{R}$.

Using Theorem 2.2.13, the reader can verify the following assertion.

Theorem 2.2.18 *The topology carried over from the space Ξ to $(\mathcal{A}/\mathcal{K}L_2(\mathcal{M}))^\wedge$ coincides with the Jacobson topology.*

Let us consider the composition series

$$0 \subset \mathcal{K}L_2(\mathcal{M}) \subset \operatorname{com} \mathcal{A} \subset \mathcal{A}, \quad (2.2.55)$$

where $\operatorname{com} \mathcal{A}$ denotes the ideal in $\mathcal{A}$ generated by the commutators of the elements in $\mathcal{A}$. It is obvious that all one-dimensional representations vanish on the ideal $\operatorname{com} \mathcal{A}$. From Theorem 2.2.17, it follows that $\operatorname{com} \mathcal{A}/\mathcal{K}L_2(\mathcal{M}) \simeq C_0(\mathbb{R}) \otimes \mathcal{K}L_2(S^{n-1})$. Finally, $\mathcal{A}/\operatorname{com}\mathcal{A} \simeq C(X)$, where X is a compact manifold with a boundary obtained by gluing

$S(\mathcal{M})_{x^0} \times S^*(\mathcal{M})_{x^0}$ to $S^*(\mathcal{M})|(\mathcal{M} \setminus x^0)$. Thus, composition series (2.2.55) is solving, and the length of the algebra $\mathcal{A}$ is equal to 2.

Series (2.2.55) is stratified and, consequently, coincides with the maximal radical series (see Sect. 1.3.7). Indeed, by Theorem 2.2.17, all irreducible representations but the identity one annihilate $\mathcal{K}L_2(\mathcal{M})$. Among the nonzero irreducible representations of the ideal com$\mathcal{A}$, in addition to the identity one, there are representations of the form $\pi(\lambda)$, $\lambda \in \mathbb{R}$. However, $\pi(\lambda)(\text{com}\mathcal{A}) = \mathcal{K}L_2(S^{n-1})$, while the algebra $\pi(\lambda)\mathcal{A}$ contains some noncompact operators (for example, the identity operator on $L_2(S^{n-1})$). Summing up these facts, we obtain the following result.

Theorem 2.2.19 *Composition series (2.2.55) is solving and*

$$\text{com}\mathcal{A}/\mathcal{K}L_2(\mathcal{M}) \simeq C_0(\mathbb{R}) \otimes \mathcal{K}L_2(S^{n-1}),$$

$$\mathcal{A}/\text{com}\mathcal{A} \simeq C(X),$$

where X is a compact manifold with a boundary obtained by gluing $S(\mathcal{M})_{x^0} \times S^*(\mathcal{M})_{x^0}$ to $S^*(\mathcal{M})|(\mathcal{M} \setminus x^0)$. The length of the algebra $\mathcal{A}$ is equal to 2. Series (2.2.55) coincides with the maximal radical series.

2.3 Algebras of Pseudodifferential Operators with Discontinuities in Symbols Along a Submanifold

2.3.1 The Statement of Basic Theorems

Let $\mathcal{M}$ be a smooth m-dimensional compact Riemannian manifold without boundary and let $\mathcal{N}$ be a submanifold of codimension n, where $1 \le n \le m-1$. We denote by $\mathfrak{M}_{\mathcal{N}}$ the set of smooth functions f on $\mathcal{M} \setminus \mathcal{N}$ that have a limit $f^0(z, \varphi) = \lim_{x \to z} f(x)$ smoothly depending on $z \in \mathcal{N}$ and on $\varphi \in \nu(\mathcal{N})_z$, where $\nu(\mathcal{N})$ is the bundle of the unit vectors normal to $\mathcal{N}$; thus, the functions f are continuous on the compact obtained by gluing the boundary of a tubular neighborhood of $\mathcal{N}$ to $\mathcal{M} \setminus \mathcal{N}$. We consider the algebra $\mathcal{A}$ generated on $L_2(\mathcal{M})$ by the operators of the class $\bar{\Psi}^0(\mathcal{M})$ (see Definition 2.1.4) and the multiplication operators $f\cdot$ for $f \in \mathfrak{M}_{\mathcal{N}}$. This algebra is irreducible and contains the ideal $\mathcal{K}L_2(\mathcal{M})$.

Let $A \in \bar{\Psi}^0(\mathcal{M})$ and let σ^0 be the principal symbol of the operator A. We introduce the maps:

$$\pi(x, \omega) : A \mapsto \sigma^0(\omega), \quad \pi(x, \omega) : f \mapsto f(x), \tag{2.3.1}$$

where $\omega \in S^*(\mathcal{M})_x$ and $x \in \mathcal{M} \setminus \mathcal{N}$;

$$\pi(z, \varphi, \omega) : A \mapsto \sigma^0(\omega), \quad \pi(z, \varphi, \omega) : f \mapsto f^0(z, \varphi), \tag{2.3.2}$$

where $z \in \mathcal{N}$, $\varphi \in \nu(\mathcal{N})_z$, and $\omega \in S^*(\mathcal{M})_z$;

$$\pi(z,\theta) : A \mapsto F^{-1}_{\eta\to x}\sigma^0(\eta,\theta)F_{y\to\eta} \in \mathcal{B}L_2(\mathbb{R}^n), \tag{2.3.3}$$

$$\pi(z,\theta) : f \mapsto f^0(z,\cdot)\cdot \in \mathcal{B}L_2(\mathbb{R}^n),$$

where F is the Fourier transform on $\mathbb{R}^n = T(\mathcal{N})^{\perp}_z \subset T(\mathcal{M})_z$, T is the tangent bundle, $\theta \in S^*(\mathcal{N})_z$, and $f^0(z,\cdot)\cdot$ denotes the operator of multiplication by the function $\mathbb{R}^n \setminus 0 \ni x \mapsto f^0(z, x/|x|)$;

$$\pi(z,\lambda) : A \mapsto \mathfrak{A}(z,\lambda) := E_{\omega\to\varphi}(\lambda)^{-1}\sigma^0(\omega,0)E_{\psi\to\omega}(\lambda) \in \mathcal{B}L_2(\Sigma_z), \tag{2.3.4}$$

$$\pi(z,\lambda) : f \mapsto f^0(z,\cdot)\cdot \in \mathcal{B}L_2(\Sigma_z),$$

where $\Sigma_z = \nu(\mathcal{N})_z$ is the unit sphere in the space $T(\mathcal{N})^{\perp}_z$, $\omega \in \Sigma_z$.

Theorem 2.3.1 *The maps (2.3.1)–(2.3.4) extend to irreducible pairwise nonequivalent representations of the algebra $\mathcal{A}/\mathcal{K}L_2(\mathcal{M})$. Every irreducible representation of the algebra $\mathcal{A}$ is equivalent either to the identity representation or to one of those in (2.3.1)–(2.3.4). The algebra $\mathcal{A}$ is a type I-algebra.*

If $\operatorname{codim}\mathcal{N} = 1$, representations (2.3.4) turn out to be two-dimensional, see (2.2.48)–(2.2.50). As a rule, in what follows, we do not emphasize this fact by special remarks.

We now proceed to describing the spectral topology. Let $\mathcal{M}_{\mathcal{N}}$ be the compact manifold with the boundary obtained by gluing $\nu(\mathcal{N})$ to $\mathcal{M}\setminus\mathcal{N}$ and $p : \mathcal{M}_{\mathcal{N}} \to \mathcal{M}$ the projection. We denote by $p^*(S^*(\mathcal{M}))$ the induced bundle over $\mathcal{M}_{\mathcal{N}}$ and by Ξ the disjoint union of the sets $p^*(S^*(\mathcal{M}))$, $S^*(\mathcal{N})$, and $\mathcal{N}\times\mathbb{R}$. According to Theorem 2.3.1, the correspondence

$$\pi(x,\omega) \mapsto (x,\omega) \in p^*(S^*(\mathcal{M})), \;\; x \in \mathcal{M}\setminus\mathcal{N}, \tag{2.3.5}$$

$$\pi(z,\varphi,\omega) \mapsto (z,\varphi,\omega) \in p^*(S^*(\mathcal{M})), \;\; (z,\varphi) \in \nu(\mathcal{N})_z \subset \mathcal{M}_{\mathcal{N}},$$

$$\pi(z,\theta) \mapsto (z,\theta) \in S^*(\mathcal{N})_z, \;\; z \in \mathcal{N},$$

$$\pi(z,\lambda) \mapsto (z,\lambda) \in \mathcal{N}\times\mathbb{R}$$

is a bijection of the set $(\mathcal{A}/\mathcal{K}L_2(\mathcal{M}))^\wedge$ onto Ξ. We introduce a topology on Ξ by indicating typical neighborhoods of the points, i.e., neighborhoods forming a fundamental system.

A neighborhood in Ξ of a point $\omega \in p^*(S^*(\mathcal{M}))_z$, $z \notin \mathcal{N}$, is a usual neighborhood $\mathcal{V}(\omega)$ of this point in $S^*(\mathcal{M})$, $\mathcal{V}(\omega) \cap (S^*(\mathcal{M})|\mathcal{N}) = \emptyset$. A neighborhood in Ξ of a point $\theta \in S^*(\mathcal{N})_z$, $z \in \mathcal{N}$, is its neighborhood in the bundle $S^*(\mathcal{N})$. A neighborhood of a point $(z,\lambda) \in \mathcal{N}\times\mathbb{R}$ is defined as the union of an open set $\mathcal{U} \subset \mathcal{N}\times\mathbb{R}$ containing (z,λ) and the part of $S^*(\mathcal{N})$ over the projection of $\mathcal{U}$ into $\mathcal{N}$.

Let $z^0 \in \mathcal{N}$, $\varphi_0 \in \nu(\mathcal{N})_{z^0}$, and $\omega_0 \in S^*(\mathcal{M})_{z^0}$. We can assume that, near z^0, the manifold $\mathcal{M}$ coincides with $\mathbb{R}^m = \{x = (x^{(1)}, x^{(2)}) : x^{(1)} = (x_1, \ldots, x_n), x^{(2)} = (x_{n+1}, \ldots, x_m)\}$ and $\mathcal{N} = \mathbb{R}^{m-n} = \{x \in \mathbb{R}^m : x = (x^{(1)}, x^{(2)}), x^{(1)} = 0\}$. Setting $\omega_0 = (\omega_0^{(1)}, \omega_0^{(2)})$, we first consider the case $\omega_0^{(2)} \neq 0$. A typical neighborhood of a point $(z^0, \varphi_0, \omega_0)$ in Ξ is the union

$$\mathcal{U}(z^0, \varphi_0, \omega_0) \cup (\mathcal{V}(z^0) \times \mathcal{W}(\theta_0)),$$

where $\mathcal{U}(z^0, \varphi_0, \omega_0)$ is a usual neighborhood of $(z^0, \varphi_0, \omega_0)$ in the bundle $p^*(S^*(\mathcal{M}))$ not containing the points (z, φ, ω) with $\omega^{(2)} = 0$; $\mathcal{V}(z^0)$ is a neighborhood of z^0 in $\mathcal{N}$; $\theta_0 = \omega_0^{(2)}/|\omega_0^{(2)}|$ and $\mathcal{W}(\theta_0)$ is a neighborhood of θ_0 in S^{m-n-1}.

For $\omega_0 = (\omega_0^{(1)}, 0)$, $\varphi_0\omega_0^{(1)} = 0$, a typical neighborhood of $(z^0, \varphi_0, \omega_0)$ is of the form

$$\mathcal{U}(z^0, \varphi_0, \omega_0) \cup (\mathcal{V}(z^0) \times S^{m-n-1}) \cup (\mathcal{V}(z^0) \times \mathbb{R}),$$

where $\mathcal{U}(z^0, \varphi_0, \omega_0)$ is an arbitrary neighborhood of $(z^0, \varphi_0, \omega_0)$ in $p^*(S^*(\mathcal{M}))$. If $\varphi_0\omega_0^{(1)} \gtrless 0$, then the line $\mathbb{R}$ must be changed for the set $\{\lambda \in \mathbb{R} : \lambda \gtrless Q\}$ with any real number Q. In case of $\operatorname{codim} \mathcal{N} = 1$, the points subject to the conditions $\varphi_0\omega_0^{(1)} = 0$ are absent, and any point $(z^0, \varphi_0, \omega_0)$ has a neighborhood containing only a part of the line $\mathbb{R}$.

Theorem 2.3.2 *The topology carried over from the space Ξ to $(\mathcal{A}/\mathcal{K}L_2(\mathcal{M}))^\wedge$ by the bijection (2.3.5) coincides with the Jacobson topology. The spectrum $(\mathcal{A}/\mathcal{K}L_2(\mathcal{M}))^\wedge$ is a T_0-space.*

Theorem 2.3.3 *Let I denote the ideal in the algebra $\mathcal{A}$ equal to the intersection of the kernels of all one-dimensional representations and all representations $\pi(z, \lambda)$ for $(z, \lambda) \in \mathcal{N} \times \mathbb{R}$. Then, the composition series $0 \subset \mathcal{K}L_2(\mathcal{M}) \subset I \subset \operatorname{com} \mathcal{A} \subset \mathcal{A}$ is solving and*

$$I/\mathcal{K}L_2(\mathcal{M}) \simeq C(S^*(\mathcal{N})) \otimes \mathcal{K}L_2(\mathbb{R}^n),$$

$$\operatorname{com} \mathcal{A}/I \simeq C_0(\mathcal{N} \times \mathbb{R}) \otimes \mathcal{K}L_2(S^{n-1}),$$

$$\mathcal{A}/\operatorname{com} \mathcal{A} \simeq C(p^*(S^*(\mathcal{M}))).$$

The length of $\mathcal{A}$ is equal to 3. The composition series is the maximal radical series.

2.3.2 Algebras $\mathcal{L}(\theta)$: Irreducibility

Let us introduce Cartesian coordinates in the tangent plane to $\mathcal{M}$ at a point $z^0 \in \mathcal{N}$ choosing axes $x_1, \ldots, x_n$ to be orthogonal to the submanifold $\mathcal{N}$. A point x in the tangent plane $\mathbb{R}^m$ will be written in the form $x = (x^{(1)}, x^{(2)})$, where $x^{(1)} = (x_1, \dots, x_n)$ and $x^{(2)} = (x_{n+1}, \dots, x_m)$. We also introduce the operators

$$A = F^{-1}_{\xi \to x} \Phi(\varphi, \xi) F_{y \to \xi}, \tag{2.3.6}$$

where F is the Fourier transform in $\mathbb{R}^m$, $\Phi(\varphi, \xi) = f^0(\varphi)\sigma^0(\xi)$, $\varphi = x^{(1)}/|x^{(1)}|$, and σ^0 is a degree zero homogeneous function, while $f^0 \in C^\infty(S^{n-1})$ and $\sigma^0 \in C^\infty(S^{m-1})$.

Denote by $\mathcal{L}$ the algebra spanned by the operator-functions

$$S^{m-n-1} \ni \theta \mapsto A(\theta) = F^{-1}_{\eta \to x} \Phi(\varphi, \eta, \theta) F_{y \to \eta} \in \mathcal{B}L_2(\mathbb{R}^n), \tag{2.3.7}$$

with $\|A(\cdot); \mathcal{L}\| = \sup\{\|A(\theta); \mathcal{B}L_2(\mathbb{R}^n)\|, \theta \in S^{m-n-1}\}$ (here, F denotes the Fourier transform in $\mathbb{R}^n$). Let $\mathcal{L}(\theta)$ with $\theta \in S^{m-n-1}$ be the algebra generated by the operators $A(\theta)$ on $L_2(\mathbb{R}^n)$.

To apply the localization principle (Proposition 1.3.24) to the algebra $\mathcal{A}$, we choose $C(\mathcal{M})$ as a localizing subalgebra and $J = \mathcal{K}L_2(\mathcal{M})$. Let $\mathcal{A}_{z^0}$ be a local algebra obtained by localizing $\mathcal{A}$ at a point z^0.

Proposition 2.3.4

(1) *The algebra $\mathcal{A}_{z^0}$ for $z^0 \in \mathcal{N}$ is generated on $L_2(\mathbb{R}^m)$ by the operators of the form (2.3.6).*

(2) *The algebras $\mathcal{A}_{z^0}$ and $\mathcal{L}$ are isomorphic. The equalities*

$$\widehat{\mathcal{L}} = \cup_{\theta \in S^{m-n-1}} \widehat{\mathcal{L}}(\theta) \tag{2.3.8}$$

hold, where $\widehat{\mathcal{L}}$ and $\widehat{\mathcal{L}}(\theta)$ are the spectra of the algebras $\mathcal{L}$ and $\mathcal{L}(\theta)$.

Proof

(1) Let V be a coordinate neighborhood of a point z^0 in $\mathcal{M}$ and let $\varkappa$ be a coordinate diffeomorphism projecting V into the tangent space to $\mathcal{M}$ at z^0 so that $\varkappa(z^0) = 0$, $\varkappa'(z^0) = I$, and $V \ni z \mapsto \varkappa(z) = (x^{(1)}, x^{(2)})$. Moreover, let B denote a product of finitely many generators of the algebra $\mathcal{A}$ and let $B_\varkappa$ denote the operator B written in local coordinates. For $t > 0$, we introduce the unitary operator $U_t = t^{m/2}u(t\cdot)$ on $L_2(\mathbb{R}^m)$. Finally, we choose $\chi \in C_c^\infty(V)$ and consider $U_t(\chi B \chi)_\varkappa U_t^{-1}$, which is a product of finitely many operators of the form

$$v \mapsto \int e^{ix\xi} f(tx)\sigma_\varkappa(tx, \xi/t)\hat{v}(\xi)\, d\xi,$$

where $\sigma_\varkappa$ is a symbol of operator in $\bar{\Psi}^0(\mathcal{M})$ and f is a function in $\mathfrak{M}_\mathcal{N}$ written in local coordinates. Therefore, in $L_2(\mathbb{R}^m)$, there exists a strong limit

$$Q_0 := \lim_{t \to 0} U_t(\chi B \chi)_\varkappa U_t^{-1}.$$

The operator Q_0 is a product of finitely many operators of the form (2.3.6) and

$$\|Q_0\| \leq |\chi(0)|^2 \|Q\|,$$

where $Q = (\psi B \psi)_\varkappa$, $\psi \in C_c^\infty(V)$, and $\psi \chi = \chi$.

Let $\mathfrak{Q}$ be the algebra spanned in $L_2(\mathbb{R}^m)$ by operators of the form $(\chi B \chi)_\varkappa$ and let $\mathfrak{J}$ be the ideal in $\mathfrak{Q}$ generated by the operators of multiplication by the functions $\zeta \in C_c^\infty(\mathbb{R}^m)$ such that $\zeta(0) = 0$. From the inequality $\|Q_0\| \leq |\chi(0)|^2 \|Q\|$, it follows that the map $Q \mapsto Q_0$ extends to an epimorphism $q : \mathfrak{Q}/\mathfrak{J} \to \mathfrak{Q}_0$, where $\mathfrak{Q}_0$ is the algebra spanned by operators of the form (2.3.6). We show that q is an isomorphism. Let $\phi \in C_c^\infty(\mathbb{R}^m)$ and $\phi(0) = 1$. Then, $\phi Q_0 \phi \in \mathfrak{Q}$ and $q : [\phi Q_0 \phi] \mapsto Q_0$, where $[\phi Q_0 \phi]$ is the residue class of the operator $\phi Q_0 \phi$ in the algebra $\mathfrak{Q}/\mathfrak{J}$. The fact that q is a monomorphism follows from the relations $Q - \phi Q_0 \phi \in \mathfrak{J}$ and $\|\phi Q_0 \phi\| \leq |\phi|^2 \|Q_0\|$. Let $\mathcal{J}_{z^0}$ be the ideal in $\mathcal{A}$ generated by the operators of multiplication by the functions $\eta \in C(\mathcal{M})$ with $\eta(z^0) = 0$. An isomorphism $\mathcal{A}_{z^0} \simeq \mathfrak{Q}_0$ is established by the map

$$\mathcal{A}_{z^0} = \mathcal{A}/\mathcal{J}_{z^0} \ni [A] \mapsto [(\varkappa^{-1})^*(\chi A \chi)(\varkappa)^*] \in \mathfrak{Q}/\mathfrak{J} \to \mathfrak{Q}_0;$$

the latter arrow indicates the map q, $\chi \in C_c^\infty(V)$, and $\chi(z) = 1$ near z^0.

(2) We denote by $\hat{u}(y^{(1)}, \xi^{(2)}) = F_{y^{(2)} \to \xi^{(2)}} u(y^{(1)}, y^{(2)})$ the partial Fourier transform of a function u. Then, from (2.3.6), it follows that

$$(Au)\hat{}(x^{(1)}, \xi^{(2)}) = F^{-1}_{\xi^{(1)} \to x^{(1)}} \Phi(\varphi, \xi^{(1)}, \xi^{(2)}) F_{y^{(1)} \to \xi^{(1)}} \hat{u}(y^{(1)}, \xi^{(2)}). \tag{2.3.9}$$

Setting $X = x^{(1)}|\xi^{(2)}|$ and $Y = y^{(1)}|\xi^{(2)}|$, we rewrite (2.3.9) as

$$(Au)\hat{}(X|\xi^{(2)}|^{-1}, \xi^{(2)}) = (2\pi)^{-n} \int_{\mathbb{R}^n} e^{iX\eta} \Phi(\varphi, \eta, \theta)\, d\eta \int_{\mathbb{R}^n} e^{-iY\eta} \hat{u}(Y|\xi^{(2)}|^{-1}, \xi^{(2)})\, dY \tag{2.3.10}$$

because Φ is a homogeneous function. Let us write x, y ($\in \mathbb{R}^n$) instead of X, Y and $v(y)$ and $A(\theta)v(x)$ instead of $\hat{u}(Y|\xi^{(2)}|^{-1}, \xi^{(2)})$ and $(Au)\hat{}(X|\xi^{(2)}|^{-1}, \xi^{(2)})$. Equality (2.3.10) now coincides with the formula

$$A(\theta)v(x) = F^{-1}_{\eta \to x} \Phi(\varphi, \eta, \theta) F_{y \to \eta} v(y), \quad \theta \in S^{m-n-1}. \tag{2.3.11}$$

Since the transform $F_{y^{(2)}\to\xi^{(2)}}$ is unitary on $L_2(\mathbb{R}^{m-n})$, we find that $\mathcal{A}_{z^0}$ and $\mathcal{L}$ are isomorphic. Formula (2.3.8) follows from Proposition 1.3.16.

□

Proposition 2.3.5 *The algebra $\mathcal{L}(\theta)$ is irreducible, i.e., every subspace in $L_2(\mathbb{R}^n)$ invariant with respect to $\mathcal{L}(\theta)$ is either* 0 *or $L_2(\mathbb{R}^n)$.*

Proof We first obtain a new representation for operators (2.3.11). According to Proposition 1.2.2, the Fourier transform is a composition of the Mellin transform, $E(\lambda)$, and the inverse Mellin transform. We substitute the corresponding expressions for $F^{\pm 1}$ in (2.3.11) and, after the change of variable $\rho = e^t$, arrive at

$$A(\theta)u(x) = \frac{1}{\sqrt{2\pi}} \int_0^\infty r^{i\lambda - n/2} E_{\omega\to\varphi}(\lambda)^{-1} \Phi_{\mu\to\lambda}(\varphi, \omega, \theta) E_{\psi\to\omega}(\mu) \times \tilde{u}(\mu + in/2, \psi)\, d\mu, \quad (2.3.12)$$

where $\tilde{u}$ denotes the Mellin transform of u and

$$\Phi_{\mu\to\lambda}(\varphi, \omega, \theta) g(\mu) = \frac{1}{2\pi} \int_{-\infty}^{+\infty} e^{i\lambda t} \Phi(\varphi, \omega, e^{-t}\theta)\, dt \int_{-\infty}^{+\infty} e^{-it\mu} g(\mu)\, d\mu. \quad (2.3.13)$$

Let $H \subset L^2(\mathbb{R}^n)$ be a subspace invariant with respect to $\mathcal{L}(\theta)$. We show that it is also invariant with respect to all operators

$$Bu(x) = (2\pi)^{-1/2} \int_0^\infty r^{i\lambda - n/2} E_{\omega\to\varphi}(\lambda)^{-1} \Psi(\varphi, \omega) E_{\psi\to\omega}(\lambda) \tilde{u}(\lambda + in/2, \psi)\, d\lambda, \quad (2.3.14)$$

where $\Psi \in C^\infty(S^{n-1} \times S^{n-1})$. It suffices to verify that every operator (2.3.14) is the limit of a sequence of operators in $\mathcal{L}(\theta)$ in the strong topology.

We first consider the case where the function Ψ depends only on $\omega \in S^{n-1}$. Let $\chi \in C^\infty(\mathbb{R})$, let $\chi(t) = 1$ for $t > 0$, and let $\chi(t) = 0$ for $t < -1$. We smoothly extend the function $\omega \mapsto \Psi(\omega)$ inside the sphere S^{n-1} and introduce $\Phi_k(\xi) = \Psi(\xi^1/|\xi|)\chi_k(|\xi^2|/|\xi|)$, where $k \in \mathbb{N}$, $\xi \in \mathbb{R}^m \setminus 0$, and the function χ_k is defined by $\chi_k(e^{-t}/\sqrt{1+e^{-2t}}) = \chi(t+k)$. Denote by $A_k(\theta)$ operator (2.3.12) assuming that $\Phi(\omega, e^{-t}\theta) \equiv \Phi_k(\omega, e^{-t}\theta) = \Psi(\omega)\chi_k(e^{-t}/\sqrt{1+e^{-2t}})$.

The sequence of the operators $h \mapsto \Phi_k h$ on $L_2(\mathbb{R}, L_2(S^{n-1}))$ strongly converges to the operator $h \mapsto \Psi h$ as $k \to \infty$. Therefore, the sequence $\{A_k(\theta)\}$ strongly converges to the operator B because the Mellin transform $M : L_2(\mathbb{R}^n) \to L_2(\mathbb{R}, L_2(S^{n-1}))$ and $E(\lambda) : L_2(S^{n-1}) \to L_2(S^{n-1})$ are unitary as well as the Fourier transform on $L^2(\mathbb{R}, L_2(S^{n-1}))$

sending a function h to the map

$$(t, \omega) \mapsto \frac{1}{\sqrt{2\pi}} \int_{-\infty}^{+\infty} e^{-it\lambda} h(\lambda, \omega)\, d\lambda.$$

We have assumed that the function Ψ does not depend on $\varphi \in S^{n-1}$. To eliminate this restriction, it suffices to observe that the operator (2.3.14) with an arbitrary function $\Psi \in C^\infty(S^{n-1} \times S^{n-1})$ can be approximated in the norm by operators in which the role of $\Psi(\varphi, \omega)$ is played by the finite sums $\sum a_j(\varphi) b_j(\omega)$.

The orthogonal complement $\mathcal{H}^\perp$ of the subspace $\mathcal{H}$ is also an invariant subspace for the algebra $\mathcal{L}(\theta)$. It follows from the foregoing that $\mathcal{H}^\perp$ is invariant under all the operators (2.3.14). The subspace $\mathcal{H}$ is thereby an invariant subspace for the algebra $\mathcal{L}(0)$ generated by operators of the form (2.3.14) on $L_2(\mathbb{R}^n)$. The algebra $\mathcal{L}(0)$ is isomorphic to the algebra $\mathfrak{S}$ spanned by the operator-valued functions $\lambda \mapsto \mathfrak{A}(\lambda) = E(\lambda)^{-1}\Psi(\varphi, \omega) E(\lambda) \in \mathcal{B}L_2(S^{n-1})$, see Sect. 2.2.1. According to Proposition 2.2.3, we have $C_0(\mathbb{R}) \otimes \mathcal{K}L_2(S^{n-1}) \subset \mathfrak{S}$. This means that the algebra $\mathcal{L}(0)$ contains all operators of the form

$$L_2(\mathbb{R}^n) \ni f \mapsto M^{-1}\Big(h(\lambda) \sum_{k=1}^{N} \langle (Mf)(\lambda + in/2\,, \cdot), e_k \rangle e_k\Big), \tag{2.3.15}$$

where $\langle\ ,\ \rangle$ and $\{e_k\}$ are the inner product and a basis in $L^2(S^{n-1})$, and $h \in C_0(\mathbb{R})$. It is clear that the sequence of operators (2.3.15) converges strongly as $N \to \infty$ to the operator $f \mapsto M^{-1}(h(\lambda)(Mf)(\lambda + in/2\,, \cdot))$. Hence, this operator carries the subspace $\mathcal{H}$, which is invariant under $\mathcal{L}(\theta)$, into itself.

Let f and g be elements of $\mathcal{H}$ and $\mathcal{H}^\perp$, respectively. It follows from the above that

$$\int_{-\infty}^{+\infty} h(\lambda) \langle \tilde{f}(\lambda + n/2\,, \cdot)\,, \tilde{g}(\lambda + in/2\,, \cdot)\rangle\, d\lambda = 0.$$

Since $h \in C_0(\mathbb{R})$ is arbitrary, we obtain

$$\langle \tilde{f}(\lambda + in/2\,, \cdot)\,, \tilde{g}(\lambda + in/2\,, \cdot)\rangle = 0$$

for almost all $\lambda \in \mathbb{R}$. Every operator in $\mathcal{L}(0)$ can be written in the form $M^{-1}\mathfrak{A}(\lambda)M$, where $\mathfrak{A} \in \mathfrak{S}$. Therefore, $\tilde{f}(\lambda + in/2\,, \cdot)$ can be replaced by $\mathfrak{A}(\lambda)\tilde{f}(\lambda + in/2\,, \cdot)$. Thus, for almost all $\lambda \in \mathbb{R}$

$$\langle \mathfrak{A}(\lambda)\tilde{(}\lambda + in/2\,, \cdot)\,, \tilde{g}(\lambda + in/2\,, \cdot)\rangle = 0. \tag{2.3.16}$$

The algebra $\mathfrak{S}(\lambda)$ generated on $L_2(S^{n-1})$ by the operators $\mathfrak{A}(\lambda)$ is irreducible for each $\lambda \in \mathbb{R}$ (Proposition 2.2.1). Every nonzero vector $\tilde{f}(\lambda + in/2\,, \cdot) \in L_2(S^{n-1})$ is thereby totalizing for $\mathfrak{S}(\lambda)$, i.e., the set $\{\mathfrak{A}(\lambda)\tilde{f}(\lambda+in/2\,,\cdot) : \mathfrak{A}(\lambda) \in \mathfrak{S}(\lambda)\}$ is dense in $L_2(S^{n-1})$. Consequently, if $f \in \mathcal{H}$, $g \in H^{\perp}$, and $\lambda \in \mathbb{R}$ are such that (2.3.16) holds and $\tilde{g}(\lambda + in/2\,,\cdot) \neq 0$, then $\tilde{f}(\lambda + in/2\,,\cdot) = 0$.

Assume that there is a nonzero element g in $\mathcal{H}^{\perp}$. Then, any function $\lambda \mapsto \tilde{f}(\lambda+in/2\,,\cdot)$ with $f \in \mathcal{H}$ vanishes almost everywhere on the set $\{\lambda \in \mathbb{R} : \tilde{g}(\lambda + in/2\,,\cdot) \neq 0\}$ of positive measure. We show that this is possible only if $\mathcal{H} = 0$. This concludes the proof of the proposition.

Choose a sequence $\{\chi_j\} \subset C_0^{\infty}(\mathbb{R})$ converging to the function $t \to \delta(t - t_0)$. We define the functions $\Phi_j \in C^{\infty}(S^{m-1})$ by $\Phi_j(\omega(1 + e^{-2t})^{-1/2}, e^{-t}\theta(1 + e^{-2t})^{-1/2}) = \chi_j(t)$, where $\omega \in S^{n-1}$, $\theta \in S^{m-n-1}$, and $t \in \mathbb{R}$, and extend Φ_j to $\mathbb{R}^m \setminus 0$ as a homogeneous function of zero degree. Let $A_j(\theta)$ denote the right-hand side of (2.3.12) with $\Phi = \Phi_j$. The equality $v_j = A_j(\theta)f$ can be rewritten in the form

$$E(\lambda)v_j(\lambda + in/2\,,\cdot) = \frac{1}{2\pi}\int_{-\infty}^{+\infty} e^{i\lambda t}\chi_j(t)\,dt \int_{-\infty}^{+\infty} e^{-it\mu}E(\mu) \times \tilde{f}(\mu + in/2\,,\cdot)\,d\mu. \tag{2.3.17}$$

Assume that $\mathcal{H} \neq 0$ and $f \in \mathcal{H}$ is a nonzero element. Taking into account that the operator of multiplication by a function $h \in C_0(\mathbb{R})$ does not lead out of the subspace $M\mathcal{H}$, we can take $f \in \mathcal{H}$ such that the function $\lambda \mapsto \tilde{f}(\lambda + in/2\,,\cdot)$ has compact support and

$$\int_{-\infty}^{+\infty} e^{-it_0\mu}E(\mu)\tilde{f}(\mu + in/2\,,\cdot)\,d\mu \neq 0.$$

Clearly, the operator $\tilde{v}(\lambda+in/2\,,\cdot) \mapsto E(\lambda)v(\lambda+in/2\,,\cdot)$ does not change the support of a function in $L_2(\mathbb{R}, L_2(S^{n-1}))$. Therefore, choosing the number j in (2.3.17) sufficiently large, we obtain that the function $\lambda \mapsto \tilde{v}_j(\lambda + in/2, \cdot)$ has no zero values on any given finite interval. Take this interval so that the measure of its intersection with the set $\{\lambda : \tilde{g}(\lambda + in/2\,,\cdot) \neq 0\}$ is positive. Then, the function $\lambda \mapsto \tilde{g}(\lambda + in/2\,,\cdot)$ does not vanish on the indicated intersection, which contradicts the inclusion $v_j \in \mathcal{H}$. □

2.3.3 Localization in the Algebra $\mathcal{L}(\theta)$

Let $\bar{\mathbb{R}}^n$ be the compact set obtained by adjoining the $(n-1)$-dimensional sphere S_{∞}^{n-1} to the space $\mathbb{R}^n$ at infinity. We are going to apply Proposition 1.3.26 for localization in the algebra $\mathcal{L}(\theta)$ choosing $C(\bar{\mathbb{R}}^n)$ as a localizing algebra $\mathcal{C}$. We first prove some technical results for verification of the conditions of this proposition.

Lemma 2.3.6 *Let $\mathbb{R}^m \setminus 0 \ni \xi \mapsto \Psi(\xi) = |\xi|^a f(\xi/|\xi|)$ be a homogeneous function of a complex degree a with $\operatorname{Re} a \leq 0$, $f \in C^{\infty}(S^{m-1})$, and*

$$B(\theta) = F_{\eta\to x}^{-1}\Psi(\eta, \theta)F_{y\to\eta}, \tag{2.3.18}$$

where $\eta \in \mathbb{R}^n$ *and* $\theta \in S^{m-n-1}$. *Then,* $(B(\theta)v)\tilde{}(\lambda + ia + in/2, \varphi)$ *is equal to*

$$E_{\omega\to\varphi}(\lambda + ia)^{-1}\Psi_{\mu\to\lambda}(\omega, \theta)E_{\psi\to\omega}(\mu)\tilde{v}(\mu + in/2, \psi) \tag{2.3.19}$$
$$= \int_{S^{n-1}} d\psi \int_{\mathrm{Im}\mu=0} \tilde{H}(\varphi, \psi; \lambda, \lambda - \mu, \theta)\tilde{v}(\mu + in/2, \psi)\, d\mu;$$

here, $\mathrm{Im}\,\lambda = 0$, *the operator* $\Psi_{\mu\to\lambda}$ *is defined by (2.3.13) (with* $\Phi(\varphi, \omega, e^{-t}\theta)$ *replaced by* $\Psi(\omega, e^{-t}\theta)$*),* $v \in C_0^\infty(\mathbb{R}^n \setminus 0)$, *and*

$$\tilde{H}(\varphi, \psi; \lambda, \nu, \theta) = (2\pi)^{-1/2} \int_0^\infty \rho^{-i\nu-1}\, d\rho \int_0^\infty t^{-i(\lambda+ia+in/2)-1}\hat{H}(t\varphi - \psi, \rho\theta)\, dt, \tag{2.3.20}$$
$$\hat{H}(x, \theta) = (2\pi)^{(n-m)/2} \int_{\mathbb{R}^{m-n}} e^{-i\theta z} H(x, z)\, dz,$$

where $H = F^{-1}\Psi$ *(in the last equality,* F *is the Fourier transform on* $\mathbb{R}^m$, *and in (2.3.18), it denotes the Fourier transform on* $\mathbb{R}^n$*). In (2.3.20)* $\varphi \neq \psi$, *the Mellin transform with respect to* ρ *is understood in the sense of the theory of generalized functions, and the inside integral (with respect to the variable* t*) is defined by means of analytic extension with respect to the parameter* a *(an explicit formula is written below for this extension). The relation (2.3.19) holds everywhere in the half-plane* $\mathrm{Re}a \le 0$ *except at the poles* $a = -n/2 - k + i\lambda$, *where* $k = 0, 1, \ldots$

Proof Let us first assume that $0 \ge \mathrm{Re}\, a > -n/2$. We write operator (2.3.18) in the form

$$B(\theta)v(x) = \int_{\mathbb{R}^n} \hat{H}(x - y, \theta)v(y)\, dy.$$

Applying the Mellin transform to this equality, we obtain

$$(B(\theta)v)\tilde{}(\lambda + ia + in/2, \varphi) \tag{2.3.21}$$
$$= (2\pi)^{-1/2} \int_0^\infty r^{-i(\lambda+ia+in/2)-1}\, dr \int_{\mathbb{R}^m} \hat{H}(x - y, \theta)v(y)\, dy$$
$$= (2\pi)^{-1/2} \int_{S^{n-1}} d\psi \int_0^\infty v(\rho, \psi)\rho^{n-1}\, d\rho \int_0^\infty \hat{H}(r\varphi - \rho\psi, \theta)r^{-i(\lambda+ia+in/2)-1}\, dr$$
$$= (2\pi)^{-1/2} \int_{S^{n-1}} d\psi \int_0^\infty v(\rho, \psi)\rho^{-i(\lambda+in/2)-1}\, d\rho$$
$$\times \int_0^\infty \rho^{i(\lambda-in/2)} r^{-i(\lambda+ia+in/2)-1}\hat{H}(r\varphi - \rho\psi, \theta)\, dr.$$

The quantity $\hat{H}(x, \rho\theta)$ tends to zero faster than any power of ρ as $\rho \to \infty$ for $x \neq 0$ and $0 \geq \operatorname{Re} a > -n/2$. Since H is a homogeneous function of degree $-m - a$, it follows that $\hat{H}(tx, \theta) = \hat{H}(x, t\theta)t^{-a-n}$ for any $t > 0$. Therefore,

$$\int_0^\infty \rho^{i(\lambda - in/2)} r^{-i(\lambda + ia + in/2) - 1} \hat{H}(r\varphi - \rho\psi, \theta)\, dr \qquad (2.3.22)$$

$$= \int_0^\infty t^{-(\lambda + ia + in/2) - 1} \hat{H}(t\varphi - \psi, \rho\theta)\, dt.$$

We take account of the fact that

$$\int_0^\infty \rho^{-i(\lambda + in/2) - 1} v(\rho) w(\rho)\, d\rho = \int_{\operatorname{Im}\mu = 0} \tilde{w}(\lambda - \mu)\tilde{v}(\mu + in/2)\, d\mu.$$

From this, (2.3.21), and (2.3.22), we obtain

$$(B(\theta)v)\tilde{\ }(\lambda + ia + in/2, \varphi) \qquad (2.3.23)$$

$$= \int_{S^{n-1}} d\psi \int_{\operatorname{Im}\mu = 0} \tilde{H}(\varphi, \psi; \lambda, \lambda - \mu, \theta)\tilde{v}(\mu + in/2, \psi)\, d\mu.$$

However, for $0 \geq \operatorname{Re} a > -n/2$,

$$(B(\theta)v)\tilde{\ }(\lambda + ia + in/2) = E_{\omega \to \varphi}(\lambda + ia)^{-1} \Psi_{\mu \to \lambda}(\omega, \theta) E_{\psi \to \omega}(\mu)\tilde{v}(\mu + in/2, \psi)$$

(cf. the derivation of (2.3.12)). Thus, (2.3.23) can be rewritten in the form (2.3.19).

The function $\lambda \mapsto E(\lambda)^{-1}$ is meromorphic on the whole plane (the poles are located at the points $\lambda = -i(n/2 + k)$, $k = 0, 1, \dots$). Therefore, the left-hand side of (2.3.19) can be extended analytically to the half-plane $\operatorname{Re} a \geq 0$ except for the points $a = -n/2 - k + i\lambda$. Obviously, the right-hand side of (2.3.19) has the same property. The analytic extension of the inside integral in (2.3.20) on the right is implemented in the strip $0 \geq \operatorname{Re} a > -n/2 - p$ by the formula

$$\int_0^\infty t^{-is-1} \chi_0(t)\hat{H}(t\varphi - \psi, \rho\theta)\, dt \quad (2.3.24)$$

$$= \frac{(-1)^p}{(-is)\dots(-is + p - 1)} \int_0^\infty t^{-is+p-1} \frac{d^p}{dt^p}(\chi_0(t)\hat{H}(t\varphi - \psi, p\theta))\, dt,$$

where $s = \lambda + ia + in/2$, $\chi \in C^\infty(\bar{R}_+, \chi(t) = 1$ for $t < 1/4$, and $\chi(t) = 0$ for $t > 1/2$.
□

Lemma 2.3.7 *Let $A(\theta)$ be the operator defined by (2.3.11) with a function Φ independent of φ and homogeneous of degree 0. Furthermore, suppose that $G(x) = F^{-1}_{\xi \to x}\Phi(\xi)$, a zero* *degree homogeneous function σ belongs to $C^\infty(\mathbb{R}^n \setminus 0)$, and let ζ_1 and ζ_2 be arbitrary functions in $C^\infty(S^{n-1})$ such that the union* $\operatorname{supp}\zeta_1 \cup \operatorname{supp}\zeta_2$ *of their supports lies in a semisphere. Then, for all $\lambda \in \mathbb{R}$ except perhaps for $\lambda = 0$,*

$$(\zeta_1 A(\theta)\zeta_2\sigma u)\tilde{}(\lambda + in/2, \varphi) = \sum_{|\gamma|=0}^{N} \frac{(-1)^{|\gamma|}}{\gamma!}(\zeta_1 \partial_\varphi^\gamma \sigma)(\varphi) E_{\omega\to\varphi}(\lambda - i|\gamma|)^{-1} \tag{2.3.25}$$

$$\times \Phi^{(\gamma)}_{\mu\to\lambda}(\omega, \theta) E_{\psi\to\omega}(\mu)\zeta_2(\psi)\tilde{u}(\mu + in/2, \psi) + (\zeta_1 R_N \zeta_2 \tilde{u})(\lambda + in/2, \varphi),$$

where $\partial_\varphi^\gamma = \partial^{|\gamma|}/\partial\varphi_1^{\gamma_1}\dots\partial\varphi_n^{\gamma_n}$, *the operator* $\Phi^{(\gamma)}_{\mu\to\lambda}$ *is defined by (2.3.13) with* $\Phi(\varphi, \omega, e^{-t}\theta)$ *replaced by* $(-1)^{|\gamma|}\partial_\omega^\gamma \Phi(\omega, e^{-t}\theta)$, *and*

$$(\zeta_1 R_N \zeta_2 \tilde{u})(\lambda + in/2, \varphi) = \int_{S^{n-1}} d\psi \int_0^\infty \zeta_1(\varphi)\tilde{G}_N(\varphi, \psi; \lambda, \lambda - \mu, \theta)$$

$$\times \zeta_2(\psi)\tilde{u}(\mu + in/2, \psi)\, d\mu,$$

$$\zeta_1(\varphi) G_N(\varphi, \psi\,; \lambda, \nu, \theta)\zeta_2(\psi) = \frac{1}{\sqrt{2\pi}} \int_0^\infty \rho^{-i\nu-1}\, d\rho \int_0^\infty t^{-i(\lambda - i(N+1) + in/2) - 1}$$

$$\times \sum_{|\gamma|=N+1} \hat{G}(t\varphi - \psi, \rho\theta)(t\varphi - \psi)^\gamma \varkappa_\gamma(\varphi, \psi, t)\, dt,$$

$$\varkappa_\gamma(\varphi, \psi, t) = \frac{(N+1)(-1)^{N+1}}{\gamma!}\zeta_1(\varphi)\zeta_2(\psi) \int_0^1 (1 - s)^N (\partial^\gamma \sigma)(\varphi + s(t^{-1}\psi - \varphi))\, ds.$$

(The inside integral with respect to t is understood in the sense of analytic extension with respect to λ; the extension is implemented by a formula of the form (2.3.24).)

Proof By Taylor's formula,

$$\zeta_1(\varphi)\sigma(\psi)\zeta_2(\psi) = \sum_{|\gamma|=0}^{N} \frac{(-1)^{|\gamma|}}{\gamma!}\zeta_1(\varphi)\zeta_2(\psi)(\partial^\gamma\sigma)(\varphi)(\varphi - t^{-1}\psi)^\gamma$$

$$+ \sum_{|\gamma|=N+1} (\varphi - t^{-1}\psi)^\gamma \varkappa_\gamma(\varphi, \psi, t).$$

According to Lemma 2.3.6,

$$(\zeta_1 A(\theta)\zeta_2\sigma u)\tilde{}(\lambda + in/2, \varphi) = \sum_{|\gamma|=0}^{N} \frac{(-1)^{|\gamma|}}{\gamma!}(\zeta_1 \partial_\varphi^\gamma \sigma)(\varphi) \tag{2.3.26}$$

$$\times \int_{S^{n-1}} d\psi \int 0^\infty \zeta_2(\psi)\tilde{v}(\mu + in/2, \psi)$$

$$\times \tilde{G}_\gamma(\varphi, \psi; \lambda, \lambda - \mu, \theta)\, d\mu + (\zeta_1 R_N \zeta_2 \tilde{v})(\lambda + in/2, \varphi),$$

and

$$\tilde{G}_\gamma(\varphi, \psi; \lambda, \nu, \theta) = \frac{1}{\sqrt{2\pi}} \int_0^\infty \rho^{-i\nu-1}\, d\rho \tag{2.3.27}$$

$$\int_0^\infty t^{-i(\lambda - i|\gamma| + in/2)} \hat{G}(t\varphi - \psi, \rho\theta)(t\varphi - \psi)^\gamma\, dt.$$

Using (2.3.19) for the terms on the right-hand side of (2.3.26) and taking into account (2.3.27), we conclude the proof. □

Lemma 2.3.8 *Let T be the operator on $L_2(\mathbb{R}, L_2(S^{n-1}))$ defined by*

$$(Tv)(\lambda) = \int_{-\infty}^{+\infty} g(\lambda, \mu) v(\mu)\, d\mu. \tag{2.3.28}$$

Assume that the kernel g is continuous on $\mathbb{R} \times \mathbb{R}$ with respect to the norm of operators on $L_2(S^{n-1})$, its values belong to $\mathcal{K}L_2(S^{n-1})$, and

$$\iint_{\mathbb{R}\times\mathbb{R}} \|g(\lambda, \mu)\|^2 d\lambda d\mu < \infty.$$

Then, T is a compact operator.

The proof is left to the reader.

Proposition 2.3.9 *Suppose that $\Phi \in C^\infty(S^{n-1} \times (\mathbb{R}^m \setminus 0))$, $\xi \mapsto \Phi(\varphi, \xi)$ is a homogeneous function of degree zero, and $\Phi(\varphi, \omega, 0) \equiv f(\varphi)$, where $\omega \in S^{n-1}$ and f is an arbitrary element of $C^\infty(S^{n-1})$. Moreover, let $A(\theta)$ be the operator defined by (2.3.11). Then, for $\sigma \in C^\infty(S^{n-1})$, the commutator $[A(\theta), \sigma] = A(\theta)\sigma - \sigma A(\theta)$ is compact on $L_2(\mathbb{R}^n)$.*

Proof Without loss of generality, it can be assumed that $\Phi(\varphi, \omega, 0) \equiv 0$ (if not, then $\Phi(\varphi, \omega, \theta)$ is replaced by the difference $\Phi(\varphi, \omega, \theta) - f(\varphi)$). The compactness of the commutator $[A(\theta), \sigma]$ on $L_2(\mathbb{R}^n)$ is equivalent to the compactness of the commutator $[MA(\theta)M^{-1}, \sigma]$ on $L_2(\mathbb{R}, L_2(S^{n-1}))$.

We first assume that Φ is independent of φ. Denote by $\{\eta_j\}$ a partition of unity on S^{n-1} such that the union $\operatorname{supp}\eta_j \cup \operatorname{supp}\eta_k$ of any two intersecting supports lies in some hemisphere. It is clear that $A(\theta) = \sum_{j,k} \eta_j A(\theta)\eta_k$. Let j and k be such that $\operatorname{supp}\eta_j \cap \operatorname{supp}\eta_k \neq \emptyset$. We use (2.3.25) setting $\zeta_1 = \eta_j$ and $\zeta_2 = \eta_k$. The operator

$$T_\gamma \equiv E_{\omega\to\varphi}(\lambda - i|\gamma|)^{-1} \Phi_{\mu\to\lambda}(\omega, \theta) E_{\psi\to\omega}(\mu)$$

is compact on $L^2(\mathbb{R}, L_2(S^{n-1}))$ for $|\gamma| \geq 1$. Indeed, we write T_γ in the form (2.3.28), where $g(\lambda, \mu) = E(\lambda - i|\gamma|)^{-1}h(\lambda - \mu)E(\mu)$ and $h(\nu) = F^{-1}_{t\to\nu} D^\gamma_\omega \Phi(\omega, e^{-t}\theta)$. For each $\nu \in \mathbb{R}$, we have $h(\nu) \in C^\infty(S^{n-1})$ and $h(\nu)$ decreases rapidly as $\nu \to \infty$. The operator $E(\mu)$ is unitary, while $E(\lambda - i|\gamma|)^{-1}$ is compact on $L_2(S^{n-1})$ and $\|E(\lambda - i|\gamma|)^{-1}\| \leq c(1 + |\lambda|)^{-|\gamma|}$, $\lambda, \mu \in \mathbb{R}$ (see Sect. 1.2.1). The compactness of T_γ, and hence that of $\eta_j T_\gamma \eta_k$, now follows from Lemma 2.3.8. The compactness of the operator $\tilde{u} \mapsto \eta_j R_N \eta_k \tilde{u}$ on $L_2(\mathbb{R}, L_2(S^{n-1}))$ is ensured by the smoothness of its kernel for sufficiently large N(see Lemma 2.3.7) and the same Lemma 2.3.8. Using (2.3.25), we obtain that the commutator $[\eta_j M A(\theta) M^{-1} \eta_k, \sigma]$ is compact on the indicated space. The arguments only simplify in the case $\mathrm{supp}\eta_j \cap \mathrm{supp}\eta_k = \emptyset$.

If Φ depends on φ, then we expand Φ in a series $\Phi(\varphi, \xi) = \sum_{j,k} Y^{(n)}_{jk}(\varphi) a_{jk}(\xi)$, where $Y^{(n)}_{jk}$ are spherical harmonics, and the coefficients a_{jk} satisfy the condition $a_{jk}(\omega, 0) \equiv 0$. According to this, $A(\theta)$ can be represented as a series $\sum Y^{(n)}_{jk}(\varphi) A_{jk}(\theta)$, where $A_{jk}(\theta)$ is the operator (2.3.11) with $\Phi = a_{jk}$. It remains to use the first part of the proof in connection with the operators $A_{jk}(\theta)$. □

Recall that $\bar{\mathbb{R}}^n$ denotes the compact obtained by adjoining the $(n-1)$-dimensional sphere S^{n-1}_∞ to the space $\mathbb{R}^n$ at infinity. The next assertion follows from [2], Theorem C.

Proposition 2.3.10 *Let $A(\theta)$ be an operator of the form (2.3.7) and $a \in C(\bar{\mathbb{R}}^n)$. Then, the commutator $[A(\theta), a]$ is compact on $L_2(\mathbb{R}^n)$.*

Now, we are ready to describe localization in the algebra $\mathcal{L}(\theta)$ directly. The assumptions of Proposition 1.3.26 are fulfilled for $\mathcal{L}(\theta)$ and the localizing algebra $\mathcal{C} = C(\bar{R}^n)$. Indeed, if the operator $A(\theta)$ is subject to the requirements of Proposition 2.3.9 and $\sigma \in C^\infty(S^{n-1})$, then $[A(\theta), \sigma] \in \mathcal{K}L_2(\mathbb{R}^n)$. Since the algebra $\mathcal{L}(\theta)$ is irreducible, it contains the ideal $\mathcal{K}L_2(\mathbb{R}^n)$. Therefore, the first condition of Proposition 1.3.26 is fulfilled. The second condition is obvious, and the third one is ensured by Proposition 2.3.10. Let $\mathcal{L}(\theta)_z$ denote the local algebra of $\mathcal{L}(\theta)$ at a point $z \in \bar{R}^n$.

For the operators $A(\theta) = F^{-1}_{\eta\to x} \Phi(\varphi, \eta, \theta) F_{y\to\eta}$ in (2.3.7), we introduce the maps:

(i) $p(z) : A(\theta) \mapsto \Phi(\varphi, \cdot, 0) \in C(S^{n-1})$, $z \in \mathbb{R}^n \setminus 0$, and $\varphi = z/|z|$.
(ii) $p(0) : A(\theta) \mapsto \mathfrak{A}$, where $\mathfrak{A}$ is the function

$$\mathbb{R} \ni \lambda \mapsto \mathfrak{A}(\lambda) = E_{\omega\to\varphi}(\lambda)^{-1} \Phi(\varphi, \omega, 0) E_{\psi\to\omega}(\lambda) \in \mathcal{B}L_2(S^{n-1}).$$

(iii) $p(\varphi, \theta) : A(\theta) \mapsto \Phi(\varphi, \cdot, \theta) \in C(\bar{R}^n)$.

Proposition 2.3.11 *The maps (i)–(iii) extend to isomorphisms*
$p(z) : \mathcal{L}(\theta)_z \simeq C(S^{n-1})$ for $z \in \mathbb{R}^n \setminus 0$;

$p(0) : \mathcal{L}(\theta)_0 \simeq \mathfrak{S}$, *where as in Sect. 2.2.1,* $\mathfrak{S}$ *is the algebra generated by the operator-functions* $\mathfrak{A}$;

$p(\varphi, \theta) : \mathcal{L}(\theta)_z \simeq C(\bar{R}^n)$, *where* $z \in S_\infty^{n-1}$ *and the vector* φ *is directed to the point* z.

Proof Let $L(\theta)$ be the algebra spanned by $\mathcal{L}(\theta)$ and $C(\bar{\mathbb{R}}^n)$. We denote by $\mathcal{J}_z$ the ideal in $L(\theta)$ generated by the functions $c \in C(\bar{\mathbb{R}}^n)$ that vanish at $z \in \bar{\mathbb{R}}^n$). We have $\mathcal{K}L_2(\mathbb{R}^n) \subset \mathcal{J}_z$ for all $z \in \bar{R}^n$. The assertions about $p(z)$ and $p(0)$ can be verified as in the proof of Proposition 2.1.5.

Let us consider the map (iii). We set

$$A_{jk}(\theta) = a_{jk}(\varphi) F^{-1}_{\eta\to x} \Phi_{jk}(\eta, \theta) F_{y\to\eta},$$

where $\varphi = x/|x|$, $a_{jk} \in C^\infty(S^{n-1})$, and $\Phi_{jk} \in C^\infty(S^{m-1})$, and choose $\chi_{jk} \in C(\bar{\mathbb{R}}^n)$, where the indices j and k run over finite sets. Moreover, let $z^0 \in S_\infty^{n-1}$ and let $\varphi^0 \in S^{n-1}$ be the vector directed to z^0. On $L_2(\mathbb{R}^n)$, we introduce the unitary operator $U_t u(x) = u(x + t\varphi^0)$ for $t \in \mathbb{R}$. Since the operators $A(\theta) = F^{-1}_{\eta\to x} \Phi(\eta, \theta) F_{y\to\eta}$ and U_t commute, the equality

$$U_t \left(\sum_j \prod_k \chi_{jk} a_{jk} A_{jk}(\theta) \right) U_t^{-1} u = \left(\sum_j \prod_k U_t(\chi_{jk} a_{jk}) A_{jk}(\theta) \right) u \tag{2.3.29}$$

holds. It is clear that

$$U_t(\chi_{jk}(x) a_{jk}(x)) = U_t(\chi_{jk}(x)) U_t(a_{jk}(x)) = \chi_{jk}(x + t\varphi^0) a_{jk}(\varphi + t\varphi^0) \longrightarrow \chi_{jk}(z^0) a_{jk}(\varphi^0)$$

as $t \to +\infty$. Therefore, there exists a strong limit

$$\sum_j \prod_k U_t(\chi_{jk} a_{jk}) A_{jk}(\theta) \longrightarrow \sum_j \prod_k \chi_{jk}(z^0) a_{jk}(\varphi^0)) A_{jk}(\theta).$$

This and (2.3.29) imply that the operator $U_t \left(\sum_j \prod_k \chi_{jk} a_{jk} A_{jk}(\theta) + J \right) U_t^{-1}$ strongly tends to the same limit as $t \to +\infty$, J being any element of the ideal $\mathcal{J}_{z^0}$. Taking into account the unitarity of U_t and the property of strong limit "the norm of strong limit does not exceed the low limit of the norms," we obtain the inequality

$$\| \sum_j \prod_k \chi_{jk}(z^0) a_{jk}(\varphi^0)) A_{jk}(\theta); \mathcal{B}L_2(\mathbb{R}^n) \| \leq \tag{2.3.30}$$

$$\leq \inf_J \{ \| \sum_j \prod_k \chi_{jk} a_{jk} A_{jk}(\theta) + J; \mathcal{B}L_2(\mathbb{R}^n) \|; J \in \mathcal{J}_{z^0} \}.$$

Obviously, the operators $a_{jk}(\varphi^0))A_{jk}(\theta)$ and $a_{jk}(\varphi))A_{jk}(\theta)$ are in the same class in the quotient algebra $L(\theta)/\mathcal{J}_{z^0}$. Therefore, the inequality converse for (2.3.30) holds. Thus,

$$\inf_J\{\|\sum_j\prod_k \chi_{jk}a_{jk}A_{jk}(\theta)+J;\mathcal{B}L_2(\mathbb{R}^n)\|;\ J\in\mathcal{J}_{z^0}\}= \tag{2.3.31}$$

$$=\|\sum_j\prod_k \chi_{jk}(z^0)a_{jk}(\varphi^0))A_{jk}(\theta);\mathcal{B}L_2(\mathbb{R}^n)\|=$$

$$=\|\sum_j\prod_k \chi_{jk}(z^0)a_{jk}(\varphi^0))\Phi_{jk}(\theta);C(\mathbb{R}^n)\|.$$

□

2.3.4 The Spectrum of Algebra $\mathcal{L}(\theta)$

1^0. **Representations.** The following theorem contains a list of all equivalence classes of irreducible representations of the algebra $\mathcal{L}(\theta)$ generated on $L_2(\mathbb{R}^n)$ by operators of the form $A(\theta) = F^{-1}_{\eta\to x}\Phi(\varphi,\eta,\theta)F_{y\to\eta}$, where $\Phi(\varphi,\eta,\theta) = f(\varphi)\sigma(\eta,\theta)$, $\varphi = x/|x|$, and $f\in C^\infty(S^{n-1})$; the function $(\mathbb{R}^m\setminus 0)\ni\xi\mapsto\sigma(\xi)$ is homogeneous of zero degree, $\sigma\in C^\infty(S^{m-1})$, and $\theta\in S^{m-n-1}$.

Introduce the maps:

(1) $\pi(\varphi,\eta,\theta): A(\theta)\mapsto\Phi(\varphi,\eta,\theta)$ for $\varphi\in S^{n-1}$ and $\eta\in\mathbb{R}^n$.
(2) $\pi(\varphi,\omega): A(\theta)\mapsto\Phi(\varphi,\omega,0)$ for $\varphi,\omega\in S^{n-1}$.
(3) $\pi(\lambda): A(\theta)\mapsto\mathfrak{A}(\lambda)$, where $\mathfrak{A}(\lambda)=E_{\omega\to\varphi}(\lambda)^{-1}\Phi(\varphi,\omega,0)E_{\psi\to\omega}(\lambda)\in\mathcal{B}L_2(S^{n-1})$ and $\lambda\in\mathbb{R}$.

Theorem 2.3.12 *The algebra $\mathcal{L}(\theta)$ is irreducible and contains the ideal $\mathcal{K}L_2(\mathbb{R}^n)$. The maps (1)–(3) extend to representations of the quotient algebra $\mathcal{L}(\theta)/\mathcal{K}L_2(\mathbb{R}^n)$ that are irreducible and nonequivalent. Any irreducible representation of $\mathcal{L}(\theta)$ is equivalent either to one of those listed or to the identity representations $\iota(\theta)$.*

Proof The irreducibility of the algebra $\mathcal{L}(\theta)$ is established by Proposition 2.3.5. Together with Proposition 2.3.9, this provides the inclusion $\mathcal{K}L_2(\mathbb{R}^n)\subset\mathcal{L}(\theta)$. It remains to apply the localization principle (Proposition 1.3.26) and to recall the description of the local algebras (Proposition 2.3.11) and the list of the irreducible representations of the algebra $\mathfrak{S}$ given in Theorem 2.2.11. □

We will also write representations (1) and (2) in a somewhat different form. Let $\mathbb{R}^{n+1}(\theta)$ be the subspace in $\mathbb{R}^m$ spanned by the plane $\{\xi\in\mathbb{R}^m : \xi =$

$(\eta_1, \dots, \eta_n, 0, \dots, 0)\}$ and the vector $\theta = (0, \dots, 0, \theta_1, \dots, \theta_{m-n})$, $|\theta| = 1$. Denote by $S^n_+(\theta)$ the open hemisphere with pole θ in $\mathbb{R}^{n+1}(\theta)$. Since $\mathbb{R}^m \setminus 0 \mapsto \Phi(\varphi, \xi)$ is a homogeneous function of zero degree, we have $\Phi(\varphi, \eta, \theta) = \Phi(\varphi, \xi)$, where $\xi = (\eta/(|\eta|^2 + 1)^{1/2}, \theta/(|\eta|^2 + 1)^{1/2})$. Therefore, the representations in (1) can be written as $\pi(\varphi, \xi) : A(\theta) \mapsto \Phi(\varphi, \xi)$ for $(\varphi, \xi) \in S^{n-1} \times S^n_+(\theta)$ and those in (2) as $\pi(\varphi, \xi)$ for $(\varphi, \xi) \in S^{n-1} \times \partial S^n_+(\theta)$.

2^0. **The topology on the spectrum.** Denote by $\Lambda(\theta)$ the disjoint union of the sets $S^{n-1} \times \overline{S^n_+(\theta)}$, $\mathbb{R}$, and the point ι. We introduce a topology on $\Lambda(\theta)$. As before, neighborhoods making up a fundamental system will be called typical. The point ι is an open set in the space $\Lambda(\theta)$. The closure of ι coincides with $\Lambda(\theta)$. A typical neighborhood in $\Lambda(\theta)$ of a point $(\varphi, \xi) \in S^{n-1} \times S^n_+(\theta)$ is taken to be the union of an ordinary neighborhood of it in $S^{n-1} \times S^n_+(\theta)$ with the point ι.

Let $(\varphi, \xi) \in S^{n-1} \times \partial S^n_+(\theta)$ and let $\varphi\xi = 0$ (here ξ is regarded as a vector in R^n, since its last $m - n$ coordinates are equal to zero). A neighborhood of (φ, ξ) is the union of an ordinary neighborhood of it in $S^{n-1} \times \overline{S^n_+(\theta)}$ with the point ι and the line $\{\lambda : \lambda \in \mathbb{R}\}$. In the case $\varphi\xi \gtrless 0$, the line is replaced by a set $\{\lambda \in \mathbb{R} : \lambda \gtrless N\}$, where N is an arbitrary real number. Finally, a neighborhood of a point $\lambda \in \mathbb{R}$ in $\Lambda(\theta)$ is the union of an interval on $\mathbb{R}$ containing λ with the point ι.

Theorem 2.3.13 *The correspondence* $\iota(\theta) \mapsto \iota$, $\pi(\varphi, \xi) \mapsto (\varphi, \xi) \in S^{n-1} \times \overline{S^n_+(\theta)}$, $\pi(\lambda) \mapsto \lambda \in \mathbb{R}$ *determines a bijection from the spectrum* $\widehat{\mathcal{L}}(\theta)$ *onto the space* $\Lambda(\theta)$. *The topology carried over by means of this bijection from* $\Lambda(\theta)$ *to* $\widehat{\mathcal{L}}(\theta)$ *coincides with the Jacobson topology.*

Proof The correspondence indicated in the theorem is denoted by h. The fact that h is a bijection follows from Theorem 2.3.12 and the remark after its proof. It remains to see that the maps $h : \widehat{\mathcal{L}}(\theta) \to \Lambda(\theta)$ and $h^{-1} : \Lambda(\theta) \to \widehat{\mathcal{L}}(\theta)$ are continuous (we mean that the space $\widehat{\mathcal{L}}(\theta)$ is endowed with the Jacobson topology).

Denote by $J(\theta)$ the intersection of the kernels of all representations (2) and (3) in Theorem 2.3.12. Thus, the representations $\pi(\varphi, \xi)$ for $(\varphi, \xi) \in S^{n-1} \times \partial S^n_+(\theta)$ and $\pi(\lambda)$ for $\lambda \in \mathbb{R}$ vanish on $J(\theta)$. The ideal $J(\theta)$ is irreducible, $\mathcal{K}L_2(\mathbb{R}^n) \subset J(\theta)$, and by Theorem 2.3.12, the spectrum $\hat{J}(\theta)$ consists of the equivalence classes of the representations $\iota(\theta)$ and $\pi(\varphi, \xi)$ for $(\varphi, \xi) \in S^{n-1} \times S^n_+(\theta)$; such representations $\pi(\varphi, \xi)$ constitute the spectrum $(J(\theta)/\mathcal{K}L_2(\mathbb{R}^n))^\wedge$. Recall that $\|B\| = \sup\{\|\pi B\|, \pi \in \widehat{\mathcal{B}}\}$. It follows that the algebras $J(\theta)/\mathcal{K}L_2(\mathbb{R}^n)$ and $C_0(S^{n-1} \times S^n_+(\theta))$ are isomorphic, where $C_0(S^{n-1} \times S^n_+(\theta))$ denotes the algebra of the continuous functions on $S^{n-1} \times S^n_+(\theta)$ that vanish on $S^{n-1} \times \partial S^n_+(\theta)$. The spectrum $(\mathcal{L}(\theta)/J(\theta))^\wedge$ consists of the equivalence classes of the representations $\pi(\lambda)$ for $\lambda \in \mathbb{R}$ and the representations $\pi(\varphi, \xi)$ for $(\varphi, \xi) \in S^{n-1} \times \partial S^n_+(\theta)$. This and Theorem 2.2.11 imply that the algebras $\mathcal{L}(\theta)/J(\theta)$ and $\mathfrak{S}$ are isomorphic.

Let I be a closed two-sided ideal of the algebra $\mathcal{B}$. By virtue of Proposition 1.3.7, we have $\widehat{\mathcal{B}} = \widehat{I} \cup (\mathcal{B}/I)^\wedge$, and the set $\widehat{I}$ is open in $\widehat{\mathcal{B}}$, while $(\mathcal{B}/I)^\wedge$ is closed. Therefore, $\widehat{\mathcal{L}}(\theta) = \widehat{\mathcal{K}} \cup (J(\theta)/\mathcal{K})^\wedge \cup (\mathcal{L}(\theta)/J(\theta))^\wedge$, where $\mathcal{K} := \mathcal{K}L_2(\mathbb{R}^n)$. The sets $\widehat{\mathcal{K}}$ and $\widehat{\mathcal{K}} \cup (J(\theta)/\mathcal{K})^\wedge = \widehat{J}(\theta)$ are open in $\widehat{\mathcal{L}}(\theta)$ (being the spectra of ideals).

Recall that $\widehat{\mathcal{K}} = \iota(\theta)$. Thus, the points $\iota(\theta)$ and $\iota = h(\iota(\theta))$ are open in $\widehat{\mathcal{L}}(\theta)$ and $\Lambda(\theta)$, respectively. The maps h and h^{-1} are thereby continuous at these points. The spectrum $(J(\theta)/\mathcal{K})^\wedge$ is homeomorphic to the set $S^{n-1} \times S^n_+(\theta) \subset \Lambda(\theta)$, and a homeomorphism is implemented by the map h. Since any neighborhood of a point $\pi(\varphi, \xi) \in \widehat{\mathcal{L}}(\theta)$ with $(\varphi, \xi) \in S^{n-1} \times S^n_+(\theta)$ contains $\iota(\theta)$, while the neighborhoods of (φ, ξ) contain ι, we find that $h : \widehat{J}(\theta) \to (S^{n-1} \times S^n_+(\theta)) \cup \iota$ is a homeomorphism.

We now consider a point $(\varphi_0, \xi_0) \in S^{n-1} \times \partial S^n_+(\theta)$. Let $l = \mathbb{R}$ if $\varphi_0\xi_0 = 0$, and let $l = \{\lambda \in \mathbb{R} : \lambda \gtrless N\}$, where N is a number if $\varphi_0\xi_0 \gtrless 0$. Let Ω be some open subset of $S^{n-1} \times \overline{S^n_+(\theta)}$ containing the point (φ_0, ξ_0). Then, $\Omega \cup l$ is a neighborhood of (φ_0, ξ_0) in $\Lambda(\theta)$. From Theorem 2.2.13, it follows that the algebra $\mathfrak{S}$ contains an element $\mathfrak{A} = \sum_j \prod_k \mathfrak{A}_{jk}$ (j and k run through finite sets of values), where

$$\mathfrak{A}_{jk}(\lambda) = E(\lambda)^{-1}\Psi_{jk}(\varphi, \omega)E(\lambda), \quad \Psi_{jk} \in C^\infty(S^{n-1} \times \partial S^n_+(\theta)),$$

having the properties:

(1) $\{\lambda : \|\mathfrak{A}(\lambda)\| > 1\} \subset l$.
(2) $\{(\varphi, \omega) : \left|\sum_j \prod_k \Psi_{jk}(\varphi, \omega)\right| > 1\} \subset \Omega \cap (S^{n-1} \times \partial S^n_+(\theta))$.

(A union of sets of the form (1) and (2) is a typical neighborhood of (φ_0, ω_0) in the spectrum $\widehat{\mathfrak{S}}$.) Let Φ_{jk} be functions in $C^\infty(S^{n-1} \times S^{m-1})$ such that $\Phi_{jk}(\varphi, \omega, 0) = \Psi_{jk}(\varphi, \omega)$ on $S^{n-1} \times \partial S^n_+(\theta)$ and $(S^{n-1} \times \overline{S^n_+(\theta)}) \cap \operatorname{supp} \Phi_{jk} \subset \Omega$. We consider the operator $A(\theta) = \sum \prod A_{jk}(\theta)$ in the algebra $\mathcal{L}(\theta)$, where $A_{jk}(\theta) = F^{-1}_{\eta\to x}\Phi_{jk}(\varphi, \eta, \theta)F_{y\to\eta}$ and the functions $\xi \mapsto \Phi_{jk}(\varphi, \xi)$ are extended as homogeneous of zero degree to $\mathbb{R}^m \setminus 0$. It is clear that

$$h(\{\pi \in \widehat{\mathcal{L}}(\theta) : \|\pi(A(\theta))\| > 1\}) \subset \Omega \cup l \cup \iota.$$

Thus, the map h is continuous at $\pi(\varphi_0, \xi_0)$.

We verify that the inverse map h^{-1} is continuous at (φ_0, ξ_0). Let

$$A_{jk}(\theta) = F^{-1}_{\eta\to x}\Phi_{jk}(\varphi, \eta, \theta)F_{y\to\eta}$$

be arbitrary elements in $\mathcal{L}(\theta)$ and $A(\theta) = \sum \prod A_{jk}(\theta)$. We assume that

$$\|\pi(\varphi_0, \xi_0)A(\theta)\| = \left|\sum \prod \Phi_{jk}(\varphi_0, \omega_0, 0)\right| > 1$$

and denote by $\mathfrak{A}$ the element of $\mathfrak{S} \simeq \mathcal{L}(\theta)/J(\theta)$ corresponding to the residue class $[A(\theta)] \in \mathcal{L}(\theta)/J(\theta)$. The set $V := \{\pi \in \widehat{\mathcal{L}}(\theta) : \|\pi(A(\theta))\| > 1\}$ is a neighborhood of $\pi(\varphi_0, \xi_0)$ in $\widehat{\mathcal{L}}(\theta)$. Clearly, this set contains the identity representation $\iota(\theta)$ and all the representations $\pi(\varphi, \xi)$ with $(\varphi, \xi) \in S^{n-1} \times S^n_+(\theta)$ sufficiently close to (φ_0, ξ_0). Again, using Theorem 2.2.13, we obtain that if $\varphi_0\omega_0 = 0$, then $\|\mathfrak{A}(\lambda)\| > 1$ for all $\lambda \in \mathbb{R}$, while if $\varphi_0\omega_0 \gtrless 0$, then $\|\mathfrak{A}(\lambda)\| > 1$ for $\lambda \gtrless N$, where N is some number. It follows from the foregoing that the set $h(V)$ contains a neighborhood of (φ_0, ξ_0) in $\Lambda(\theta)$. This shows that the map $h^{-1} : \Sigma(\theta) \to \widehat{\mathcal{L}}(\theta)$ is continuous at $(\varphi_0, \omega_0) \in S^{n-1} \times \partial S^n_+(\theta)$.

It remains to be proven that h and h^{-1} are continuous at the respective points $\pi(\lambda)$ and λ. By Theorem 2.2.13, there exists an element $\mathfrak{A} = \sum \prod \mathfrak{A}_{jk} \in \mathfrak{S}$ subject to the following conditions:

(1) $\mathfrak{A}_{jk}(\lambda) = E(\lambda)^{-1}\Psi_{jk}(\varphi, \omega)E(\lambda)$.
(2) $\{\lambda \in \mathbb{R} : \|\mathfrak{A}(\lambda)\| > 1\} \subset (\lambda_0 - \varepsilon\,, \lambda + \varepsilon)$.
(3) $\max\left|\sum \prod \Psi_{jk}(\varphi, \omega)\right| < 1, \quad (\varphi, \omega) \in S^{n-1} \times \partial S^n_+(\theta)$.

Suppose that $\Phi_{jk} \in C^\infty(S^{n-1} \times S^{m-1})$, $\Phi_{jk}(\varphi, \omega, 0) = \Psi_{jk}(\varphi, \omega)$ for $(\varphi, \omega) \in S^{n-1} \times \partial S^n_+(\theta)$, and $\max\left|\sum \prod \Phi_{jk}\right| < 1$ on $S^{n-1} \times S^{m-1}$. As earlier, we introduce the operators $A_{jk}(\theta) \in \mathcal{L}(\theta)$ and $A(\theta) = \sum \prod A_{jk}(\theta)$. It is clear that $h(\{\pi \in \widehat{\mathcal{L}}(\theta) : \|\pi(A(\theta))\| > 1\} \subset (\lambda_0 - \varepsilon\,, \lambda_0 + \varepsilon)$. This gives the continuity of h at $\pi(\lambda_0)$. Conversely, if $A(\theta) \in \mathcal{L}(\theta)$ and $\|\pi(\lambda_0)A(\theta)\| = \|\mathfrak{A}(\lambda_0)\| > 1$, then $\|\mathfrak{A}(\lambda)\| > 1$ at all points λ close to λ_0. Thus, h^{-1} is continuous at λ_0. □

3^0. **Solvability and the length of the algebra $\mathcal{L}(\theta)$.**

Theorem 2.3.14

(1) $\mathcal{K}L_2(\mathbb{R}^n) \subset \mathrm{com}\mathcal{L}(\theta)$.
(2) *The composition series* $0 \subset \mathcal{K}L_2(\mathbb{R}^n) \subset \mathrm{com}\,\mathcal{L}(\theta) \subset \mathcal{L}(\theta)$ *is solving and*

$$\mathrm{com}\,\mathcal{L}(\theta)/\mathcal{K}L_2(\mathbb{R}^n) \simeq C_0(\mathbb{R}) \otimes \mathcal{K}L_2(S^{n-1}),$$
$$\mathcal{L}(\theta)/\mathrm{com}\,\mathcal{L}(\theta) \simeq C(S^{n-1} \times \overline{S^n_+(\theta)}).$$

(3) *The length of* $\mathcal{L}(\theta)$ *is equal to 2.*
(4) *The above composition series is the maximal radical series.*

Proof

(1) Since the algebra $\mathcal{L}(\theta)$ is irreducible, the same is true for the ideal $\mathrm{com}\,\mathcal{L}(\theta)$. If $A_1(\theta)$ and $A_2(\theta)$ are elements of the ideal $J(\theta)$, then $[A_1(\theta), A_2(\theta)] \in \mathcal{K}L_2(\mathbb{R}^n)$. Therefore, $\mathcal{K}L_2(\mathbb{R}^n) \subset \mathrm{com}\,\mathcal{L}(\theta)$.

(2) We show that the algebras $\operatorname{com}\mathcal{L}(\theta)/\mathcal{K}L_2(\mathbb{R}^n)$ and $C_0(\mathbb{R})\otimes\mathcal{K}L_2(S^{n-1})$ are isomorphic. If $A\in\operatorname{com}\mathcal{L}(\theta)$ and $p:\mathcal{L}(\theta)\to\mathcal{L}(\theta)/J(\theta)\simeq\mathfrak{S}$ is the projection, then $p(A)\in\operatorname{com}\mathfrak{S}$. Moreover, $\operatorname{com}\mathfrak{S}\simeq C_0(\mathbb{R})\otimes\mathcal{K}L_2(S^{n-1})$ (see Remark 2.2.16). The map $q:\operatorname{com}\mathcal{L}(\theta)/\mathcal{K}L_2(\mathbb{R}^n)\to C_0(\mathbb{R})\otimes\mathcal{K}L_2(S^{n-1})$ is defined by the equality $q([A])=p(A)$, where $[A]$ is the residue class of the element A in $\operatorname{com}\mathcal{L}(\theta)/\mathcal{K}L_2(\mathbb{R}^n)$ (this definition is unambiguous in view of the inclusion $\mathcal{K}L_2(\mathbb{R}^n)\subset J(\theta)$). Let us verify that q is an isomorphism. Since the spectrum of the algebra $C_0(\mathbb{R})\otimes\mathcal{K}L_2(S^{n-1})$ coincides with $\mathbb{R}$, we have

$$\|q([A])\|=\sup\{\|\mathfrak{A}(\lambda):\mathcal{B}L_2(S^{n-1})\|;\ \lambda\in\mathbb{R}\},\tag{2.3.32}$$

where $\mathfrak{A}=q([A])$. The spectrum of $\operatorname{com}\mathcal{L}(\theta)/\mathcal{K}L_2(\mathbb{R}^n)$ consists of the representations of $\mathcal{L}(\theta)/\mathcal{K}L_2(\mathbb{R}^n)$ that do not annihilate $\operatorname{com}\mathcal{L}(\theta)/\mathcal{K}L_2(\mathbb{R}^n)$. Taking into account Theorem 2.3.12, we see that the spectrum $\operatorname{com}\mathcal{L}(\theta)/\mathcal{K}L_2(\mathbb{R}^n)^\wedge$ consists of representations of the form $\pi(\lambda)$. Hence,

$$\big\|[A]\big\|=\sup_{\lambda\in\mathbb{R}}\|\pi(\lambda)A\|=\sup_{\lambda\in\mathbb{R}}\|\mathfrak{A}(\lambda)\|.$$

Together with (2.3.32), this gives us that q is a monomorphism. It is obviously an epimorphism. Thus,

$$\operatorname{com}\mathcal{L}(\theta)/\mathcal{K}L_2(\mathbb{R}^n)\simeq C_0(\mathbb{R})\otimes\mathcal{K}L_2(S^{n-1}).$$

The spectrum of $\mathcal{L}(\theta)/\operatorname{com}\mathcal{L}(\theta)$ consists of all the representations of $\mathcal{L}(\theta)$ vanishing on $\operatorname{com}\mathcal{L}(\theta)$, i.e., of all the one-dimensional representations. By Theorem 2.3.12,

$$(\mathcal{L}(\theta)/\operatorname{com}\mathcal{L}(\theta))^\wedge=S^{n-1}\times\overline{S^n_+(\theta)}.$$

In other words, $\mathcal{L}(\theta)/\operatorname{com}\mathcal{L}(\theta)\simeq C(S^{n-1}\times\overline{S^n_+(\theta)})$.

(3) It follows from (2) that the length of $\mathcal{L}(\theta)$ is no greater than 2. Since the algebra $\mathcal{L}(\theta)$ has irreducible representations of different dimensions (see Theorem 2.3.12), $\mathcal{L}(\theta)$ cannot be isomorphic to an algebra of the form $C_0(X)\otimes\mathcal{K}(H)$. Hence, the length of $\mathcal{L}(\theta)\geq 1$. Assume that the length is equal to 1. Then, there exists an ideal $I\neq 0$ such that $I\simeq C_0(X_1)\otimes\mathcal{K}(H_1)$ and $\mathcal{L}(\theta)/I\simeq C_0(X_2)\otimes\mathcal{K}(H_2)$, where X_1 and X_2 are locally compact spaces. The spectrum $\widehat{I}$ is an open subset of $\widehat{\mathcal{L}}(\theta)$. Every open subset of the spectrum $\widehat{\mathcal{L}}(\theta)$ contains the point $\iota(\theta)$ corresponding to the identity representation (Theorem 2.3.13). Since the point $\iota(\theta)$ is not closed, the spectrum $\widehat{I}$ would not be a Hausdorff space in the case $\iota(\theta)\neq\widehat{I}$, and that is impossible because $\widehat{I}$ is homeomorphic to X_1. Hence, $\widehat{I}=\iota(\theta)$ and $I=\mathcal{K}L_2(\mathbb{R}^n)$. By Theorem 2.3.12, the quotient algebra $\mathcal{L}(\theta)/\mathcal{K}L_2(\mathbb{R}^n)$ has representations of different dimensions and thus

cannot be isomorphic to an algebra $C_0(X_2) \otimes \mathcal{K}H_2$. Thus, the length of $\mathcal{L}(\theta)$ is equal to 2.

(4) This assertion is immediately verified by means of Theorem 2.3.12.

□

2.3.5 The Spectrum of the Algebra of Pseudodifferential Operators with Symbols Discontinuous Along a Submanifold

Here, we prove the theorems stated in Sect. 2.3.1 for the algebra $\mathcal{A}$ (we keep the same notation). Let us first describe the spectrum of the local algebra $\mathcal{A}_{z^0}$ at a point $z^0 \in \mathcal{N}$. This algebra is generated on $L_2(\mathbb{R}^m)$ by the operators (2.3.6) whose symbols may have discontinuities on an $(m-n)$-dimensional plane. According to Proposition 2.3.4, $\mathcal{A}_{z^0} \simeq \mathcal{L}$, where $\mathcal{L}$ is the algebra spanned by the functions

$$S^{m-n-1} \ni \theta \mapsto A(\theta) = F^{-1}_{\eta \to x} \Phi(\varphi, \eta, \theta) F_{y \to \eta} \in \mathcal{B}L_2(\mathbb{R}^n) \tag{2.3.33}$$

with $\varphi = x^{(1)}/|x^{(1)}|$ and $\Phi \in C^\infty(S^{n-1} \times (\mathbb{R}^m \setminus 0))$, while $\xi \mapsto \Phi(\varphi, \xi)$ is a homogeneous function of degree zero.

1^0. **The spectrum of the algebra** $\mathcal{L}$. Let $A(\cdot)$ be an algebra of the form (2.3.33). Introduce the notations:

(1) $\pi(\theta) : A \mapsto A(\theta) \in \mathcal{B}L_2(\mathbb{R}^n)$ for $\theta \in S^{m-n-1}$.

(2) $\pi(\lambda) : A(\cdot) \mapsto \mathfrak{A}(\lambda) \in \mathcal{B}L_2(S^{n-1})$, where $\mathfrak{A}(\lambda) = E_{\omega \to \varphi}(\lambda)^{-1} \Phi(\varphi, \omega, 0) E_{\psi \to \omega}(\lambda)$ and $\lambda \in \mathbb{R}$.

(3) $\pi(\varphi, \xi) : A(\cdot) \mapsto \Phi(\varphi, \xi)$ for $(\varphi, \xi) \in S^{n-1} \times S^{m-1}$.

Theorem 2.3.15 *The mappings (1)–(3) extend to irreducible pairwise nonequivalent representations of $\mathcal{L}$. Every irreducible representation of $\mathcal{L}$ is equivalent to one of those listed.*

Proof It suffices to take into account Proposition 2.3.4, Theorem 2.3.12, and the remark after its proof.

□

Let us discuss a spectral topology. Denote by Λ the disjoint union of the sets S^{m-n-1}, $S^{n-1} \times S^{m-1}$, and $\mathbb{R}$. We introduce a topology on Λ. The typical neighborhoods in Λ of a point $\theta \in S^{m-n-1}$ coincide with its neighborhoods in S^{m-n-1}. If $\xi = (\xi^{(1)}, \xi^{(2)})$, $\xi^{(1)} = (\xi_1, \ldots, \xi_n)$, $\xi^{(2)} = (\xi_{n+1}, \ldots, \xi_m)$, and $\xi^{(2)} \neq 0$, then a neighborhood in Λ of the point $(\varphi, \xi) \in S^{n-1} \times S^{m-1}$ is defined as a union $(\mathcal{U}(\varphi) \times \mathcal{V}(\xi)) \cup \mathcal{W}(\theta)$, where $\theta = \xi^{(2)}/|\xi^{(2)}|$, while $\mathcal{U}(\varphi)$, $\mathcal{V}(\xi)$, and $\mathcal{W}(\theta)$ are neighborhoods of the points φ, ξ, and θ in S^{n-1}, S^{m-1}, and S^{m-n-1}, respectively, and $\mathcal{V}(\xi) \cap \{\xi : \xi^{(2)} = 0\} = \emptyset$. For $\xi = (\xi^{(1)}, 0)$, $\varphi\xi^{(1)} = 0$, a typical neighborhood of a point (φ, ξ) is of the form $S^{m-n-1} \cup (\mathcal{U}(\varphi) \times \mathcal{V}(\xi)) \cup \mathbb{R}$, and for $\varphi\xi^1 \gtrless 0$, the line $\mathbb{R}$ is replaced by the set $\{\lambda \in \mathbb{R} : \lambda \gtrless N\}$, where N is an arbitrary real number. A neighborhood of a point $\lambda \in \mathbb{R}$ is defined as a union $(\lambda - \varepsilon, \lambda + \varepsilon) \cup S^{m-n-1}$, $\varepsilon > 0$. With its topology, Λ is a T_0-space.

Theorem 2.3.16 *The correspondence*

$$\pi(\theta) \mapsto \theta \in S^{m-n-1}, \ \pi(\lambda) \mapsto \lambda \in \mathbb{R}, \ \pi(\varphi, \xi) \in S^{n-1} \times S^{m-1}$$

is a bijection of the spectrum $\widehat{\mathcal{L}}$ onto the set Λ. The topology carried over via this bijection from Λ to $\widehat{\mathcal{L}}$ coincides with the Jacobson topology.

Proof Denote by J the closed two-sided ideal in $\mathcal{L}$ spanned by the functions $\theta \mapsto A(\theta) \in J(\theta)$, where $J(\theta)$ is the ideal in $\mathcal{L}(\theta)$ introduced in the proof of Theorem 2.3.13. We show that any continuous function $S^{m-n-1} \ni \theta \mapsto A(\theta) \in \mathcal{B}L_2(\mathbb{R}^n)$ for $A(\theta) \in J(\theta)$ belongs to J. Let J_0 be a closed two-sided ideal in $\mathcal{L}(\theta)$ generated by operators of the form (2.3.33), where the function $\eta \mapsto \Phi(\varphi, \eta, \theta)$ vanishes for $|\eta| > R$ and all $\varphi \in S^{n-1}$, while R is a sufficient large number. From Theorem 2.3.12, it follows that $J_0(\theta)$ is a rich subalgebra in $J(\theta)$ and, therefore, $J_0(\theta) = J(\theta)$. This property of J means that the algebra J is generated by the continuous field $\{J(\theta), J\}$ of the algebras $J(\theta)$. Therefore, a topology on the spectrum $\widehat{J} = \cup_\theta \widehat{J}(\theta)$ can be described with the help of Proposition 1.3.17.

We now verify that the algebras $\mathcal{L}/J$ and $\mathfrak{S}$ are isomorphic. Let a map $r : \mathcal{L} \to \mathfrak{S}$ be the composition of the morphisms

$$\mathcal{L} \to \mathcal{L}(\theta) \to \mathcal{L}(\theta)/J(\theta) \cong \mathfrak{S},$$

where θ is a fixed point, the first arrow denotes the calculation of a value at θ, $\mathcal{L} \ni A \mapsto A(\theta) \in \mathcal{L}(\theta)$, the second arrow is a projection, and $\cong$ stands for the isomorphism established in the proof of Theorem 2.3.13. It is clear that r vanishes on J; therefore, $r : \mathcal{L}/J \to \mathfrak{S}$ is a well-defined epimorphism. Moreover, from Theorems 2.3.12 and 2.3.15, it follows that $\|[A]; \mathcal{L}/J\| = \|r(A); \mathfrak{S}\|$ for all $A \in \mathcal{L}$, i.e., $r : \mathcal{L}/J \to \mathfrak{S}$ is a monomorphism. Thus, $\mathcal{L}/J \simeq \mathfrak{S}$.

The set $\widehat{J}$ is open in the spectrum $\widehat{\mathcal{L}}$. Therefore, the typical neighborhoods of the points (φ, ξ) for $\xi = (\xi^{(1)}, \xi^{(2)})$ with $\xi^{(2)} \neq 0$ in the space $\widehat{J}$ can be taken as neighborhoods of these points also in $\widehat{\mathcal{L}}$. Neighborhoods in $\widehat{\mathcal{L}}$ of the points (φ, ξ) for $\xi = (\xi^{(1)}, 0)$ and those of the points λ (i.e., the points in $\widehat{\mathfrak{S}}$) can be obtained by evident modifications in the proof of Theorem 2.3.13. □

Theorem 2.3.17 *Let $I \subset \mathcal{L}$ be the ideal in $\mathcal{L}$ equal to the intersections of the kernels of all representations of the form $\pi(\lambda)$ and $\pi(\varphi, \xi)$ (defined in Theorem 2.3.15). Then, the composition series $\{0\} \subset I \subset \mathrm{com}\mathcal{L} \subset \mathcal{L}$ is solving, and*

$$I \cong C(S^{m-n-1}) \otimes \mathcal{K}L_2(\mathbb{R}^n),$$

$$\mathrm{com}\mathcal{L}/I \cong C_0(\mathbb{R}) \otimes \mathcal{K}L_2(S^{n-1}),$$

$$\mathcal{L}/\mathrm{com}\mathcal{L} \cong C(S^{n-1} \times S^{m-1}).$$

The length of $\mathcal{L}$ is equal to 2. The composition series is the maximal radical series.

Proof We saw earlier that $\mathcal{L}$ is isomorphic to a certain algebra of operator-valued functions $S^{m-n-1} \ni \theta \mapsto A(\theta) \in \mathcal{L}(\theta)$ that contains all continuous functions $S^{m-n-1} \ni \theta \mapsto A(\theta) \in J(\theta)$ (see the proof of Theorem 2.3.16). In particular, $C(S^{m-n-1}) \otimes \mathcal{K}L_2(\mathbb{R}^n)$ is a subalgebra of $\mathcal{L}$. It follows from the definition of the representations $\pi(\varphi, \xi)$ and $\pi(\lambda)$ that each of them is a representation of $\mathcal{L}(\theta)$ for some θ. The intersections of the kernels of such representations of $\mathcal{L}(\theta)$ coincide with the ideal $\mathcal{K}L_2(\mathbb{R}^n) \subset \mathcal{L}(\theta)$. Therefore, $C(S^{m-n-1}) \otimes \mathcal{K}L_2(\mathbb{R}^n)$ is the intersection of the kernels of $\pi(\varphi, \xi)$ and $\pi(\lambda)$, regarded as representations of $\mathcal{L}$, i.e., $I \cong C(S^{m-n-1}) \otimes \mathcal{K}L_2(\mathbb{R}^n)$.

As in Theorem 2.3.16, let r be the composition of the morphisms $\mathcal{L} \to \mathcal{L}/J \xrightarrow{\sim} \mathfrak{S}$. It is clear that $r(\mathrm{com}\,\mathcal{L}) \subset \mathrm{com}\mathfrak{S}$. Define a morphism $q : \mathrm{com}\,\mathcal{L}/I \to \mathrm{com}\,\mathfrak{S}$ by the equality $q([A]) = r(A)$, where $[A]$ is the residue class of an element A in $\mathrm{com}\,\mathcal{L}/I$; this definition is unambiguous by virtue of the inclusion $I \subset J$. Recall that $\mathrm{com}\,\mathfrak{S} \cong C_0(\mathbb{R}) \otimes \mathcal{K}L_2(S^{n-1})$. As in the proof of Theorem 2.3.14, we see that q is an isomorphism. The relation $\mathcal{L}/\mathrm{com}\,\mathcal{L} \cong C(S^{n-1} \times S^{m-1})$ follows from Theorem 2.3.15. Thus, the composition series indicated in the theorem is solving.

We show that the length of $\mathcal{L}$ equals 2. Assume that there exists a solving composition series of the form $\{0\} \subset Q \subset \mathcal{L}$, i.e., it is shorter than that in the formulation of the theorem. It follows from Theorem 2.3.15 that the spectrum of the ideal Q must contain either all infinite-dimensional representations or all one-dimensional representations (the spectrum $\hat{Q}$ must consist of representations of a single dimension). The first case is impossible since $\hat{Q}$ is then not a Hausdorff space, and the composition series is not solving. The second case cannot be because the set of all one-dimensional representations is not open (Theorem 2.3.16), while the spectrum of an ideal is always an open part of the spectrum of an algebra. □

Proof of Theorem 2.3.1 It suffices to compare the localization principle (Proposition 1.3.26), Theorem 2.3.15, and Proposition 2.1.5. □

Proof of Theorem 2.3.2 We identify the sets $(\mathcal{A}/\mathcal{K}L_2(\mathcal{M}))^\wedge$ and Ξ with the help of bijection (2.3.5). It must be verified that the same subsets of Ξ play the role of the typical neighborhoods of points both in the Ξ-topology (defined before the statement of Theorem 2.3.2) and in the Jacobson topology. This is clear for the points $(x, \omega) \in p^*(S^*(\mathcal{M}))$ with $x \in \mathcal{M} \setminus \mathcal{N}$.

Let $z^0 \in \mathcal{N}$ and $\theta^0 \in S^*(\mathcal{N})_{z^0}$. We choose an element $A \in \text{com}\mathcal{A}$ that is annihilated by all representations of the form $\pi(z, \lambda)$ (and, of course, by all one-dimensional representations of $\mathcal{A}$). Moreover, we assume that the set $\{\theta \in S^*(\mathcal{N})_{z^0} : \|\pi(z^0, \theta)A\| > 1\}$ coincides with a given neighborhood of θ^0 in $S^*(\mathcal{N})_{z^0}$. (The needed operator A can be related to a neighborhood of z^0 in $\mathcal{M}$ and constructed in local coordinates.) Multiplication by functions in $C^\infty(\mathcal{M})$ does not remove from the algebra $\mathcal{A}$. Therefore, it is possible to choose such a function $\chi \in C^\infty(\mathcal{M})$ that the set $\{\pi \in (\mathcal{A}/\mathcal{K}L_2(\mathcal{M}))^\wedge : \|\pi(\chi A)\| > 1\}$ is a sufficiently small neighborhood of (z^0, θ^0) in Ξ. On the other hand, the same set is a neighborhood of (z^0, θ^0) in the Jacobson topology as well. Thus, the same sets are neighborhoods of (z^0, θ^0) in both topologies.

Let $(z^0, \lambda^0) \in \mathcal{N} \times \mathbb{R}$ and let $A_0 \in \mathcal{A}_{z^0}$ be such that $\{\pi \in \hat{\mathcal{A}}_{z^0} : \|\pi(A_0)\| > 1\}$ is a union of an interval in $\mathbb{R}$ and the sphere $S^*(\mathcal{N})_{z^0}$ (the existence of A_0 follows from Theorem 2.3.16). Choosing $\chi \in C^\infty(\mathcal{M})$ and $A \in \text{com}\mathcal{A}$ so that $p_{z^0}(A) = A_0$, where $p_z \mathcal{A} \to \mathcal{A}_z$ is the canonical localizing map, one can argue as in the preceding case.

Let us now consider a neighborhood of a point $(z, \varphi_0, \omega_0) \in p^*(S^*(\mathcal{M}))$, where $(z, \varphi_0) \in \nu(\mathcal{N})_z \subset \mathcal{M}_\mathcal{N}$. Assume, for example, that $\omega_0 = (\omega_0^1, 0)$ and $\varphi_0 \omega_0^1 = 0$. Let A be an operator in $\mathcal{A}$ whose symbol Φ satisfies $|\Phi(z, \varphi, \omega)| < 1$ outside a small neighborhood V of (z, φ_0, ω_0) on $p^*(S^*(\mathcal{M}))$ (see notation before (2.3.5)) and $|\Phi(z, \varphi, \omega)| > 1$ in another neighborhood U of this point, $\overline{U} \subset V$. Furthermore, let $\mathcal{U}$ be the projection of U in $\mathcal{N}$. Theorem 2.3.16 gives us the inequalities $\|\pi(z, \lambda)A\| > 1$ for $(z, \lambda) \in \mathcal{U} \times \mathbb{R}$ and $\|\pi(z, \theta)A\| > 1$ for $\theta \in S^*(\mathcal{N})_z$, $z \in \mathcal{U}$. From this, it is easy to deduce that every neighborhood of (z, φ_0, ω_0) in the Ξ- topology (the Jacobson topology) contains a neighborhood of this point in the Jacobson topology (the Ξ-topology), which is what was needed. The cases $\varphi_0 \omega_0^1 \neq 0$ and $\omega_0 = (\omega_0^1, \omega_0^2)$ with $\omega_0^2 \neq 0$ are handled similarly.

□

Proof of Theorem 2.3.3 Let $p_z : \mathcal{A} \to \mathcal{A}_z$ be the canonical localizing map; as usual, $\mathcal{A}_z$ is the local algebra at a point $z \in \mathcal{M}$, and $C(\mathcal{M})$ is a localizing algebra. We consider the algebra $\mathcal{D}$ generated by functions of the form

$$\mathcal{M} \ni z \mapsto \alpha(z) := p_z(A) \in \mathcal{A}_z,$$

Where $a \in \mathcal{A}$. The operations in $\mathcal{D}$ are pointwise, and the norm is defined by

$$\|\alpha; \mathcal{D}\| = \sup\{\|\alpha(z); \mathcal{A}_z\|, z \in \mathcal{M}\}.$$

Taking into account Theorem 2.3.1, one can see that there is an isomorphism $\mathcal{D} \simeq \mathcal{A}/\mathcal{K}L_2(\mathcal{M})$.

Let I be the ideal in the statement of Theorem 2.3.3 and let $J := I/\mathcal{K}L_2(\mathcal{M})$. We first show that the composition series $0 \subset J \subset \text{com}\mathcal{D} \subset \mathcal{D}$ is solving. The elements of $\mathcal{D}$ can be multiplied by functions in $C^\infty(\mathcal{M})$. This implies that the algebra $\text{com}\mathcal{D}|\mathcal{N}$ of restrictions to $\mathcal{N}$ of elements in $\text{com}\mathcal{D}$ is defined by the continuous field of algebras $\{\text{com}\mathcal{D}_z, \text{com}\mathcal{D}|\mathcal{N}\}$, while $J|\mathcal{N}$ is defined by the continuous field $\{J_z, J|\mathcal{N}\}$, where, for example, J_z is the algebra of values at z of the functions in J (see [3], 10.4.2). By Theorem 2.3.17, for $z \in \mathcal{N}$, we have $J_z \simeq C(S^*(\mathcal{N})_z) \otimes \mathcal{K}L_2(\mathbb{R}^n_z)$ and $\text{com}\mathcal{D}_z/J_z \simeq C_0(\mathbb{R}) \otimes \mathcal{K}L_2(\Sigma_z)$; here $\mathbb{R}^n_z = T(\mathcal{N})^\perp_z$ is the subspace of the tangent space $T(\mathcal{M})_z$ orthogonal to the submanifold $\mathcal{N}$, and Σ_z is the unit sphere in $\mathbb{R}^n_z$. This means that $J|\mathcal{N}$ is defined by the continuous field on $S^*(\mathcal{N})$ of the elementary algebras $\mathcal{K}L_2(\mathbb{R}^n_z)$, while $\text{com}\mathcal{D}|\mathcal{N}$ is defined by the continuous field on $\mathcal{N} \times \mathbb{R}$ of the elementary algebras $\mathcal{K}L_2(\Sigma_z)$. The fields $J|\mathcal{N}$ and $\text{com}\mathcal{D}|\mathcal{N}$ are locally trivial; the local trivializations of the fields come from the local trivializations of the normal bundle over the submanifold $\mathcal{N}$. According to Proposition 1.3.18, these fields are trivial. Therefore, $J|\mathcal{N} \simeq C(S^*(\mathcal{N})) \otimes \mathcal{K}L_2(\mathbb{R}^n)$ and $(\text{com}\mathcal{D}|\mathcal{N})/(J|\mathcal{N}) \simeq C_0(\mathcal{N} \times \mathbb{R}) \otimes \mathcal{K}L_2(S^{n-1})$. The elements of $\text{com}\mathcal{D}$ are identically equal to zero outside $\mathcal{N}$. Hence, in the last two relations, we can replace $\text{com}\mathcal{D}|\mathcal{N}$ and $J|\mathcal{N}$ by $\text{com}\mathcal{D}$ and J, respectively. Finally, the quotient algebra $\mathcal{D}/\text{com}\mathcal{D}$ is commutative, and its spectrum coincides with $p^*(S^*(\mathcal{M}))$ (see (2.3.5)), i.e., $D/\text{com}\mathcal{D} \simeq C(p^*(S^*(\mathcal{M})))$. Thus, the composition series $0 \subset J \subset \text{com}\mathcal{D} \subset \mathcal{D}$ is solving.

Let us now consider the composition series

$$0 \subset \mathcal{K}L_2(\mathcal{M}) \subset I \subset \text{com}\mathcal{A} \subset \mathcal{A}.$$

We have the isomorphisms $I/\mathcal{K}L_2(\mathcal{M}) \simeq J$, $\text{com}\mathcal{A}/\mathcal{K}L_2(\mathcal{M}) \simeq \text{com}\mathcal{D}$, and $\mathcal{A}/\mathcal{K}L_2(\mathcal{M}) \simeq \mathcal{D}$. Therefore,

$$\begin{aligned} I/\mathcal{K}L_2(\mathcal{M}) &\simeq C(S^*(\mathcal{N})) \otimes \mathcal{K}L_2(\mathbb{R}^n), \\ \text{com}\mathcal{A}/I &\simeq C_0(\mathcal{N} \times \mathbb{R}) \otimes \mathcal{K}L_2(S^{n-1}), \\ \mathcal{A}/\text{com}\mathcal{A} &\simeq C(p^*(S^*(\mathcal{M}))), \end{aligned}$$

i.e., the composition series is solving.

We show that the length of $\mathcal{A}$ is equal to 3. It follows from the last paragraph that the length is at most 3. It must be shown that if a composition series $0 \subset I_0 \subset I_1 \subset \cdots \subset I_N = \mathcal{A}$ is solving, then $N \geq 3$. According to Theorem 2.3.2, the spectrum $\widehat{\mathcal{A}}$ is homeomorphic to the space obtained by adjoining to Ξ a single point ι corresponding to the identity representation; this point is an open set, and its closure coincides with the whole space. The spectra $\hat{I}_j$ of the ideals I_j form an increasing sequence of open subsets of $\widehat{\mathcal{A}}$. Any open subset contains the point ι; in particular, $\iota \in \hat{I}_0$. In the case $\iota \neq \hat{I}_0$, the spectrum $\hat{I}_0$ would not be a Hausdorff space; the isomorphism $I_0 \simeq C_0(X) \otimes \mathcal{K}H$ would be impossible. Hence, $\hat{I}_0 = \iota$, $I_0 = \mathcal{K}L_2(\mathcal{M})$. Assume that $N = 1$. Then, $I_1/I_0 = \mathcal{A}/\mathcal{K}L_2(\mathcal{M})$. However, the spectrum $(\mathcal{A}/\mathcal{K}L_2(\mathcal{M}))^\wedge$ contains representations of different dimensions (Theorem 2.3.1), and therefore, the isomorphism $\mathcal{A}/\mathcal{K}L_2(\mathcal{M}) \simeq C_0(X)\otimes\mathcal{K}H$ is impossible. Consequently, $N > 1$. Finally, assume that $N = 2$. Each open subset of $\hat{\mathcal{A}}$ intersects $S^*(\mathcal{N})$ (Theorem 2.3.2). In particular, $S^*(\mathcal{N}) \cap \hat{I}_1 \neq \emptyset$. In the case $\hat{I}_1 \not\subset S^*(\mathcal{N})$, the spectrum $(I_1/I_0)^\wedge$ is not a Hausdorff space, and a relation of the form $I_1/I_0 \simeq C_0(X)\otimes \mathcal{K}H$ does not hold. However, if $\hat{I}_1 \subset S^*(\mathcal{N})$, then the space $(\mathcal{A}/I_1)^\wedge$ turns out not to be Hausdorff, and again the isomorphism $\mathcal{A}/I_1 \simeq C_0(X) \otimes \mathcal{K}H$ cannot be. Therefore, the smallest possible number N in a solving series is equal to 3. $\square$

3 Algebra of Pseudodifferential Operators with Piecewise Smooth Symbols on a Smooth Manifold

On a smooth compact m-dimensional manifold $\mathcal{M}$ without boundary, we consider the C^*-algebra $\mathcal{A}$ generated on $L_2(\mathcal{M})$ by the operators of two classes. One of the classes consists of zero order pseudodifferential operators with smooth symbols. The other class comprises the operators of multiplication by functions ("coefficients") that may have discontinuities along a given collection of submanifolds (with boundary) of various dimensions; the intersections of the submanifolds under nonzero angles are admitted. The situation is formally described by a stratification of the manifold $\mathcal{M}$. All the equivalence classes of irreducible representations of $\mathcal{A}$ are listed; the topology on the spectrum is described; solving composition series are presented.

3.1 Algebra $\mathcal{A}$ and Its Irreducible Representations

3.1.1 Stratification of Manifold $\mathcal{M}$. Algebra $\mathcal{A}$

Let $\mathcal{T} = \{s_\alpha^p\}$ be a finite partition of manifold $\mathcal{M}$ into subsets (strata). In what follows, the discontinuities in coefficients will be supported by the strata of positive codimension. A stratum s_α^p is a connected p-dimensional submanifold of $\mathcal{M}$ (possibly, nonclosed). The partition consisting of a single element (the manifold $\mathcal{M}$ itself) is not ruled out. The boundary $\bar{s}_\alpha^p \setminus s_\alpha^p$ of a stratum is the union of strata of smaller dimension (or the empty set). The collection of all strata whose boundary contains s_α^p is called the star $\mathrm{st}(s_\alpha^p)$ of s_α^p.

Let us subject a stratification $\mathcal{T}$ of $\mathcal{M}$ to some additional requirements. Namely, we suppose that for any point $x \in s_\alpha^p$ there exist a neighborhood U in $\mathcal{M}$ and a diffeomorphism $\varkappa$ of U onto a neighborhood of the origin in $\mathbb{R}^m$ that locally rectifies s_α^p and all $s_\beta^q \in \mathrm{st}(s_\alpha^p)$; this means that $\varkappa(s_\alpha^p \cap U)$ is a neighborhood of the origin in $\mathbb{R}^p = \{y = (y_1, \dots, y_m) \in \mathbb{R}^m : y_1 = \dots = y_{m-p} = 0\}$ and $\varkappa(s_\beta^q \cap U)$ coincides near

© The Author(s), under exclusive license to Springer Nature Switzerland AG 2023

B. Plamenevskii, O. Sarafanov, *Solvable Algebras of Pseudodifferential Operators*, Pseudo-Differential Operators 15, https://doi.org/10.1007/978-3-031-28398-7_3

$\varkappa(s^p_\alpha \cap U)$ with the product $\mathbb{R}^p \times K^{q-p}$, where K^{q-p} is a $(q-p)$-dimensional cone in the subspace $\mathbb{R}^{m-p}$ orthogonal to $\mathbb{R}^p$. It will be assumed further that $\varkappa(x) = 0$ and $\varkappa'(x) = 1$. Let S^{m-p-1} be the unit sphere in this subspace $\mathbb{R}^{m-p}$. It is assumed that the projections $\mathrm{pr}(s^q_\beta)$ of the sets $\varkappa(s^q_\beta \cap U)$ onto S^{m-p-1}, where $s^q_\beta \in \mathrm{st}(s^p_\alpha)$, in turn, form a partition of S^{m-p-1} satisfying the above requirements; denote the partition by $\mathcal{T}(s^p_\alpha)$ (the point x is not indicated in the notation for simplicity). Under the given conditions, the partition $\mathcal{T}$ of $\mathcal{M}$ is said to be admissible.

We give to the partition $\mathcal{T}(s^p_\alpha)$ and the stratum $\mathrm{pr}(s^q_\beta)$, $q \le m-1$, the roles of $\mathcal{T}$ and s^p_α. This determines a partition of the sphere S^{m-q-1}; denote it by $\mathcal{T}(s^p_\alpha, s^q_\beta)$. In general, let $s_0, \dots, s_k$ be a chain of strata such that $\dim s_p = d_p$, $d_k \le m-1$, $s_i \in \mathrm{st}(s_j)$ for $i > j$ (simplifying the notation, we write strata with a single subscript). Continuing the procedure, we obtain partitions $\mathcal{T}(s_0), \mathcal{T}(s_0, s_1), \dots, \mathcal{T}(s_0, \dots, s_k)$. If $\mathcal{T}(s_0, \dots, s_k)$ consists only of a single element, i.e., the sphere S^{m-d_k-1} (for $d_k < m-1$), or of two points, i.e., the sphere S^0 (for $d_k = m-1$), then the chain $s_0, \dots, s_k$ is said to be complete.

Example 3.1.1 A partition $\mathcal{T}$ of S^2 is formed by a zero-dimensional stratum consisting of the point $P_+ = (0, 0, 1))$ and by the two-dimensional stratum $s = S^2 \setminus P_+$. It is clear that $\mathcal{T}(P_+) = S^1$. A complete chain is formed by P_+.

Example 3.1.2 A partition $\mathcal{T}$ of S^2 is formed by zero-dimensional strata: the points $P_\pm = (0, 0, \pm 1)$, a one-dimensional stratum $\mu = \{x = (x_1, x_2, x_3) : |x| = 1, x_1 > 0, x_2 = 0\}$ (the meridian), and the two-dimensional stratum $s = S^2 \setminus \bar{\mu}$. A locally rectifying diffeomorphism in a neighborhood of the points $P_\pm$ or a point $x \in \mu$ acts as the orthogonal projection onto the tangent plane at the corresponding point. There are three complete chains at all: P_+, μ; P_-, μ; and μ. The chain P_+, μ generates a partition $\mathcal{T}(P_+)$ consisting of two strata: a zero-dimensional one $p(\mu)$ and a one-dimensional $S^1 \setminus p(\mu)$, and a partition $\mathcal{T}(P_+, \mu) = S^0$. The partition $\mathcal{T}(\mu) = S^0$ corresponds to the complete chain μ.

Example 3.1.3 A partition $\mathcal{T}$ of the sphere S^{m-1} consists of an arbitrary k-dimensional $(k < m-1)$ smooth closed submanifold γ and a stratum $S^{m-1} \setminus \gamma$. Then, $\mathcal{T}(\gamma) = S^{m-k-2}$.

To determine the generators of $\mathcal{A}$, we first describe the coefficients. On a manifold $\mathcal{M}$ endowed with an admissible partition $\mathcal{T}$, we introduce a class $\mathfrak{M}(\mathcal{T}) \equiv \mathfrak{M}(\mathcal{T}, \mathcal{M})$ of functions. The operators aI with $a \in \mathfrak{M}(\mathcal{T})$ form one type of the generators of $\mathcal{A}$. We define the classes $\mathfrak{M}(\mathcal{T}, \mathcal{M})$ using induction on $\dim \mathcal{M}$. For $\dim \mathcal{M} = 0$, such a class is the set of all functions on $\mathcal{M}$ (by $\mathcal{M}$ is meant a finite collection of points with discrete topology). Assume that the classes have been defined for $k = \dim \mathcal{M} \le m-1$ and turn to the case $\dim \mathcal{M} = m$. We suppose that $\mathcal{T}$ is an admissible partition of $\mathcal{M}$ and $|\mathcal{T}|$ is the union of all strata (as subsets of the manifold $\mathcal{M}$) of dimension no greater than $m-1$. By definition, a smooth function f given on $\mathcal{M} \setminus |\mathcal{T}|$ belongs to the class $\mathfrak{M}(\mathcal{T}, \mathcal{M})$ if a representation of the form

$$(f \cdot \varkappa^{-1})(y, z) = \Psi(y, z) + \Psi_1(y, z) \tag{3.1.1}$$

holds in a neighborhood U of any point of every stratum s_j, $d_j \leq m - 1$; here, $(y, z) \in \varkappa(U) \subset \mathbb{R}^m$, $y = (y_1, \ldots, y_{m-d_j})$, $z = (z_1, \ldots, z_{d_j})$, $\Psi(y, z) = \Psi(y/|y|, z)$ for $y \neq 0$, and $\Psi(\cdot, z) \in \mathfrak{M}(\mathcal{T}(s_j), S^{m-d_j-1})$ for all z; the terms in (3.1.1) depend continuously on z in the norm of L_∞ and $\Psi_1(y, z) = o(1)$ for $y \to 0$. The function $\Psi(\cdot, z)$ is called the limit value of f at the point $z \in s_j$ (instead of $\varkappa^{-1}(0, z)$, we write simply z).

The other type of the generators of the algebra $\mathcal{A}$ is formed by pseudodifferential operators of the class $\bar{\Psi}^0(\mathcal{M})$ (with smooth symbols) introduced by Definition 2.1.4.

Thus, we assume that an admissible partition $\mathcal{T}$ is given on a smooth compact $\mathcal{M}$ without boundary and let $\mathcal{A}$ stand for the algebra generated on $L_2(\mathcal{M})$ by the operators aI with a in $\mathfrak{M}(\mathcal{T}, \mathcal{M})$ and by ψDOs of the class $\bar{\Psi}^0(\mathcal{M})$.

3.1.2 The Irreducible Representations of the Algebra $\mathcal{A}$ (Formulation of a Theorem)

We suppose that the manifold $\mathcal{M}$ is endowed with the Riemannian metric. Let $s_0, \ldots, s_k$ be a chain of strata of the partition $\mathcal{T}$, i.e., $s_i \in \operatorname{st}(s_j)$ for $i > j$, $\dim s_j = d_j$, and $d_k \leq m - 1$, where $m = \dim \mathcal{M}$. For the generators of $\mathcal{A}$, we define a localization procedure along the chain of strata.

Let us start with the coefficients $a \in \mathfrak{M}(\mathcal{T}, \mathcal{M})$. Let $a(z^0; \cdot) \in \mathfrak{M}(\mathcal{T}(s_0), S^{m-d_0-1})$ be the limit value of a at a point $z^0 \in s_0$. The $(d_1 - d_0 - 1)$-dimensional stratum $\operatorname{pr}(s_1)$ of the partition $\mathcal{T}(s_0)$ corresponds to the stratum s_1. Denote by $a(z^0, z^1; \cdot)$ the limit value of $a(z^0; \cdot)$ at a point $z^1 \in \operatorname{pr}(s_1)$. Continuing the process, we arrive at the collection $a(z^0; \cdot), \ldots, a(z^0, \ldots, z^k; \cdot)$, where $a(z^0, \ldots, z^j; \cdot) \in \mathfrak{M}(\mathcal{T}(s_0, \ldots, s_j), S^{m-d_j-1})$. We say that this collection is obtained by localizing a along the chain $s_0, \ldots, s_k$. If the chain is complete, then $a(z^0, \ldots, z^k; \cdot) \in C(S^{m-d_k-1})$.

We extend the localization procedure to generators of the other type, i.e., to the ψDOs in $\bar{\Psi}^0(\mathcal{M})$. Let $A \in \bar{\Psi}^0(\mathcal{M})$ and let $\Phi(z^0; \cdot)$ be the principal symbol of A at a point $z^0 \in s_0$. The function $\xi \mapsto \Phi(z^0; \xi)$ is homogeneous of zero degree and smooth on any fiber $T^*(\mathcal{M})_{z^0} \setminus 0$ of nonzero cotangent vectors. In the tangent space $T(\mathcal{M})_{z^0}$, we introduce orthogonal coordinates $(x_1, \ldots, x_m)$, where the axes $x_{m-d_0+1}, \ldots, x_m$ are parallel to the stratum s_0. Assume that the symbol $\Phi(z^0; \cdot)$ is written in that coordinate system. We introduce operators "of the first generation"

$$A(z^0; \theta^0) = F^{-1}_{\eta \to x} \Phi(z^0; \eta, \theta^0) F_{y \to \eta} \in \mathcal{B}L_2(\mathbb{R}^{m-d_0}), \tag{3.1.2}$$

where $\eta = (\eta_1, \dots, \eta_{m-d_0})$, $\theta^0 = (\theta_1, \dots, \theta_{d_0}) \in \mathbb{R}^{d_0}$, $|\theta^0| = 1$, and F is the Fourier transform in $\mathbb{R}^{m-d_0}$;

$$\mathfrak{A}(z^0; \lambda) = E_{\omega \to \varphi}(\lambda)^{-1} \Phi(z^0; \omega, 0) E_{\psi \to \omega}(\lambda) \in \mathcal{B}L_2(S^{m-d_0-1}), \tag{3.1.3}$$

where $\varphi, \psi, \omega \in \mathbb{R}^{m-d_0}$, $|\varphi| = |\psi| = |\omega| = 1$, $\lambda \in \mathbb{R}$, and the operators $E(\lambda)^{\pm 1}$ are defined by (1.2.2) and (1.2.8) for $n = m - d_0$. The operators (3.1.2) are not defined for $d_0 = 0$.

We now choose in $T(\mathcal{M})_{z^0}$ new Cartesian coordinates $\tilde{x}_1, \dots, \tilde{x}_m$ with the same origin. The axes $\tilde{x}_{m-d_0+1}, \dots, \tilde{x}_m$ coincide with the old $x_{m-d_0+1}, \dots, x_m$. The axis $\tilde{x}_{m-d_0}$ is directed toward the point $z^1 \in \operatorname{pr}(s_1)$, $\dim \operatorname{pr}(s_1) = d_1 - d_0 - 1$; the axes $\tilde{x}_{m-d_1}, \dots, \tilde{x}_{m-d_0-1}$ are parallel to the $(d_1 - d_0 - 1)$-dimensional space tangent to the stratum $\operatorname{pr}(s_1)$ at z^1. The rest axes $\tilde{x}_1, \dots, \tilde{x}_{m-d_1-1}$ are parallel to the space tangent to S^{n-1} and orthogonal to $\operatorname{pr}(s_1)$ at z^1. The coordinate transformation (with block-diagonal matrix) will be written in the form $x = \operatorname{diag}(J(z^1), I_{m-d_0})\tilde{x}$. In what follows, the old coordinates do not appear, and the new ones are denoted by x. We introduce operators "of the second generation"

$$A(z^0, z^1; \theta^0, \theta^1) = F^{-1}_{\eta \to x} \Phi(z^0; J(z^1)(\eta, \theta^1), \theta^0) F_{y \to \eta} \in \mathcal{B}L_2(\mathbb{R}^{m-d_1}), \tag{3.1.4}$$

where $\theta^0 \in \mathbb{R}^{d_0}$, $\theta^1 \in \mathbb{R}^{d_1 - d_0}$, $|\theta^0|^2 + |\theta^1|^2 = 1$, and F is the Fourier transform in $\mathbb{R}^{m-d_1}$;

$$\mathfrak{A}(z^0, z^1; \lambda) = E_{\omega \to \varphi}(\lambda)^{-1} \Phi(z^0; J(z^1)(\omega, 0), 0) E_{\psi \to \omega}(\lambda) \in \mathcal{B}L_2(S^{m-d_1-1}), \tag{3.1.5}$$

where $\lambda \in \mathbb{R}$ and the operators $E(\lambda)^{\pm 1}$ are defined by (1.2.2) and (1.2.8) for $n = m - d_1$. If $d_0 = 0$, then in (3.1.4), the operator

$$A(z^0, z^1; \theta^1) = F_{\eta \to x} \Phi(z^0; J(z^1)(\eta, \theta^1)) F_{y \to \eta}$$

is defined instead of $A(z^0, z^1; \theta^0, \theta^1)$, and the operator

$$\mathfrak{A}(z^0, z^1; \lambda) = E_{\omega \to \varphi}(\lambda)^{-1} \Phi(z^0; J(z^1)(\omega, 0)) E_{\psi \to \omega}(\lambda)$$

is defined in (3.1.5). Continuing the process, we arrive at the collection of the operators $A(z^0, \dots, z^j; \theta^0, \dots, \theta^j)$, $\mathfrak{A}(z^0, \dots, z^j; \lambda_j)$, where $0 \le j \le k$, $\lambda_j \in \mathbb{R}$, $\theta^0 \in \mathbb{R}^{d_0}$, $\theta^j \in \mathbb{R}^{d_j - d_{j-1}}$ for $j \ge 1$, and $|\theta^0|^2 + \dots + |\theta^j|^2 = 1$. We say that such a collection is obtained by localizing $A \in \bar{\Psi}^0(\mathcal{M})$ along the chain $s_0, \dots, s_k$.

Although the operators $A(z^0, \dots, z^j; \theta^0, \dots, \theta^j)$, $\mathfrak{A}(z^0, \dots, z^j; \lambda_j)$ are constructed for special coordinate systems, they are independent of the arbitrariness in their definition. To explain this, suppose that there is chosen a chain of strata $s_0, \dots, s_j$. Let us glue the strata $s_{j-1}, \dots, s_0$ to s_j. The collection $(z^0, \dots, z^j)$ determines a d_j-dimensional subspace $T(z^0, \dots, z^j)$ of $T(\mathcal{M})_{z^0}$ tangent to s_j. The point $(\theta^0, \dots, \theta^j)$ runs over the unit sphere $S^*(z^0, \dots, z^j)$ in the cotangent space $T^*(z^0, \dots, z^j)$. Denote by $\mathbb{R}^{m-d_j}(z^0, \dots, z^j)$ the orthogonal complement of $T^*(z^0, \dots, z^j)$ in $T(\mathcal{M})_{z^0}$ and by $S^{m-d_j-1}(z^0, \dots, z^j)$ the unit sphere in the mentioned orthogonal complement. Recall that the principal symbol Φ of $A \in \bar{\Psi}^0(\mathcal{M})$ is defined on the bundle of nonzero cotangent vectors. The operator $A(z^0, \dots, z^j; \theta^0, \dots, \theta^j)$ acts in $L_2(\mathbb{R}^{m-d_j}(z^0, \dots, z^j))$ and $\mathfrak{A}(z^0, \dots, z^j; \lambda_j)$ does in $L_2(S^{m-d_j-1}(z^0, \dots, z^j))$. This implies that the operators are independent of arbitrariness in their definition.

The functions a in $\mathfrak{M}(\mathcal{T}, \mathcal{M})$ are smooth on $\mathcal{M} \setminus |\mathcal{T}|$. The set $\mathcal{M} \setminus |\mathcal{T}|$ is embedded in a compact $\mathfrak{C}$, and the functions a extend to $\mathfrak{C}$ by continuity; we now describe $\mathfrak{C}$. We first glue the boundary of a tubular neighborhood (in $\mathcal{M}$) of every $(m-1)$-dimensional stratum to $\mathcal{M} \setminus |\mathcal{T}|$ (distinct points have to be glued to distinct ones). Denote the obtained set by $\mathfrak{C}_{m-1}$. The boundary of a tubular neighborhood in $\mathcal{M}$ of each $(m-2)$-dimensional stratum s is cut by the $(m-1)$-dimensional strata of the star $\operatorname{st} s$. We glue the boundary to $\mathfrak{C}_{m-1}$ so that it be identified along the cut with the boundary of an already glued tubular neighborhood of $(m-1)$-dimensional stratum (distinct points have to be identified with distinct ones). Denote the new set by $\mathfrak{C}_{m-2}$. Descending to strata of smaller dimensions, we finally obtain the compact $\mathfrak{C} := \mathfrak{C}_0$.

Let $p : \mathfrak{C} \to \mathcal{M}$ be the natural projection and let $S^*(\mathfrak{C}) := p^* S^*(\mathcal{M})$, where $S^*(\mathcal{M})$ is the cospherical bundle over $\mathcal{M}$ and $p^* S^*(\mathcal{M})$ the induced bundle. In what follows, the one-dimensional representations of the algebra $\mathcal{A}$ will be parametrized by the points of $S^*(\mathfrak{C})$.

We now tour to listing the irreducible representations of $\mathcal{A}$. As before, let $s_0, \dots, s_k$ be a chain of strata of the partition $\mathcal{T}$, $\dim s_j = d_j$. We introduce the following series of mappings for the generators $A \in \bar{\Psi}^0(\mathcal{M})$ and $a \in \mathfrak{M}(\mathcal{T}, \mathcal{M})$ of $\mathcal{A}$.

Mapping into $\mathcal{B}L_2(\mathbb{R}^{m-d_j}(z^0, \dots, z^j))$:

$$\pi(z^0, \dots, z^j; \theta^0, \dots, \theta^j) : A \mapsto A(z^0, \dots, z^j; \theta^0, \dots, \theta^j) \in \mathcal{B}L_2(\mathbb{R}^{m-d_j}(z^0, \dots, z^j)), \tag{3.1.6}$$

$$\pi(z^0, \dots, z^j; \theta^0, \dots, \theta^j) : a \mapsto a(z^0, \dots, z^j; \cdot)I \in \mathcal{B}L_2(\mathbb{R}^{m-d_j}(z^0, \dots, z^j))I,$$

where $j = 0, \dots, k$ and the function $a(z^0, \dots, z^j; \cdot)$ is zero degree homogeneous on $\mathbb{R}^{m-d_j} \setminus 0$. For $d_0 = 0$, the mappings $\pi(z^0; \theta^0)$ are absent; given a chain of strata, the mappings $\pi(z^0, \dots, z^j; \theta^0, \dots, \theta^j)$ in (3.1.6) are changed for $\pi(z^0, z^1, \dots, z^j; \theta^1, \dots, \theta^j)$, where in particular $\pi(z^0, \dots, z^j; \theta^1, \dots, \theta^j)A = A(z^0, \dots, z^j; \theta^1, \dots, \theta^j)$. In what follows, we, as a rule, make no remarks related to the specificity of the case $d_0 = 0$.

Mapping into $\mathcal{B}L_2(S^{m-d_j-1}(z^0,\ldots,z^j))$:

$$\pi(z^0,\ldots,z^j;\lambda_j): A \mapsto \mathfrak{A}(z^0,\ldots,z^j;\lambda_j) \in \mathcal{B}L_2(S^{m-d_j-1}(z^0,\ldots,z^j)), \tag{3.1.7}$$
$$\pi(z^0,\ldots,z^j;\lambda_j): a \mapsto a(z^0,\ldots,z^j;\cdot)I \in \mathcal{B}L_2(S^{m-d_j-1}(z^0,\ldots,z^j)),$$

where $j = 0,\ldots,k$ and $\lambda_j \in \mathbb{R}$.

Mapping into $\mathbb{C}$:

$$\pi(x,\xi): A \mapsto \Phi(p(x),\xi) \in \mathbb{C}, \tag{3.1.8}$$
$$\pi(x,\xi): a \mapsto a(x),$$

where $(x,\xi) \in S^*(\mathfrak{C})_x$, while $S^*(\mathfrak{C})_x$ is the fiber over $x \in \mathfrak{C}$ of the cospherical bundle $S^*(\mathfrak{C})$, $p: \mathfrak{C} \to \mathcal{M}$ is the projection, and Φ is the principal symbol of A.

From Sect. 2.1, it follows that the algebra $\mathcal{A}$ is irreducible and contains the ideal $\mathcal{K}L_2(\mathcal{M})$. The next theorem describes all (up to equivalence) irreducible representations of the algebra $\mathcal{A}$.

Theorem 3.1.4 *The mappings (3.1.6)–(3.1.8) extend to representations of the quotient algebra $\mathcal{A}/\mathcal{K}L_2(\mathcal{M})$. The representations (3.1.6) and (3.1.7) corresponding to all complete chains of strata of the partition $\mathcal{T}$ are irreducible and nonequivalent. Any irreducible representation π of $\mathcal{A}$ with $\dim \pi > 1$ is equivalent either to one of the representations (3.1.6), (3.1.7) or to the identity representation. Any one-dimensional representation of $\mathcal{A}$ coincides with one of those in (3.1.8).*

The algebras $\pi(z^0,\ldots,z^j;\theta^0,\ldots,\theta^j)\mathcal{A}$ contain the subalgebras spanned by the operators $\pi(z^0,\ldots,z^j;\theta^0,\ldots,\theta^j)A$, where $A \in \bar{\Psi}^0(\mathcal{M})$, and by the coefficients $a(z^0,\ldots,z^j;\cdot)$ smooth on S^{m-d_j-1}. Hence, due to Theorem 2.3.12, the inclusion

$$\mathcal{K}L_2(\mathbb{R}^{m-d_j}) \subset \pi(z^0,\ldots,z^j;\theta^0,\ldots,\theta^j)\mathcal{A}$$

holds. Moreover, the algebras $\pi(z^0,\ldots,z^j;\lambda_j)\mathcal{A}$ contain the subalgebras generated by the operators $\pi(z^0,\ldots,z^j;\lambda_j)A$, where $A \in \bar{\Psi}^0(\mathcal{M})$, and by the coefficients $a(z^0,\ldots,z^j;\cdot)$ smooth on S^{m-d_j-1}. Proposition 2.2.1 leads to the inclusion

$$\mathcal{K}L_2(S^{m-d_j-1}) \subset \pi(z^0,\ldots,z^j;\lambda_j)\mathcal{A}.$$

Thus, Theorem 3.1.4 implies the following assertion.

Corollary 3.1.5 *$\mathcal{A}$ is an algebra of type I.*

3.1.3 Proof of Theorem 3.1.4

1^0. **Plan of the proof.** In what follows, if no misunderstanding is possible, we write $\mathbb{R}^{m-d_j}$ and S^{m-d_j-1} instead of $\mathbb{R}^{m-d_j}(z^0,\dots,z^j)$ and $S^{m-d_j-1}(z^0,\dots,z^j)$. We apply the localization principle (Proposition 1.3.24) to the algebra $\mathcal{A}$, taking $C(\mathcal{M})$ as a localizing algebra and $J = \mathcal{K}L_2(\mathcal{M})$. It is clear that the local algebra $\mathcal{A}_x$ at a point $x \in \mathcal{M} \setminus |\mathcal{T}|$ is isomorphic to $C(S^*(\mathcal{M})_x)$ (Proposition 2.1.6). If z^0 is a point of a d_0-dimensional stratum s_0 and $d_0 > 0$, then the same argument as in the proof of Proposition 2.3.4 shows that the local algebra $\mathcal{A}_{z^0}$ is isomorphic to the algebra of functions on S^{d_0-1} ranging in $\mathcal{B}L_2(\mathbb{R}^{m-d_0})$ spanned by the functions $\theta^0 \mapsto A(z^0;\theta^0) = \pi(z^0;\theta^0)A$ and by the operators of multiplication by $a(z^0;\cdot)I = \pi(z^0;\theta^0)a$; here $A \in \bar{\Psi}^0(\mathcal{M})$ and $a \in \mathfrak{M}(\mathcal{T},\mathcal{M})$, while $A(z^0;\theta^0)$ and $a(z^0;\cdot)$ are defined as in 3.1.2. Localization of $\mathcal{A}(z^0;\theta^0)$ leads in particular to the algebra $\mathfrak{S}(z^0)$ generated by the functions $\mathbb{R} \ni \lambda \mapsto \mathfrak{A}(z^0,\lambda)$ (see (3.1.3)) and by the operators of multiplication by the coefficients $a(z^0;\cdot)|S^{n-1}$, where $n = m - d_0$. The operations in $\mathfrak{S}(z^0)$ are pointwise, and a norm is given by

$$\|\mathfrak{A}(\cdot)\| = \sup\{\|\mathfrak{A}(\lambda);\mathcal{B}L_2(S^{n-1})\|;\ \lambda \in \mathbb{R}\}.$$

If $d_0 = 0$, then the algebra $\mathcal{A}(z^0;\theta^0)$ does not appear, and instead, $\mathfrak{S}(z^0)$ arises even at the first step of localization. The algebras $\mathcal{A}(z^0;\theta^0)$ and $\mathfrak{S}(z^0)$ are said to be algebras of the first generation. Localizing such algebras, we obtain algebras of the second generation (of the same type), and so on. The last generation consists of commutative algebras. In essence, description of the dynasty of local algebras is tantamount to listing all the equivalence classes of irreducible representations of the algebra $\mathcal{A}$. We proceed to implement this plan.

2^0. **Localization of the algebra** $\mathcal{A}(z^0,\theta^0)$. To apply the localization principle from Proposition 1.3.26, we take $C(\bar{\mathbb{R}}^n)$ as a localizing algebra, where $n = m - d_0$ and $\bar{\mathbb{R}}^n$ is a compact obtained by adding to $\mathbb{R}^n$ the sphere S^{n-1}_∞ at infinity. The conditions of the Proposition 1.3.26 are fulfilled; this can be proved in the same way as in Sect. 2.3.3 (before Proposition 2.3.11). Remind that $\mathcal{A}(z^0,\theta^0)$ is generated in $L_2(\mathbb{R}^n)$ by the operators of the form

$$A(z^0;\theta^0) = F^{-1}_{\eta\to x}\Phi(z^0;\eta,\theta^0)F_{y\to\eta} \in \mathcal{B}L_2(\mathbb{R}^n)$$

(see (3.1.2) for notations) and the operators of multiplication by the coefficients

$$\mathbb{R}^n \setminus 0 \ni x \mapsto a(z^0;x/|x|),\quad a(z^0;\cdot)|S^{n-1} \in \mathfrak{M}(\mathcal{T}(s_0),S^{n-1}).$$

On such generators, we define some mappings that, as shown below, arise under localization at a point $z \in \bar{\mathbb{R}}^n$; in items (i)–(v), the mapping $\mathfrak{l}(z)$ is acting on the algebra l_z.

(i) For $z \in \mathbb{R}^n \setminus 0$, $z/|z| \in S^{n-1} \setminus |\mathcal{T}(s_0)|$, we set $l_z = C(S^{n-1})$ and introduce the mapping

$$\begin{aligned} \mathfrak{l}(z) &: a(z^0; \cdot) \mapsto a(z^0; z/|z|), \\ \mathfrak{l}(z) &: A(z^0; \theta^0) \mapsto \Phi(z^0; \cdot, 0). \end{aligned} \tag{3.1.9}$$

(ii) For $z = 0$, we denote by l_z the algebra generated on $L_2(\mathbb{R}^n)$ by operators of the form

$$A(z^0; 0) = F^{-1}_{\eta \to x} \Phi(z^0; \eta, 0) F_{y \to \eta}$$

and the operators of multiplication by the functions $\mathbb{R}^n \setminus 0 \ni x \mapsto a(z^0; x/|x|)$. Introduce the mapping

$$\begin{aligned} \mathfrak{l}(z) &: a(z^0; \cdot) \mapsto a(z^0; \cdot) I, \\ \mathfrak{l}(z) &: A(z^0; \theta^0) \mapsto A(z^0; 0). \end{aligned} \tag{3.1.10}$$

(iii) For $z \in S^{n-1}_\infty$ and $\varphi_z \in S^{n-1} \setminus |\mathcal{T}(s_0)|$, where φ_z is the vector directed toward z, we set $l_z = C(\bar{\mathbb{R}}^n)$ and introduce the mapping

$$\begin{aligned} \mathfrak{l}(z) &: a(z^0; \cdot) \mapsto a(z^0; \varphi_z), \\ \mathfrak{l}(z) &: A(z^0; \theta^0) \mapsto \Phi(z^0; \cdot, \theta^0). \end{aligned} \tag{3.1.11}$$

We pass on to the mappings corresponding to the points $z \in \bar{\mathbb{R}}^n \setminus 0$ for $\varphi_z \in |\mathcal{T}(s_0)|$. Let $s_1 \in \mathrm{st}(s_0)$, $\dim s_1 = d_1$, so that $\mathrm{pr}(s_1)$ is a $(d_1 - d_0 - 1)$-dimensional stratum of the partition $\mathcal{T}(s_0)$ of the sphere S^{n-1}. We choose new coordinates $\tilde{x}_1, \dots, \tilde{x}_n$ in $\mathbb{R}^n$ with the same origin. The axis $\tilde{x}_n$ is directed toward $\varphi_z \in \mathrm{pr}(s_1)$; the axes $\tilde{x}_{n-d_1+d_0}, \dots, \tilde{x}_{n-1}$ are parallel to the space tangent to the stratum $\mathrm{pr}(s_1)$ at φ_z; the remaining axes are orthogonal to the stratum $\mathrm{pr}(s_1)$ at the point φ_z and parallel to the plane tangent to the sphere S^{n-1} at that point. We write the coordinate transformation in the form $x = J(z)\tilde{x}$. Further, the old coordinates do not appear, and the new ones are denoted by x. A point $x = (x_1, \dots, x_n)$ will be written in the form $x = (x^1, x^2)$, where $x^1 = (x_1, \dots, x_{n-(d_1-d_0)-1})$ and $x^2 = (x_{n-(d_1-d_0)}, \dots, x_n)$.

(iv) For $z \in \mathbb{R}^n \setminus 0$ and $z/|z| \in \mathrm{pr}(s_1)$, we denote by l_z the algebra generated on $L_2(\mathbb{R}^n)$ by operators of the form

$$B(z^0, z; 0) = F^{-1}_{\xi \to x} \Phi(z^0; J(z)\xi, 0) F_{y \to \xi}$$

and the operators of multiplication by the functions

$$\mathbb{R}^n \setminus \{x = (x^1, x^2) : x^1 = 0\} \ni x \mapsto a(z^0, z; x) = a(z^0, z; x^1/|x^1|).$$

(Remind that the function $\mathbb{R}^n \setminus 0 \ni x \mapsto a(z^0; x)$ is zero degree homogeneous; hence, $a(z^0, z; x^1/|x^1|) = a(z^0, z/|z|; x^1/|x^1|)$.) Introduce the mapping

$$\mathfrak{l}(z) : a(z^0; \cdot) \mapsto a(z^0, z; \cdot)I, \qquad (3.1.12)$$
$$\mathfrak{l}(z) : A(z^0; \theta^0) \mapsto B(z^0, z; 0).$$

(v) For $z \in S^{n-1}_\infty$ and $\varphi_z \in \mathrm{pr}(s_1)$, by l_z, we denote the algebra generated on $L_2(\mathbb{R}^n)$ by operators of the form

$$B(z^0, z; \theta^0) = F^{-1}_{\xi\to x}\Phi(z^0; J(z)\xi, \theta^0)F_{y\to\xi}$$

and the operators of multiplication by the functions $x \mapsto= a(z^0, z; x^1/|x^1|)$; here

$$a(z^0, z; x^1/|x^1|) = a(z^0, \varphi_z; x^1/|x^1|).$$

Introduce the mapping

$$\mathfrak{l}(z) : a(z^0; \cdot) \mapsto a(z^0, z; \cdot)I, \qquad (3.1.13)$$
$$\mathfrak{l}(z) : A(z^0; \theta^0) \mapsto B(z^0, z; \theta^0).$$

Proposition 3.1.6 *Let $\mathcal{A}(z^0, \theta^0)_z$ be the local algebra at $z \in \bar{\mathbb{R}}^n$ obtained by the localization of $\mathcal{A}(z^0, \theta^0)$ with commutative algebra $C(\bar{\mathbb{R}}^n)$. The mapping $\mathfrak{l}(z)$ defined on the generators of $\mathcal{A}(z^0; \theta^0)$ by (3.1.9)–(3.1.13) extends to an isomorphism $\mathfrak{l}(z)$: $\mathcal{A}(z^0; \theta^0)_z \to l_z$ for every $z \in \bar{\mathbb{R}}^n$.*

Proof Items (i)–(iii) can be treated in the same way as in Proposition 2.3.11. Let us turn to the mapping (iv). Introduce the unitary operator $(U_t u)(x) = t^{m/2}u(z + t(x - z))$ on $L_2(\mathbb{R}^n)$ for $t > 0$ and consider the composition $U_t Q U_t^{-1}$, where Q is a product of finitely many generators of $\mathcal{A}(z^0; \theta^0)$ and functions in $C(\bar{\mathbb{R}}^n)$. In view of the properties of the coefficients, there exists a strong limit

$$Q_0 = \lim_{t\to 0} U_t Q U_t^{-1}. \qquad (3.1.14)$$

Since

$$\lim_{t\to 0}(U_t a(z^0; \cdot)U_t^{-1}u)(x) = a(z^0, z; x^1/|x^1|)u(x), \qquad (3.1.15)$$
$$\lim_{t\to 0}(U_t F^{-1}_{\xi\to x}\Phi(z^0; J(z)\xi, \theta^0)F_{y\to\xi}U_t^{-1}u)(x) = (F^{-1}_{\xi\to x}\Phi(z^0; J(z)\xi, 0)F_{y\to\xi}u)(x),$$

the operators Q_0 generate on $L_2(\mathbb{R}^n)$ the algebra l_z defined in (iv) before Proposition 3.1.6. Let $\mathfrak{Q}$ be the algebra spanned by $\mathcal{A}(z^0;\theta^0)$ and $C(\bar{\mathbb{R}}^n)$. From (3.1.13) and (3.1.14), it follows that the mapping $Q \mapsto Q_0$ extends to an epimorphism $q : \mathfrak{Q}/\mathfrak{J} \to l_z$, where $\mathfrak{J}$ is the ideal generated in $\mathfrak{Q}$ by the functions χ in $C(\bar{\mathbb{R}}^n)$, $\chi(z) = 0$. We show that q is an isomorphism. If the support of $\chi \in C_c^\infty(\mathbb{R}^n)$ belongs to a sufficiently small neighborhood of z, then χQ_0 is in $\mathfrak{Q}$, and if $\chi(z) = 1$, then the operators Q and χQ_0, where $Q_0 = \lim_{t\to 0} U_t Q U_t^{-1}$, represent the same residue class $[Q]$ in $\mathfrak{Q}/\mathfrak{J}$. Hence, $\|[Q]\| \le \|\chi Q_0\| \le \sup|\chi|\|Q_0\|$, which means that q is a monomorphism. Thus, q is an isomorphism.

Let us finally turn to the mapping (v). We can argue as in the part of the proof of Proposition 2.3.11 relating to the case (iii). Namely, as $A_{jk}(\theta)$, we take operators of the form

$$B_{jk}(z^0, z; \theta^0) = F^{-1}_{\xi\to x}\Phi_{jk}(z^0; J(z)\xi, \theta^0)F_{y\to\xi},$$

and as coefficients, we choose $a_{jk}(z^0;\cdot)$, which are zero degree homogeneous functions satisfying $a_{jk}(z^0;\cdot)|S^{n-1} \in \mathfrak{M}(\mathcal{T}(s_0), S^{n-1})$. Then, there holds an equality similar to (2.3.29),

$$\begin{aligned} &U_t(\sum_j\prod_k \chi_{jk}a_{jk}(z^0;\cdot)B_{jk}(z^0,z;\theta^0))U_t^{-1}u \\ &= \{\sum_j\prod_k (U_t(\chi_{jk}a_{jk}(z^0;\cdot))B_{jk}(z^0,z;\theta^0)\}u, \end{aligned} \tag{3.1.16}$$

where $(U_t u)(x) = u(x + t\varphi_z)$. We have

$$\lim_{t\to+\infty}(U_t a_{jk}(z^0;\cdot))(x) = a_{jk}(z^0, z;\cdot).$$

Therefore, instead of (2.3.31), we obtain

$$\begin{aligned} &\inf\{\|\sum_j\prod_k \chi_{jk}a_{jk}(z^0;\cdot)B_{jk}(z^0,z;\theta^0) + K; \mathcal{B}L_2(\mathbb{R}^n)\|; K \in \mathcal{J}_z\} \\ &= \|\sum_j\prod_k \chi_{jk}(z)a_{jk}(z^0,z;\cdot)B_{jk}(z^0,z;\theta^0); \mathcal{B}L_2(\mathbb{R}^n)\|. \end{aligned}$$

□

3^0. **Localization of the algebra $\mathfrak{S}(z^0)$.** If a point z^0 belongs to a zero-dimensional stratum s_0 of the partition $\mathcal{T}$ of $\mathcal{M}$, then under localization of the initial algebra $\mathcal{A}$ there arises the algebra spanned by operators of the form $a(z^0;x)F^{-1}_{\xi\to x}\Phi(z^0;\xi)F_{y\to\xi}$; here, F is the

Fourier transform on $\mathbb{R}^m$. Such an algebra is isomorphic to the algebra $\mathfrak{S}(z^0)$ generated by the operator-valued functions

$$\mathbb{R} \ni \lambda \to a(z^0; \varphi) E_{\omega\to\varphi}(\lambda)^{-1} \Phi(z^0; \omega) E_{\psi\to\omega}(\lambda) \in \mathcal{B}L_2(S^{m-1});$$

the norm in $\mathfrak{S}(z^0)$ is defined by $\|\mathfrak{B}(\cdot); \mathfrak{S}(z^0)\| = \sup\{\|\mathfrak{B}(\lambda); \mathcal{B}L_2(S^{m-1})\|; \lambda \in \mathbb{R}\}$ (see 2.2.1). If $z^0 \in s_0$ and $\dim s_0 = d_0 > 0$, then $\mathfrak{S}(z^0)$ arises under localization of the algebra $\mathcal{A}(z^0, \theta^0)$ at $z = 0$. In more detail, first comes the algebra l_0 of the operators $a(z^0; x) A(z^0; 0)$ (see item (ii) before Proposition 3.1.6), which is isomorphic to the algebra $\mathfrak{S}(z^0)$ spanned by the functions

$$\mathbb{R} \ni \lambda \to a(z^0; \varphi) E_{\omega\to\varphi}(\lambda)^{-1} \Phi(z^0; \omega, 0) E_{\psi\to\omega}(\lambda) \in \mathcal{B}L_2(S^{n-1}),$$

where $n = m - d_0$ (compared with Sect. 2.2.1 and Proposition 2.3.11).

To apply the localization principle (Proposition 1.3.24) to $\mathfrak{S}(z^0)$, we take $C_0(\mathbb{R}) \otimes \mathcal{K}L_2(S^{n-1})$ as the ideal J and $C^{n-1}(S^{n-1})$ as the localizing algebra. The fulfillment of the hypotheses of Proposition 1.3.24 is in fact verified in the proof of Proposition 2.2.9. Indeed, in contrast with the algebra $\mathfrak{S}(z^0)$, the role of coefficients in $\mathfrak{S}$ from 2.2.1 is played by functions in $C^\infty(S^{n-1})$. Hence, $J = C_0(\mathbb{R}) \otimes \mathcal{K}L_2(S^{n-1}) \subset \mathfrak{S} \subset \mathfrak{S}(z^0)$ (the inclusion $J \subset \mathfrak{S}$ is proved in 2.2.2). This easily implies that the proof of Proposition 2.2.9 verifies the applicability of Proposition 1.3.24 to the algebra $\mathfrak{S}(z^0)$, too. Therefore, the equality

$$\widehat{\mathfrak{S}}(z^0) = \cup_{\varphi\in S^{n-1}} \widehat{\mathfrak{S}}(z^0)_\varphi \cup \widehat{J} \tag{3.1.17}$$

holds; here, $\mathfrak{S}(z^0)_\varphi = \mathfrak{S}(z^0)/\mathcal{J}_\varphi$ is the local algebra at a point φ, and $\mathcal{J}_\varphi$ is the ideal in $\mathfrak{S}(z^0)$ spanned by the functions $\chi \in C(S^{n-1})$ vanishing at φ.

We then describe the local algebras $\mathfrak{S}(z^0)/\mathcal{J}_\varphi$. Assume that $z^0 \in s_0$, $\dim s_0 = d_0 \geq 0$, and $T(\mathcal{M})_{z^0} = T^{(m)}(\mathcal{M})_{z^0}$ is the tangent space of $\mathcal{M}$ at z^0. Denote by $T^{(n)}(\mathcal{M})_{z^0}$, where $n = m - d_0$, the subspace of $T^{(m)}(\mathcal{M})_{z^0}$, orthogonal to the stratum s_0. We choose arbitrarily a unit vector z^1 in $T^{(n)}(\mathcal{M})_{z^0}$ and introduce in $T^{(n)}(\mathcal{M})_{z^0}$ new orthogonal coordinates $\tilde{x}_1, \dots, \tilde{x}_n$ (in the same way as in a similar situation before). Namely, the axis $\tilde{x}_n$, $n = m - d_0$, is directed toward z^1; if $s_1 \in \mathrm{st}(s_0)$, $\dim s_1 = s_0$, and $z^1 \in \mathrm{pr}(s_1)$, then the axes $\tilde{x}_{m-d_1}, \dots, \tilde{x}_{m-d_0-1}$ are parallel to the $(d_1 - d_0 - 1)$-dimensional space tangent to the stratum $\mathrm{pr}(s_1)$ at z^1. The remaining axes $\tilde{x}_1, \dots, \tilde{x}_{m-d_1-1}$ are orthogonal to $\mathrm{pr}(s_1)$ at z_1. We write the coordinate transformation in the form $x = J(z^1)\tilde{x}$ with orthogonal matrix $J(z^1)$. In what follows, we use the new coordinates, which will be given the notation $x_1, \dots, x_n$. The algebra $\mathfrak{S}(z^0)$ is spanned by the operators of multiplication $a(z^0; \cdot)I \in \mathcal{B}L_2(S^{n-1})$ and the functions $\mathbb{R} \ni \lambda \mapsto \mathfrak{A}(z^0; \lambda) \in \mathcal{B}L_2(S^{n-1})$; here

$$\mathfrak{A}(z^0; \lambda) = E_{\omega\to\varphi}(\lambda)^{-1} \Phi(z^0; J(z^1)\omega, 0) E_{\psi\to\omega}(\lambda), \tag{3.1.18}$$

while 0 is the row $(0, \dots, 0)$ containing d_0 elements (which is absent for $d_0 = 0$).

Denote by $\mathcal{L}(z^0, z^1)$ the algebra generated on $L_2(\mathbb{R}^{n-1})$ by the operators $a(z^0, z^1; \cdot)I$ and functions of the form

$$\mathbb{R} \ni \lambda \mapsto \mathfrak{D}(z^0, z^1; \lambda) = F^{-1}_{\eta \to x'} \Phi(z^0; J(z^1)(\eta, \lambda), 0) F_{y' \to \eta} \in \mathcal{B}L_2(\mathbb{R}^{n-1}); \quad (3.1.19)$$

as usual, $a(z^0, z^1; \cdot)$ denotes a zero degree homogeneous function $\mathbb{R}^{n-1} \setminus 0 \ni x' \mapsto a(z^0, z^1; x')$ and a norm in $\mathcal{L}(z^0, z^1)$ is introduced by

$$\|\mathfrak{D}(\cdot); \mathcal{L}\| = \sup\{\|\mathfrak{D}(\lambda); \mathcal{B}L_2(\mathbb{R}^{n-1})\|; \lambda \in \mathbb{R}\}.$$

The next assertion follows from Proposition 2.2.10.

Proposition 3.1.7 *Let $\mathfrak{A}$ and $\mathfrak{D}$ be the same as in (3.1.18) and (3.1.19). Then, the mapping*

$$a(z^0; \cdot) \mapsto a(z^0, z^1; \cdot), \quad \mathfrak{A}(z^0; \cdot) \mapsto \mathfrak{D}(z^0, z^1; \cdot),$$

defined for generators of the algebra $\mathfrak{S}(z^0)$ extends to an isomorphism

$$\mathfrak{S}(z^0)_{z^1} \simeq \mathcal{L}(z^0, z^1),$$

where $\mathfrak{S}(z^0)_{z^1}$ is the local algebra $\mathfrak{S}(z^0)/\mathcal{J}_{z^1}$ at $z^1 \in S^{n-1}$.

4^0. **The dynasty of local algebras.** In this section, with every (complete) chain of strata $s_0, \ldots, s_k$ of the partition $\mathcal{T}$ of $\mathcal{M}$, we associate the dynasty of $k+1$ generations of some local algebras. These algebras arise under iterations of the localization procedure. In the next section, any irreducible representation π of $\mathcal{A}$ (up to equivalence) with $\dim \pi > 1$ will be implemented in terms of such local algebras. The successive generations consist of algebras of the same type, however, increasingly simple ones: the operators of the algebras act in the spaces $L_2(\mathbb{R}^p)$ or $L_2(S^q)$ with certain p and q; as a generation is succeeded, the p and q decrease. The last generation of a dynasty is degenerate: localization of the algebras of the last generation provides only commutative algebras.

We apply the localization principle (Proposition 1.3.24) to $\mathcal{A}$ with the localizing algebra $C(\mathcal{M})$ and $J = \mathcal{K}L_2(\mathcal{M})$. The following three types of local algebras $\mathcal{A}_{z^0}$ can arise: 1) if $z^0 \in \mathcal{M} \setminus |\mathcal{T}|$, then $\mathcal{A}_{z^0}$ is commutative; 2) if $z^0 \in s_0$, where s_0 is a stratum of $\mathcal{T}$ with $\dim s_0 = d_0 > 0$, then the algebra $\mathcal{A}_{z^0}$ is generated by the functions

$$S^{d_0-1} \ni \theta^0 \mapsto a(z^0; \cdot) A(z^0; \theta^0) \in \mathcal{B}L_2(\mathbb{R}^{m-d_0})$$

(the notations are explained in 1^0); 3) if $z^0 \in s_0$ and $d_0 = \dim s_0 = 0$, then $\mathcal{A}_{z^0}$ is spanned by the functions

$$\mathbb{R} \ni \lambda \mapsto a(z^0; \cdot)\mathfrak{A}(z^0; \lambda) \in \mathcal{B}L_2(S^{m-1}).$$

First, we consider case 2). As in 1^0, by $\mathcal{A}(z^0;\theta^0)$, we denote the algebra generated on $L_2(\mathbb{R}^{m-d_0})$ by the operators of the form $a(z^0;\cdot)A(z^0;\theta^0)$ (for a fixed θ^0). The algebra $\mathcal{A}(z^0;\theta^0)$ is irreducible. The equality

$$\widehat{\mathcal{A}}_{z^0} = \bigcup_{\theta^0 \in S^{d_0-1}} \widehat{\mathcal{A}}(z^0;\theta^0) \tag{3.1.20}$$

is valid (compared with Proposition 2.3.4). Now, we apply Proposition 3.1.6. The local algebras $\mathcal{A}(z^0;\theta^0)_z$ in (i) and (iii) are commutative (see items (i)–(v) before Proposition 3.1.6); we postpone for a while their discussion and turn to items (ii), (iv), and (v). The algebra $\mathcal{A}(z^0;\theta^0)_z$ for $z=0$ (item (ii)) is isomorphic to the algebra $\mathfrak{S}(z^0)$ of functions that range in $\mathcal{B}L_2(S^{n-1})$, where $n=m-d_0$; see 1^0 and 3^0. In (iv), the local algebra $\mathcal{A}(z^0;\theta^0)_z$ is isomorphic to the algebra of the functions

$$S^{d_1-d_0-1} \ni \theta^1 \mapsto a(z^0,z;\cdot)A(z^0,z;0,\theta^1), \tag{3.1.21}$$

where

$$A(z^0,z;0,\theta^1) = F^{-1}_{\eta\to x^{(1)}}\Phi(z^0;J(z)(\eta,\theta^1),0)F_{y^{(1)}\to\eta} \in \mathcal{B}L_2(\mathbb{R}^{m-d_1}). \tag{3.1.22}$$

We pass on to item (v). Denote by $\hat{u}(y^{(1)},\xi^{(2)}) = F_{y^{(2)}\to\xi^{(2)}}u(y^{(1)},y^{(2)})$ the partial Fourier transform of u. Then,

$$(B(z^0,z;\theta^0)u)\hat{}(x^{(1)},\xi^{(2)}) = F^{-1}_{\xi^{(1)}\to x^{(1)}}\Phi(z^0;J(z)(\xi^{(1)},\xi^{(2)}),\theta^0)F_{y^{(1)}\to\xi^{(1)}}\hat{u}(y^{(1)},\xi^{(2)}).$$

We set $p=(|\xi^{(2)}|^2+1)^{1/2}$, $X=x^{(1)}p$, $Y=y^{(1)}p$ and obtain

$$\begin{aligned} &(B(z^0,z;\theta^0)u)\hat{}(Xp^{-1},\xi^{(2)}) \\ &= (2\pi)^{-(m-d_0)}\int_{\mathbb{R}^{m-d_1}} \exp(iXp^{-1}\xi^{(1)})\Phi(z^0;J(z)(\xi^{(1)},\xi^{(2)}),\theta^0)\,d\xi^{(1)} \\ &\quad\times \int_{\mathbb{R}^{m-d_1}} \exp(-i\xi^{(1)}Yp^{-1})\hat{u}(Yp^{-1},\xi^{(2)})p^{-1}\,dY. \end{aligned} \tag{3.1.23}$$

We write x, y ($\in\mathbb{R}^{m-d_1}$) instead of X, Y, and $v(y)$ and $w(x)$ instead of $u(Yp^{-1},\xi^{(2)})$ and $(B(z^0,z;\theta^0)u)\hat{}(Xp^{-1},\xi^{(2)})$. Then, (3.1.23) takes the form

$$w(x) = F^{-1}_{\eta\to x}\Phi(z^0;J(z)(\eta,\xi^{(2)}/p),\theta^0/p)F_{y\to\eta}v(y).$$

Thus, the algebra $\mathcal{A}(z^0, \theta^0)_z$ is isomorphic to the algebra of the functions

$$\mathbb{R}^{d_1-d_0} \ni \xi^{(2)} \mapsto a(z^0, z; x/|x|) \times \tag{3.1.24}$$
$$\times F^{-1}_{\eta\to x}\Phi(z^0; J(z)(\eta, \xi^{(2)}/p), \theta^0/p)F_{y\to\eta}v(y) \in \mathcal{B}L_2(\mathbb{R}^{m-d_1}).$$

Denote by $\mathcal{A}(z^0, z; \theta^0, \theta^1)$ the algebra generated on $L_2(\mathbb{R}^{m-d_1})$ by operators of the form

$$a(z^0, z; x/|x|)F^{-1}_{\eta\to x}\Phi(z^0; J(z)(\eta, \theta^1), \theta^0)F_{y\to\eta}$$

for fixed $\theta^1 \in \mathbb{R}^{d_1-d_0}$ and $\theta^0 \in \mathbb{R}^{d_0}$ such that $\theta^0 \neq 0$ and $|\theta^0|^2+|\theta^1|^2 = 1$. From (3.1.24), it follows that

$$\widehat{\mathcal{A}}(z_0; \theta_0)_z = \bigcup_{\theta^1\in\mathbb{R}^{d_1-d_0}} \widehat{\mathcal{A}}(z^0, z; \theta^0/(1 + |\theta^1|^2)^{1/2}, \theta^1/(1 + |\theta^1|^2)^{1/2}) \tag{3.1.25}$$

(compared with Proposition 2.3.4).

We turn to case 3). Let z^0 be a point of a zero-dimensional stratum s_0. As the local algebra $\mathcal{A}_{z^0}$, there arises the algebra $\mathfrak{S}(z^0)$ for $n = m - 1$, see 1^0 and 3^0; remind that such an algebra had already arisen in 4^0 with $d_0 = \dim s_0 > 0$ as the local algebra $\mathcal{A}(z^0; \theta^0)_z$ for $z = 0$. We assume that $\mathcal{A}(z^0; \theta^0)$ and $\mathfrak{S}(z^0)$ (for $d_0 \geq 0$) form the first generation of local algebras. (The commutative algebras will be considered in their own right.)

We are going to make up the second generation of local algebras. In the generation, we first include the algebras $\mathcal{A}(z^0, z^1; \theta^0, \theta^1)$ introduced after formula (3.1.24) (we changed the notation z for z^1). To clarify the contribution of the algebras $\mathfrak{S}(z^0)$ in the second generation, we turn to Proposition 3.1.7. If $z^1 \in S^{n-1} \setminus |\mathcal{T}(s_0)|$, $n = m - d_0$, then the algebra $\mathfrak{S}(z^0)_{z^1}$ is isomorphic to the algebra $C(S^{n-1})$ (of the functions $S^{n-1} \ni \omega \mapsto \Phi(z^0; J(z^1)\omega, 0)$, see (3.1.19)). If z^1 belongs to a stratum $\mathrm{pr}(s_1)$ of the partition $|\mathcal{T}(s_0)|$, then the algebra $\mathfrak{S}(z^0)_{z^1}$ is isomorphic to the algebra of operator-valued functions defined by (3.1.21) and (3.1.22) with $z = z^1$. Therefore, to the second generation, $\mathfrak{S}(z^0)$ gives the algebras $\mathcal{A}(z^0, z^1; 0, \theta^1)$. In addition, under localization of $\mathcal{A}(z^0, z^1; \theta^0, \theta^1)$, the algebras $\mathfrak{S}(z^0, z^1)$ appear (as local ones at the origin) spanned by functions of the form

$$\mathbb{R} \ni \lambda \mapsto a(z_0, z^1; \varphi)E_{\omega\to\varphi}(\lambda)^{-1}\Phi(z^0; J(z^1)(\omega, 0), 0)E_{\psi\to\omega}(\lambda) \in \mathcal{B}L_2(S^{m-d_1}).$$

The algebras $\mathfrak{S}(z^0, z^1)$ also have to be included in the second generation.

Before listing the rest of the generations, we will simplify the notation. Let $s_0, \dots, s_k$ be a chain of strata. Let us write the stratum $p(s_1)$ of the partition $\mathcal{T}(s_0)$ in the form $s_1(s_0)$; by $s_2(s_0, s_1)$, we denote the stratum of $\mathcal{T}(s_0, s_1)$ generated by s_2; the notations $s_3(s_0, \dots, s_2), \dots, s_k(s_0, \dots, s_{k-1})$ have a similar meaning.

Applying the above localization procedure to the second and succeeding generations, with a chain $s_0, \dots, s_k$, we associate the dynasty spanning $k + 1$ generations, where the j-

th generation consists of the algebras $\mathcal{A}(z^0, \dots, z^{j-1}; \theta^0, \dots, \theta^{j-1})$ and $\mathfrak{S}(z^0, \dots, z^{j-1})$ with $j \le k+1$,

$$\mathcal{A}(z^0, \dots, z^{j-1}; \theta^0, \dots, \theta^{j-1}) \subset L_2(\mathbb{R}^{m-d_{j-1}}),\ z^0 \in s_0,\ \dots,\ z^{j-1} \in s_{j-1}(s_0, \dots, s_{j-2}),$$
$$\theta^j \in \mathbb{R}^{q_j},\ |\theta^0|^2 + \dots + |\theta^{j-1}|^2 = 1,\ q_0 + \dots + q_{j-1} = d_{j-1} = \dim s_{j-1},$$
$$\mathfrak{S}(z^0, \dots, z^{j-1}) \supset C_0(\mathbb{R}) \otimes \mathcal{K}L_2(S^{m-d_{j-1}-1}).$$

The algebra $\mathcal{A}(z^0, \dots, z^{j-1}; \theta^0, \dots, \theta^{j-1})$ is generated by the operators of multiplication $a(z^0, \dots, z^{j-1}; \cdot)$ and the operators $A(z^0, \dots, z^{j-1}; \theta^0, \dots, \theta^{j-1})$ (similar types of operators appeared in (3.1.2) and (3.1.4)). The algebra $\mathfrak{S}(z^0, \dots, z^{j-1})$ is spanned by the functions

$$\mathbb{R} \ni \lambda \mapsto a(z^0, \dots, z^{j-1}; \cdot)\mathfrak{A}(z^0, \dots, z^{j-1}; \lambda) \in \mathcal{B}L_2(S^{m-d_{j-1}-1})$$

(see (3.1.5)). If the chain $s_0, \dots, s_k$ is complete (see 3.1.1), then localization of the algebras in the $(k+1)$-th generation provides only commutative algebras.

5^0. **A list of irreducible representations π of the algebra $\mathcal{A}$ with** $\dim \pi > 1$. We will show that any irreducible representation π of the algebra $\mathcal{A}$ such that $\dim \pi > 1$ is equivalent to either one of the representations (3.1.6), (3.1.7) or the identity representation.

Recall that the algebra $\mathcal{A}$ is irreducible and contains the ideal $\mathcal{K}L_2(\mathcal{M})$. Thus, the identity representation of $\mathcal{A}$ is irreducible. We start searching for other representations by localization of $\mathcal{A}$ (with Proposition 1.3.24; see 1^0). This implies, in particular, that all irreducible representations that are not equivalent to the identity one turn out to be representations of the quotient algebra $\mathcal{A}/\mathcal{K}L_2(\mathcal{M})$. The local algebras at the points $z \in \mathcal{M}\setminus|\mathcal{T}|$ are commutative; their irreducible representations are one-dimensional. Therefore, the representations $\pi \in \widehat{\mathcal{A}}$ with $\dim \pi > 1$ can be found only under localization along chains of strata.

Let $s_0, \dots, s_k$ be a complete chain of strata, $\dim s_j = d_j$. If $d_0 > 0$ and $z^0 \in s_0$, then formula (3.1.20) is valid. The algebras $\mathcal{A}(z^0; \theta^0)$ are irreducible. Hence, mappings of the form (3.1.6) for $j = 0$ extend to irreducible representations $\pi(z^0; \theta^0)$ of the algebra $\mathcal{A}/\mathcal{K}L_2(\mathcal{M})$. Such representations for distinct pairs (z^0, θ^0) are nonequivalent (they have distinct kernels). If $d_0 = 0$, then the local algebra $\mathcal{A}_{z^0}$ is isomorphic to the algebra $\mathfrak{S}(z^0)$ (see 3^0) and the equality (3.1.17) holds. The mappings (3.1.7) for $j = 0$ extend to the representations $\pi(z^0; \lambda_0)$ that arise in (3.1.17) as representations of the ideal J. These representations are irreducible and pairwise nonequivalent. If the partition $\mathcal{T}$ contains no zero-dimensional strata, then $\mathfrak{S}(z^0)$ appears all the same under localization of $\mathcal{A}(z^0; \theta^0)$, so there appear the representations (3.1.7) as well for $j = 0$.

From the description of localization for the algebras $\mathcal{A}(z^0; \theta^0)$ and $\mathfrak{S}(z^0)$, it follows that new irreducible representations of dimension > 1 can occur only in the spectrum of local algebras of succeeding generations. The same argument as applied to the algebras $\mathcal{A}(z^0, z^1; \theta^0, \theta^1)$ and $\mathfrak{S}(z^0, z^1)$ provides the representations $\pi(z^0, z^1; \theta^0, \theta^1)$

and $\pi(z^0, z^1; \lambda_1)$. Going from a generation to the succeeding generation in a dynasty of local algebras, we find all representations (3.1.6), (3.1.7). According to the localization principle, such a list contains all (up to equivalence) irreducible representations of the algebra $\mathcal{A}$ of dimension >1.

6^0. **The one-dimensional representations of the algebra** $\mathcal{A}$. We list all one-dimensional representations arising at various stages of the localization procedure. It is necessary to take their origin into account when describing the topology on the spectrum of $\mathcal{A}$.

Assume that $x \in \mathcal{M} \setminus |\mathcal{T}|$. The local algebra $\mathcal{A}_x$ is commutative; any representation of $\mathcal{A}_x$ is included in (3.1.8); in that case $p(x) = x$ and the coefficients a are continuous at the point x.

We now suppose that s_0 is a stratum of the partition $\mathcal{T}$, $d_0 = \dim s_0 > 0$, and $z^0 = s_0$. Let $T(z^0, s_0)$ $(T^*(z^0, s_0))$ be the tangent (cotangent) space to $\mathcal{M}$ that is orthogonal to s_0. Let also $z \in T(z^0, s_0)$, $z \neq 0$, and $\varphi = z/|z| \notin |\mathcal{T}(s_0)|$. Finally, we denote by ω a unit vector in $T^*(z^0, s_0)$. According to Proposition 3.1.6 and formulas (3.1.9), the mapping

$$\begin{aligned} \pi(z^0; \varphi, \omega, 0) &: a(z^0; \cdot) \mapsto a(z^0; \varphi), \\ \pi(z^0; \varphi, \omega, 0) &: A(z^0; \theta^0) \mapsto \Phi(z^0; \omega, 0) \end{aligned} \tag{3.1.26}$$

defines a one-dimensional representation of the algebra $\mathcal{A}(z^0; \theta^0)$.

Suppose that z is a point in the infinitely distant sphere of $T(z^0, s_0)$, φ is the unit vector directed toward z, and $\varphi \notin |\mathcal{T}(s_0)|$. Then, by virtue of (3.1.11), the mapping

$$\begin{aligned} a(z^0; \cdot) &\mapsto a(z^0; \varphi), \\ A(z^0; \theta^0) &\mapsto \Phi(z^0; \eta, \theta^0) \end{aligned} \tag{3.1.27}$$

defines a one-dimensional representation of $\mathcal{A}(z^0; \theta^0)$ for any $\eta \in T^*(z^0, s_0)$.

Considering that the function $\xi \mapsto \Phi(z^0; \xi)$ is homogeneous, we can combine (3.1.26) and (3.1.27). As a result, we obtain that the mapping

$$\begin{aligned} \pi(z^0; \varphi, \xi) &: a(z^0; \cdot) \mapsto a(z^0; \varphi), \\ \pi(z^0; \varphi, \xi) &: A(z^0; \theta^0) \mapsto \Phi(z^0; \xi) \end{aligned} \tag{3.1.28}$$

defines a one-dimensional representation of $\mathcal{A}(z^0; \theta^0)$ for any $\xi \in S^*(\mathcal{M})_{z^0}$ and every unit vector $\varphi \in T(z^0, s_0)$. Representations of the form (3.1.28) are contained among those in (3.1.8). Note that $\pi(z^0; \varphi, \omega, 0)$ in (3.1.26) are representations of the algebra $\mathfrak{S}(z^0)$ as well; see Theorem 2.2.11 and formulas (3.1.18). In particular, $\pi(z^0; \varphi, \omega, 0)$ for $\varphi\omega = 0$ are representations of the algebra $\mathfrak{S}^\lambda(z^0)$ spanned by multiplications by $a(z^0; \cdot)I$ and by operators (3.1.3) with a fixed $\lambda \in \mathbb{R}$. In the case $d_0 = \dim s_0 = 0$, the representations (3.1.27) are absent, and formulas (3.1.26) and (3.1.28) coincide and define representations of $\mathfrak{S}(z^0)$.

In essence, listing the one-dimensional representations under further localization along chains of strata reduces to the argument just given above. As a result, all representations in (3.1.8) appear as representations of $\mathcal{A}(z^0, \dots, z^j; \theta^0, \dots, \theta^j)$. From the localization principle, it follows that there are no other one-dimensional representations of the algebra $\mathcal{A}$. □

3.2 The Spectral Topology of Algebra $\mathcal{A}$

3.2.1 Description of the Jacobson Topology (Formulation of the Theorem)

1^0. **Parametrizing the spectrum.** We first introduce a set parametrizing the spectrum of $\mathcal{A}$. Assume that $s_0, \dots, s_k$ is a complete chain of strata of the partition $\mathcal{T}$ of $\mathcal{M}$, $z^0, \dots, z^k$ are variables related to the chain: z^0 runs over $Z_0 = s_0$, and z^j runs over the stratum Z_j of the partition $\mathcal{T}(s_0, \dots, s_{j-1})$, $1 \leq j \leq k$. The stratum s_0 is a submanifold of the manifold $\mathcal{M}$, while the stratum s_j, $1 \leq j \leq k$, is a submanifold of the corresponding sphere. We set $\mathcal{Z}_j = Z_0 \times \cdots \times Z_j$ for $j = 0, \dots, k$. A point $(z^0, \dots, z^j)$ in $\mathcal{Z}_j$ indicates the direction in $\mathcal{M}$ of approach to the point $z^0 \in s_0$ and, by doing so, defines the d_j-dimensional subspace $T(s_j; z^0, \dots, z^j)$ of $T(\mathcal{M})_{z^0}$ tangent to the stratum s_j. As before (Theorem 3.1.4), denote by $S^*(s_j; z^0, \dots, z^j)$ the unit sphere in the cotangent space $T^*(s; z^0, \dots, z^j)$. Let $S^*(\mathcal{Z}_j)$ stand for the induced bundle with base $\mathcal{Z}_j$ and fiber $S^*(s_j; z^0, \dots, z^j)$ over a point $(z^0, \dots, z^j)$. We introduce the disjoint union Σ of the sets

$$S^*(\mathfrak{C}), \quad \bigcup_{(\mathcal{T})} \bigcup_{j=0}^{k} S^*(\mathcal{Z}_j), \quad \bigcup_{(\mathcal{T})} \bigcup_{j=0}^{k} \mathcal{Z}_j \times \mathbb{R},$$

where $\cup_{(\mathcal{T})}$ is taken over all complete chains of strata in $\mathcal{T}$. From Theorem 3.1.4, it follows that the spectrum of the quotient algebra $\mathcal{A}/\mathcal{K}L_2(\mathcal{M})$ can be parametrized by the points of Σ. Assume that point e corresponds to the identity representation of $\mathcal{A}$. Then the spectrum of $\mathcal{A}$ is parametrized by the points of the set $e \cup \Sigma$.

In what follows, we compare representations obtained by localization along distinct chains of strata. Let us explain how to do that. Assume that $s_{l^0}, \dots, s^{l^q}$ is a chain consisting of some strata of an original chain $s_0, \dots, s_j$, $0 \leq l^0 < \cdots < l^q = j$. We say that the new chain is a subchain of the original one. We now describe which of the one-type representations $\pi(z^0, \dots, z^j; \theta^0, \dots, \theta^j)$ and $\pi(\zeta^0, \dots, \zeta^q; \tau^0, \dots, \tau^q)$ or $\pi(z^0, \dots, z^j; \lambda)$ and $\pi(\zeta^0, \dots, \zeta^q; \mu)$ must be considered as close ones; here, $\pi(z^0, \dots, z^j; \cdot)$ and $\pi(\zeta^0, \dots, \zeta^q; \cdot)$ are obtained by localization along the chain and subchain, respectively.

Let a point $\zeta^0 \in s_{l^0}$ belong to a sufficiently small tubular neighborhood of the stratum s_0. Then, the "cylindrical coordinates" (z^0, r^0, α^0) are defined for the point, where $z^0 \in s_0 = Z_0$, $r^0 \in \bar{\mathbb{R}}_+$, and $\alpha^0 \in S^{m-d_0-1}$. For the points α^0 in a tubular neighborhood of the stratum $\mathcal{Z}_1$, the coordinates (z^1, r^1, α^1) are defined and so on.

For a couple (ζ^0, ζ^1), we obtain a row $(z^0, r^0, \dots, z^p, r^p, \zeta^1)$, where $p = l^0 - 1$ and $\zeta^1 \in S^{m-d_p-1}$. Continuing the procedure, we rewrite the row $(\zeta^0, \dots, \zeta^q)$ in the form $(z^0, r^0, \dots, z^{j-1}, r^{j-1}, r^j)$, where $j = l^q$, $z^j = \zeta^q \in S^{m-d_{j-1}-1}$. Collections $(z_0^0, \dots, z_0^j)$ and $\zeta^0, \dots, \zeta^q) \equiv (z^0, r^0, \dots, z^{j-1}, r^{j-1}, z^j)$ are said to be close if the non-negative numbers $r^0, \dots, r^{j-1}$ are sufficiently small, while z_0^ν and z^ν are close as points of the same stratum Z_ν $(\nu = 0, \dots j)$. Furthermore, points $(\theta^0, \dots, \theta^j)$ and $(\tau^0, \dots, \tau^q)$ with $|\theta^0|^2 + \dots + |\theta^j|^2 = 1$ and $|\tau^0|^2 + \dots + |\tau^q|^2 = 1$ belong to d_j-dimensional spheres; for close $(z_0^0, \dots, z_0^j)$ and $(\zeta^0, \dots, \zeta^q)$, the spheres can be identified in a natural way with a "standard" sphere (local trivialization). Representations $\pi(z^0, \dots, z^j; \theta^0, \dots, \theta^j)$ and $\pi(\zeta^0, \dots, \zeta^q; \tau^0, \dots, \tau^q)$ are said to be close if the collections $(z_0^0, \dots, z_0^j)$ and $(\zeta^0, \dots, \zeta^q)$ are close (in the sense mentioned above), while the points $(\theta^0, \dots, \theta^j)$ and $(\tau^0, \dots, \tau^q)$ differ little in the standard sphere S^{d_j}. The proximity of representations $\pi(z^0, \dots, z^j; \lambda)$ and $\pi(\zeta^0, \dots, \zeta^q; \mu)$ can be described in a similar way.

2^0. **Fundamental system of neighborhoods.** We introduce a topology on the set $e \cup \Sigma$ by indicating typical neighborhoods of the points that form a fundamental system. (Recall that the set $e \cup \Sigma$ parametrizes the spectrum of the algebra $\mathcal{A}$.) By $\mathcal{U}(x; T)$, we denote an arbitrary neighborhood of a point x in a topological space T.

Neighborhoods of the points in $S^*(\mathcal{Z}_j)$. Let $(z^0, \dots, z^j; \theta^0, \dots, \theta^j) \in S^*(\mathcal{Z}_j)$, $0 \leq j \leq k$, where $\theta^j \in \mathbb{R}^{d_j - d_{j-1}}$, $d_{-1} = 0$, and $|\theta^0|^2 + \dots + |\theta^j|^2 = 1$. If $\theta^0 \neq 0$, then a typical neighborhood $\mathcal{U}((z^0, \dots, z^j; \theta^0, \dots, \theta^j); \Sigma)$ consists of the sets

$$\mathcal{U}((z^0, \dots, z^q); \mathcal{Z}_q) \times \mathcal{U}(\Theta^q; S^{d_q - 1}), \tag{3.2.1}$$

for $q = 0, \dots, j$, where $\Theta^q = (\theta^0, \dots, \theta^q)/(|\theta^0|^2 + \dots + |\theta^q|^2)^{1/2} \in S^{d_q - 1}$, and the neighborhood contains all sufficiently close representations of the same types obtained by localization along subchains. (We write the part of $S^*(\mathcal{Z}_j)$ over a "sufficiently small" neighborhood $\mathcal{U}((z^0, \dots, z^q); \mathcal{Z}_q)$ as the product (3.2.1).)

Suppose now that $\theta^0 = 0, \dots, \theta^i = 0, \theta^{i+1} \neq 0$. A neighborhood $\mathcal{U}((z^0, \dots, z^j; \theta^0, \dots, \theta^j); \Sigma)$ consists of the sets (3.2.1) for $q = i + 1, \dots, j$, the products

$$\mathcal{U}((z^0, \dots, z^p); \mathcal{Z}_p) \times S^{d_p - 1}, \quad p = 0, \dots, i, \tag{3.2.2}$$

$$\mathcal{U}((z^0, \dots, z^r); \mathcal{Z}_r) \times \mathbb{R}, \quad r = 0, \dots, i - 1, \tag{3.2.3}$$

and the product

$$\mathcal{U}((z^0, \dots, z^i); \mathcal{Z}_i) \times \{\lambda \in \mathbb{R} : \lambda \lessgtr N\}$$

in the case $z^{i+1}\theta^{i+1} \lessgtr 0$ or the product

$$\mathcal{U}((z^0, \dots, z^i); \mathcal{Z}_i) \times \mathbb{R}$$

for $z^{i+1}\theta^{i+1} = 0$; here, N is an arbitrary real and $z^{i+1}\theta^{i+1}$ is the projection of θ^{i+1} onto the unit vector z^{i+1}. (The vector $z^{i+1} \in \mathbb{R}^{m-d_i}$ is directed from the coordinate origin to a point in Z_{i+1}, while θ^{i+1} lies in the space spanned by z^{i+1} and by the space tangent to the stratum Z_{i+1} at the point z^{i+1}.)

Neighborhoods of the points in $\mathcal{Z}_j \times \mathbb{R}$. A typical neighborhood $\mathcal{U}((z^0, \dots, z^j; \lambda); \Sigma)$ is the union of sets (3.2.2) (3.2.3) for $i = j$, the product $\mathcal{U}((z^0, \dots, z^j); \mathcal{Z}_j) \times \mathcal{U}(\lambda; \mathbb{R})$, and close representations obtained by localization along subchains.

Neighborhoods of the points in $S^*(\mathfrak{C})$. Typical neighborhoods in Σ of the inner points of the manifold $S^*(\mathfrak{C})$ are the same as in the cospherical bundle $S^*(\mathcal{M} \setminus |\mathcal{T}|)$.

Consider the points $(z^0, \dots, z^j; \varphi^j, \xi) \in \mathcal{Z}_j \times (S^{m-d_j-1} \setminus |\mathcal{T}(s_0, \dots, s_j)|) \times S^{m-1}$ in the boundary of $S^*(\mathfrak{C})$. Let us write such a point in the form $(z^0, \dots, z^j; \varphi^j, J(z^0)(\xi^0, \theta^0))$, where $\xi^i = J(z^0, \dots, z^{i+1})(\xi^{i+1}, \theta^{i+1})$, $i = 0, \dots, j-1$; as before, $\theta^i \in \mathbb{R}^{d_i - d_{i-1}}$, $\xi^j \in \mathbb{R}^{m-d_j}$; the Jacobi matrices $J(z^0, \dots, z^{i+1})$ are defined by induction (see Sect. 3.1.2; there we have taken $J(z^0) = 1$ and written $J(z^1)$ instead of $J(z^0, z^1)$). In what follows, the points $(z^0, \dots, z^j; \varphi^j, \xi)$ are denoted by $(z^0, \dots, z^j; \varphi^j, \xi^j, \theta^0, \dots, \theta^j)$. From now on, we tacitly assume that a neighborhood, apart from explicitly listed representations, contains those of the same type obtained by localization along subchains and close to the listed representations.

If $\theta^0 \neq 0$, then a neighborhood in Σ of a point $(z^0, \dots, z^j; \varphi^j, \xi^j, \theta^0, \dots, \theta^j)$ is the union of the sets (3.2.1) for $q = 0, \dots, j$ and the product

$$\mathcal{U}((z^0, \dots, z^i); \mathcal{Z}_i) \times \mathcal{U}((\varphi^j, \xi); (S^{m-d_j-1} \setminus |\mathcal{T}(s_0, \dots, s_j)|) \times S^{m-1})$$

(in such a case, the second factor contains no points with $\theta^0 = 0$).

Assume now that $\theta^0, \dots, \theta^i = 0$, $\theta^{i+1} \neq 0$, $0 \leq i \leq 1$. Then, a typical neighborhood consists of the set (3.2.2), the set (3.2.1) for $q = i+1, \dots, j$, the set (3.2.3), and the product

$$\mathcal{U}((z^0, \dots, z^i); \mathcal{Z}_i) \times \{\lambda \in \mathbb{R} : \lambda \lesseqgtr N\}$$

in the case $z^{i+1}\theta^{i+1} \lesseqgtr 0$ or

$$\mathcal{U}((z^0, \dots, z^i); \mathcal{Z}_i) \times \mathbb{R}$$

in the case $z^{i+1}\theta^{i+1} = 0$.

Finally, suppose that $\theta^0 = 0, \dots, \theta^j = 0$. A neighborhood consists of the union of the sets (3.2.2) and (3.2.3) for $i = j$, the product $\mathcal{U}((z^0, \dots, z^i); \mathcal{Z}_i) \times \{\lambda \in \mathbb{R} : \lambda \lesseqgtr N\}$ in the case $\varphi^j \xi^j \lesseqgtr 0$, or $\mathcal{U}((z^0, \dots, z^i); \mathcal{Z}_i) \times \mathbb{R}$ in the case $\varphi^j \xi^j = 0$.

The topology defined on Σ by the above neighborhoods is inseparable; Σ is a T_0-space (at least one of any two points has a neighborhood not containing the other of the points).

Recall that the sets Σ and $(\mathcal{A}/\mathcal{K}L_2(\mathcal{M}))^\frown$ are connected by the bijection stated owing to Theorem 3.1.4).

Theorem 3.2.1 *The topology on* Σ *coincides with the Jacobson topology on the spectrum* $(\mathcal{A}/\mathcal{K}L_2(\mathcal{M}))^\wedge$.

The topology on the spectrum of $\mathcal{A}$ can be easily obtained from Theorem 3.2.1. Introduce a topology on $\Sigma \cup e$. Point e will be taken as an open set. A neighborhood in $\Sigma \cup e$ of any point $\pi \in \Sigma$ is, by definition, a neighborhood π in Σ supplemented with e.

Corollary 3.2.2 *The topology on* $\Sigma \cup e$ *coincides with the Jacobson topology on the spectrum* $\widehat{\mathcal{A}}$ *of the algebra* $\mathcal{A}$.

3.2.2 Proof of Theorem 3.2.1

Neighborhoods defining the topology on Σ were listed before Theorem 3.2.1. Moving along this list, we establish the coincidence of the topology on Σ (Σ-topology) with the Jacobson topology (J-topology) on the spectrum of $(\mathcal{A}/\mathcal{K}L_2(\mathcal{M}))^\wedge$. (The spectrum is parametrized by points of Σ.)

1^0. **Connections between representations of the algebra** $\mathcal{A}$. To verify Theorem 3.2.1, we need some relations between representations of $\mathcal{A}$. They can be seen from the localization procedure. For ease of reference, these relations are described in the following two lemmas. We use the notation of Theorem 3.1.4. Let $\theta^0 \in \mathbb{R}^{d_0}$, $\theta^i \in \mathbb{R}^{d_i - d_{i-1}}$ for $i \geq 1$. If $(\theta^0, \ldots, \theta^j) \neq 0$, then we let $\Theta^j = (\theta^0, \ldots, \theta^j)/(|\theta^1|^2 + \cdots + |\theta^j|^2)^{1/2}$.

Lemma 3.2.3

(1) *Suppose that* $(\theta^0, \ldots, \theta^{j-1}) \neq 0$. *Then for all* $A \in \mathcal{A}$, *the following inequality holds:*

$$\|\pi(z^0, \ldots, z^j; \Theta^j)A; \mathcal{B}L_2(\mathbb{R}^{m-d_j})\| \leq \|\pi(z^0, \ldots, z^{j-1}; \Theta^{j-1})A; \mathcal{B}L_2(\mathbb{R}^{m-d_{j-1}})\|.$$

(2) *Suppose that* $\theta^j \in \mathbb{R}^{d_j - d_{j-1}}$, $j \geq 1$, *and* $|\theta^j| = 1$. *Then, for all* $A \in \mathcal{A}$,

$$\|\pi(z^0, \ldots, z^j; 0, \ldots, \theta^j)A; \mathcal{B}L_2(\mathbb{R}^{m-d_j})\|$$
$$\leq \sup\{\|\pi(z^0, \ldots, z^{j-1}; \lambda)A; \mathcal{B}L_2(S^{m-d_{j-1}-1})\|, \lambda \in \mathbb{R}\}.$$

(3) *Let* $\Psi(z^0, \ldots, z^j; \xi^j, \theta^0, \ldots, \theta^j) = \Phi(z^0; J(z^0, z^1)(\xi^1, \theta^1), \theta^0)$, *where*

$$\xi^i = J(z^0, \ldots, z^{i+1})(\xi^{i+1}, \theta^{i+1})$$

for $i \leq j-1$ *(see explanation of notation before Theorem 3.2.1). Denote by* $I(z^0, \ldots, z^j; \theta^0, \ldots, \theta^j)$ *the ideal generated in the algebra* $\pi(z^0, \ldots, z^j; \theta^0, \ldots, \theta^j)$

$\mathcal{A}$ by the elements

$$F^{-1}_{\xi^j\to x}\Psi(z^1,\dots,z^j;\xi^j,\theta^0,\dots,\theta^j)F_{y\to\xi^j},$$

for which $\Psi(z^0,\dots,z^j;\xi^j;0,\dots,0)\equiv 0$. *Then, for all* $A\in\mathcal{A}$ *and* $\lambda\in\mathbb{R}$,

$$\|\pi(z^0,\dots z^j;\lambda)A;\mathcal{B}L_2(S^{m-d_j-1})\|\le\|[\pi(z^0,\dots,z^j;\theta^0,\dots,\theta^j)A]\|,$$

where the right-hand side is the norm of a residue class $[\cdot]$ *in the quotient algebra*

$$(\pi(z^0,\dots,z^j;\theta^0,\dots,\theta^j)\mathcal{A})/I(z^0,\dots,z^j;\theta^0,\dots,\theta^j).$$

Proof The representations $\pi(z^0,\dots,z^j;\Theta^j)$ of $\mathcal{A}$ arise in the localization procedure in the algebra $\pi(z^0,\dots,z^{j-1};\Theta^{j-1})\mathcal{A}$. Any morphism of C^*-algebras (considered as a morphism of involutive algebras) is continuous, and its norm does not exceed 1. This implies the first assertion of Lemma.

While localizing in $\mathfrak{S}(z^0,\dots,z^{j-1})$, the representations $\pi(z^0,\dots,z^j;0,\dots,\theta^j)$ appear (see the description of the dynasty of local algebras in Sect. 3.1.3). Hence, the second assertion of Lemma is valid.

For verification of the third assertion, one can reason in the same way as in the proof of Proposition 3.1.6 in case (iv). □

As in the list of neighborhoods before Theorem 3.2.1, the one-dimensional representations $\pi(z^0,\dots,z^j;\varphi^j,\xi)$ will also be denoted by $\pi(z^0,\dots,z^j;\varphi^j,\xi^j,\theta^0,\dots,\theta^j))$.

Lemma 3.2.4 *For all* $A\in\mathcal{A}$, *the following inequalities hold:*

$$|\pi(z^0,\dots,z^j;\varphi^j,\xi^j,\Theta^j)A|\le\|\pi(z^0,\dots,z^j;\Theta^j)A;\mathcal{B}L_2(\mathbb{R}^{m-d_j})\|, \tag{3.2.4}$$

where, as before, $\Theta^j=(\theta^0,\dots,\theta^j)$, $|\Theta^j|=1$;

$$|\pi(z^0,\dots,z^j;\varphi^j,\xi^j,0,\dots,0)A|\le\sup\{\|\pi(z^0,\dots,z^j;\lambda)A;\mathcal{B}L_2(S^{m-d_j-1})\|,\lambda\in\mathbb{R}\}. \tag{3.2.5}$$

Moreover,

$$|\pi(z^0,\dots,z^j;\varphi^j,\xi^j,0,\dots,0)A|\le\|\pi(z^0,\dots,z^j;\lambda)A;\mathcal{B}L_2(S^{m-d_j-1})\| \tag{3.2.6}$$

for any $\lambda\in\mathbb{R}$ *provided* $\varphi^j\xi^j=0$.

Proof The representations $\pi(z^0,\dots,z^j;\varphi^j,\xi^j,\Theta^j)$ of $\mathcal{A}$ appear in the localization procedure as representations of the algebra $\pi(z^0,\dots,z^j;\varphi^j,\xi^j,\Theta^j)\mathcal{A}$ (see the listing of one-dimensional representations of $\mathcal{A}$ in the proof of Theorem 3.1.4); this implies inequalities (3.2.4). The representations $\pi(z^0,\dots,z^j;\varphi^j,\xi^j,0,\dots,0)$ appear as representations of $\mathfrak{S}(z^0,\dots,z^j)$; thus, the formulas (3.2.5) are valid. Finally, we notice that if $\varphi^j\xi^j=0$, then the representations $\pi(z^0,\dots,z^j;\varphi^j,\xi^j,0,\dots,\lambda)$ are also representations of $\pi(z^0,\dots,z^j;\lambda)\mathfrak{S}(z^0,\dots,z^j)$ for any $\lambda\in\mathbb{R}$. Indeed, from Theorem 2.2.12, it follows that $\pi(z^0,\dots,z^j;\lambda)$ is a representation of the ideal in $\pi(z^0,\dots,z^j;\lambda)\mathfrak{S}(z^0,\dots,z^j)$ spanned by the coefficients $a(z^0,\dots,z^j;\cdot)$ vanishing on the set of all singularities of all the remaining coefficients $a(z^0,\dots,z^j;\cdot)$. Each representation of an ideal of any C^*-algebra extends uniquely to an irreducible representation of the whole algebra. Therefore, inequality (3.2.6) holds as well. □

2^0. **Neighborhoods of points in** $S^*(\mathcal{Z}_j)$. Assume that $s_0,\dots,s_k$ is a chain of strata of the partition $\mathcal{T}$, i.e., $s_i\in \operatorname{st}(s_j)$ for $i>j$, $\dim s_j=d_j$, and $d_k\le m-1$, where $m=\dim\mathcal{M}$.

Lemma 3.2.5 *Let $(z_0^0,\dots,z_0^j;\psi_0^j)$ be an arbitrary point in the product $\mathcal{Z}_j\times(S^{m-d_j-1}\setminus|\mathcal{T}(s_0,\dots,s_j)|)$; here, as before, $\mathcal{Z}_j=Z_0\times\cdots\times Z_j$, $0\le j\le k$. Then, there exists a function $a\in\mathfrak{M}(\mathcal{T})$ subject to the following conditions:*

1. *$a(z_0^0,\dots,z_0^j;\psi_0^j)=1$, $a(z^0,\dots,z^j;\psi^j)=0$ outside a small neighborhood of the point $(z_0^0,\dots,z_0^j;\psi_0^j)$ on $\mathcal{Z}_j\times(S^{m-d_j-1}\setminus|\mathcal{T}(s_0,\dots,s_j)|)$.*
2. *$a(z_0^0,\dots,z_0^q;\cdot)=0$ outside a neighborhood of z_0^{q+1} on S^{m-d_q-1} for $0\le q\le j-1$, and $a(\cdot)=0$ outside a neighborhood of z_0^0 on $\mathcal{M}$.*
3. *$a(z^0,\dots,z^j;\cdot)\in C(S^{m-d_j-1})$, and $a(z^0,\dots,z^p;\cdot)\equiv 0$ for $j+1\le p\le k$.*

Proof Let $\chi\in C(S^{m-d_j-1})$ be such that $\chi(\psi_0^j)=1$ and $\chi=0$ outside a neighborhood of ψ_0^j. We regard $(z^j,x)\in Z_j\times\mathbb{R}^{m-d_j-1}$ as cylindrical coordinates in a tubular neighborhood of the stratum Z_j on $S^{m-d_{j-1}-1}$, and we introduce the function $(z^j,x)\mapsto f_j(z^j,x)=\chi(x/|x|)$. Let $\chi_j\in C(S^{m-d_{j-1}-1})$, $\chi_j(z_0^j)=1$, and $\chi_j=0$ outside a neighborhood of z_0^j on $S^{m-d_{j-1}-1}$. Extending $\chi_j f_j$ by zero outside the support of χ_j, we obtain a function denned on $S^{m-d_{j-1}-1}\setminus|\mathcal{T}(s_0,\dots,s_{j-1})|$.

We give the roles of χ and $S^{m-d_{j-1}-1}$ to the function $\chi_j f_j$ and the sphere $S^{m-d_{j-2}-1}$. Repeating the argument, we define the function $\chi_{j-1}f_{j-1}$ on $S^{m-d_{j-2}-1}\setminus|\mathcal{T}(s_0,\dots,s_{j-2})|$. Continuing this process, we arrive at an element $a=\chi_0 f_0\in\mathfrak{M}(\mathcal{T})$. It follows from the construction that the function a satisfies conditions 1 and 2. Condition 3 is an obvious consequence of 1. □

Fix an arbitrary point $(z_0^0,\dots,z_0^j;\theta_0^0,\dots,\theta_0^j)$, $0\le j\le k$. We consider several cases separately.

(a) The case $\theta_0^0 \neq 0$. We show that every neighborhood $\mathcal{U}(\pi_0; \Sigma)$ of the point $\pi_0 = \pi(z_0^0, \ldots, z_0^j; \theta_0^0, \ldots, \theta_0^j)$ in hte Σ-topology contains some neighborhood $\mathcal{U}(\pi_0; J)$ of this point in the J-topology. For an arbitrary element $A \in \mathcal{A}$, the set $U_A = \{\pi \in \widehat{\mathcal{A}} : \|\pi A\| > 1\}$ is open in the J-topology. Our goal is to determine an A such that $\pi_0 \in U_A \subset \mathcal{U}(\pi_0; \Sigma)$. As in Lemma 3.2.3, let

$$\Psi(z^0, \ldots, z^j; \xi^j, \theta^0, \ldots, \theta^j) = \Phi(z^0; J(z^0, z^1)(\xi^1, \theta^1), \theta^0), \tag{3.2.7}$$

where $\xi^i = J(z^0, \ldots, z^{i+1})(\xi^{i+1}, \theta^{i+1})$ for $i \leq j - 1$. We choose a zero degree homogeneous function $\Phi(z^0; \cdot)$ on the fiber $T^*(\mathcal{M})_{z^0} \setminus 0$ of nonzero cotangent vectors such that $\Psi(z_0^0, \ldots, z_0^j; \xi_0^j, \theta_0^0, \ldots, \theta_0^j) > 1$ and $\Psi = 0$ outside a small neighborhood of the point $\zeta_0/|\zeta_0|$ on $S^*(\mathcal{M})_{z^0}$; here $\zeta^0 = (\xi_0^1, \theta_0^0, \ldots, \theta_0^j)$ and ξ_0^j is an arbitrary point in $S^{m-d_j-1} \subset S^*(\mathcal{M})_{z^0}$. We assume that this neighborhood does not intersect the set of points such that $\theta^0 = 0$. Furthermore, let P be a ψdo in the algebra $\bar{\Psi}^0(\mathcal{M})$ with principal symbol coinciding with the previously chosen function $\Phi(z^0; \cdot)$ on the fiber $T^*(\mathcal{M})_{z^0} \setminus 0$. Finally, suppose that $a \in \mathfrak{M}(\mathcal{T}, \mathcal{M})$ is the function in Lemma 3.2.5, and let $A = aP$. Then, for some $\xi_0 \in S^*(\mathcal{M})_{z^0}$, the relation

$$\pi(z_0^0, \ldots, z_0^j; \psi^0, \xi_0)A = a(z_0^0, \ldots, z_0^j; \psi_0^j)\Psi(z_0^0, \ldots, z_0^j; \xi_0^j, \theta_0^j, \ldots, \theta_0^j) > 1$$

holds. We write $\pi(z_0^0, \ldots, z_0^j; \psi^0, \xi_0)$ as $\pi(z_0^0, \ldots, z_0^j; \psi^0, \xi_0^j, \theta_0^0, \ldots, \theta_0^j)$ and use the inequality (3.2.4), obtaining the inclusion $\pi_0 \in U_A$. The representation $\pi(z^0, \ldots, z^q; \Theta^q)$ annihilates the operator A if $(z^0, \ldots, z^q; \Theta^q)$ does not belong to small neighborhood $\mathcal{U}((z_0^0, \ldots, z_0^q; \Theta_0^q); \Sigma), q = 0, \ldots, j$ (see (3.2.1). The representations $\pi(z^0, \ldots, z_0^p; \Theta^p)$ with $p \geq j + 1$ do not fall in U_A because of condition 3, which is satisfied by function a in Lemma 3.2.5.

The set U_A does not contain representations of the type $\pi(z^0, \ldots, z^q; \lambda)$, $1 \leq q \leq k$. Indeed, there are no points in the support of Ψ for which $\theta^0 = 0$. This means that all the representations of the indicated type annihilate the operator A and hence cannot belong to U_A. Thus, only those representations being considered also fall in U_A that are in $\mathcal{U}(\pi_0; \Sigma)$.

The infinite-dimensional representations in U_A obtained by localization along subchains are contained also in the neighborhood $\mathcal{U}(\pi_0; \Sigma)$; as is easy to see, this is ensured by the choice of the operator A (more precisely, by the choice of function a).

It remains to consider the one-dimensional representations. Assume that there exists an element $B \in \mathcal{A}$ such that the set $U_B = \{\pi \in \widehat{\mathcal{A}} : \|\pi B\| > 1\}$ does not contain one-dimensional representations and $\pi_0 \in U_B$. Then, $U_A \cap U_B$ is a neighborhood of π_0 in the J-topology and $U_a \cap U_B \subset \mathcal{U}(\pi_0; \Sigma)$. We determine the necessary element B. Suppose that a is the function in Lemma 3.2.5, the function Ψ is defined in (3.2.7), and

$$\mathfrak{A}(\lambda) = E_{\omega \to \varphi}(\lambda)\Psi(z^0, \ldots, z^j; \omega, 0 \ldots, 0)E_{\psi \to \omega}(\lambda) : L_2(S^{m-d_j-1}) \to L_2(S^{m-d_j-1}),$$

where $\omega = \xi_j/\|\xi^j\|$. This time, we choose Φ so that the commutator $[a(z_0^0, \dots, z_0^j; \cdot), \mathfrak{A}(\lambda)]$ does not vanish identically on the axis $-\infty < \lambda < +\infty$. (This possibility is easily seen, for example, from Proposition 1.2.11.) Assume that P is a ψdo in $\bar{\Psi}^0(\mathcal{M})$ with principal symbol coinciding on the fiber $T^*(\mathcal{M})_{z^0} \setminus 0$ with the chosen function $\Phi(z^0; \cdot)$. Let $B = [a, P] \in \mathcal{A}$. Then, $\pi(z_0^0, \dots, z_0^j; \lambda)B \neq 0$ for some $\lambda \in \mathbb{R}$; it can be assumed that $\|\pi(z_0^0, \dots, z_0^j; \lambda)B\| > 1$. According to assertion 3 in Lemma 3.2.3), we have $\|\pi_0 B\| \geq \|\pi(z_0^0, \dots, z_0^j; B\| > 1$ and $\pi_0 \in U_B$. Obviously, each one-dimensional representation annihilates the operator B and thus is not contained in U_B. Therefore, an arbitrary neighborhood $\mathcal{U}(\pi_0; \Sigma)$ of π_0 contains some neighborhood of this point in the J-topology.

We now prove that any neighborhood $\mathcal{U}(\pi_0; J)$ also contains some neighborhood of π^0 in the Σ-topology. Let $\mathcal{U}(\pi_0; J) = U_A = \{\pi \in \widehat{\mathcal{A}} : \|\pi A\| > 1\}$, where $A = \sum_l \prod_k A_{lk}$, and A_{lk} are generators of the algebra $\mathcal{A}$. Iterating the inequalities in part 1 of Lemma 3.2.3, we arrive at the estimates $\|\pi(z_0^0, \dots, z_0^q; \Theta^q)A\| > 1$, where $q = 0, \dots, j-1$. The functions $(z^0, \dots, z^q; \Theta^q) \mapsto \|\pi(z^0, \dots, z^q; \Theta^q)A\|$ are continuous. Therefore, the indicated inequalities remain valid for all representations π in a small neighborhood $\mathcal{U}(\pi_0; \Sigma)$, i.e., $\mathcal{U}(\pi_0; \Sigma) \subset \mathcal{U}(\pi_0; J)$.

(b) The case $\theta_0^0 = 0$, $\theta_0^1 \neq 0$. We show first that every neighborhood $\mathcal{U}(\pi_0; J)$ contains the half-line $\{\lambda \in \mathbb{R} : \lambda \gtrless N\}$, if $z_0^1\theta_0^1 \gtrless 0$, or the whole line, if $z_0^1\theta_0^1 = 0$. For that, we need several lemmas.

Suppose that $\mathcal{T}(S^{n-1})$ is an admissible partition of the sphere S^{n-1}, $a_{jk} \in \mathfrak{M}(\mathcal{T}, S^{n-1})$, the functions Ψ_{jk} are zero degree homogeneous on $\mathbb{R}^n$, $\Psi_{jk} \in C^\infty(S^{n-1})$, and

$$\mathfrak{A}_{jk}(\lambda) = E_{\omega\to\varphi}(\lambda)\Psi_{jk}(\omega)E_{\psi\to\omega}(\lambda), \quad \mathfrak{D}_{jk}(\lambda) = F^{-1}_{\eta\to x}\Psi_{jk}(\eta, \lambda)F_{y\to\eta}.$$

We will assume that the point z_0^1 of stratum $s^1 \in \mathcal{T}(S^{n-1})$ coincides with the north pole $(0', 1)$.

Lemma 3.2.6 *Assume that j and k run through finite sets. Then,*

$$\underline{\lim}_{\lambda\to\pm\infty}\Big\|\sum_j\prod_k a_{jk}\mathfrak{A}_{jk}(\lambda); \mathcal{B}L_2(S^{n-1})\Big\| \geq \Big\|\sum_j\prod_k a_{jk}(z_0^1; \cdot)\mathfrak{D}_{jk}(\pm 1); \mathcal{B}L_2(\mathbb{R}^{n-1})\Big\|. \tag{3.2.8}$$

Proof Suppose that the functions ζ_{jk} and η_{jk} belong to the class $C^\infty(S_+^{n-1})$ and are equal to 1 in a neighborhood of the pole $(0', 1)$. Furthermore, let $\mathfrak{A}(\lambda) = \sum_j \prod_k \zeta_{jk} a_{jk}\mathfrak{A}_{jk}(\lambda)\eta_{jk}$. In the proof of Proposition 2.2.10, inequality (2.2.42) was obtained:

$$\underline{\lim}_{t\to+\infty}\|\mathfrak{A}(\lambda t); \mathcal{B}L_2(S^{n-1})\| \geq \Big\|\sum_j\prod_k a_{jk}(z_0^1; \cdot)\mathfrak{D}_{jk}(\lambda); \mathcal{B}L_2(\mathbb{R}^{n-1})\Big\| \tag{3.2.9}$$

with $\lambda \in \mathbb{R}$. We set $\lambda = \pm 1$ in (3.2.9) and arrive at

$$\underline{\lim}_{\lambda \to \pm\infty} \|\mathfrak{A}(\lambda); \mathcal{B}L_2(S^{n-1})\| \geq \|\sum_j \prod_k a_{jk}(z_0^1; \cdot)\mathfrak{D}_{jk}(\pm 1); \mathcal{B}L_2(\mathbb{R}^{n-1})\|. \tag{3.2.10}$$

It remains to get rid of the cut-off functions ζ_{jk}, η_{jk} appearing in $\mathfrak{A}(\lambda)$. According to Proposition 1.2.11, the commutators $[\mathfrak{A}_{jk}, \sigma]$ belong to the algebra $C_0(\mathbb{R}) \otimes \mathcal{K}L_2(s^{n-1})$ for $\sigma \in S^{n-1}$. Therefore,

$$\mathfrak{A}(\lambda)\chi = \sum_j \left(\prod_k a_{jk}\mathfrak{A}_{jk}(\lambda)\right)\left(\prod_k \zeta_{jk}\eta_{jk}\right)\chi + T(\lambda)\chi = \mathfrak{B}(\lambda)\chi + T(\lambda)\chi, \tag{3.2.11}$$

where $T \in C_0(\mathbb{R}) \otimes \mathcal{K}L_2(s^{n-1})$, $\mathfrak{B}(\lambda) = \sum_j \prod_k a_{jk}\mathfrak{A}_{jk}(\lambda)$, $\chi \in C_c(S_+^{n-1})$, while $0 \leq \chi \leq 1$, $\chi = 1$ near the pole $(0', 1)$, $\chi\zeta_{jk} = \chi$, $\chi\eta_{jk} = \chi$. The operator $\mathfrak{B}(\lambda)\chi$ differs from $\mathfrak{A}(\lambda)$ only in the form of the cut-off functions. Consequently, inequality (3.2.10) holds with $\mathfrak{A}(\lambda)$ replaced by $\mathfrak{A}(\lambda)\chi$. By (3.2.11), we have $\|\mathfrak{A}(\lambda)\chi\| \leq \|\mathfrak{B}(\lambda)\| + \|T(\lambda)\|$ and $\|T(\lambda)\| \to 0$ as $\lambda \to \infty$, which implies (3.2.8). □

Let $(x = (x^{(1)}, x^{(2)}) \in \mathbb{R}^k$, $x^{(1)} = (x_1, \ldots, x_l) \in \mathbb{R}^l$, $x^{(2)} = (x_{l+1}, \ldots, x_k)$, and $l < k$. We consider the algebras $D_\pm$ generated on $L_2(\mathbb{R}^k)$ by the convolution operators $F_{\eta \to x}^{-1}\Phi(\eta, \pm 1)F_{y \to \eta}$ and the operators of multiplication by the functions $x \mapsto a(x^{(1)}/|x^{(1)}|)$; here, Φ are zero degree homogeneous functions on $\mathbb{R}^{k+1} \setminus 0$ and a are elements of some algebra of functions on S^{l-1} (for example, of type $\mathfrak{M}(\mathcal{T}, S^{l-1})$). We also introduce the algebras $\mathcal{D}_\pm$ of operator-valued functions $S_\pm^{k-l} \ni \theta \mapsto A(\theta) : L_2(\mathbb{R}^l) \to L_2(\mathbb{R}^l)$, where $A(\theta) = F_{\zeta \to t}^{-1}(\zeta, \theta)F_{s \to \zeta}$ or $A(\theta) \equiv a(\cdot)I$ (the operator of multiplication by the function $s \mapsto a(s/|s|)$), $S_\pm^l = \{\theta = (s, \sigma) : s \in \mathbb{R}^l, \sigma \in \mathbb{R}, |s|^2 + \sigma^2 = 1, \sigma \gtrless 0\}$. The algebra $\mathcal{D}_\pm$ is equipped with the norm $\|A; \mathcal{D}\| = \sup\{\|A(\theta); \mathcal{B}L_2(\mathbb{R}^l)\|, \theta \in S_\pm^l\}$.

Lemma 3.2.7 *The algebra $D_\pm$ is isomorphic to the algebra $\mathcal{D}_\pm$.*

Proof Applying the Fourier transform $F_{x^{(2)} \to \zeta}$ to the elements $A \in D_\pm$, we obtain that $D_\pm$ is isomorphic to the algebra of operator-valued functions

$$\mathbb{R}^{k-l} \ni \zeta \mapsto A_\pm(\zeta) : L_2(\mathbb{R}^l) \to L_2(\mathbb{R}^l),$$

where $A_\pm(\zeta) = F^{-1}_{\eta\to x}\Phi(\eta,\zeta,\pm 1)F_{y\to\eta}$ or $A_\pm(\zeta) \equiv a(\cdot)I$, with norm $\sup\{\|A_\pm(\zeta); \mathcal{B}L_2(\mathbb{R}^l)\|, \zeta \in \mathbb{R}^{k-l}\}$. Let $t = \sqrt{|\zeta|^2+1}$ and define a unitary operator $(U_t u)(x) = t^{l/2}u(tx)$ in $L_2(\mathbb{R}^l)$. We have

$$A_\pm(\zeta)U_t u = t^{l/2}(2\pi)^{-l}\int e^{ix\eta}\Phi(\eta,\zeta,\pm 1)\,d\eta\int e^{-i\eta y}u(ty)dy =$$

$$= t^{l/2}(2\pi)^{-l}\int e^{ix\xi t}\Phi(\xi t,\theta_\pm)\,d\xi\int e^{-i\xi z}u(z)\,dz = U_t A(\theta_\pm)u,$$

where $\theta_\pm = (\zeta,\pm 1)/\sqrt{|\zeta|^2+1} \in S^l_\pm$. Moreover, the operator U_t commutes with operators of multiplication by functions $\mathbb{R}^l\setminus 0 \ni x \mapsto a(x/|x|)$. All this leads to an isomorphism $D_\pm \simeq \mathcal{D}_\pm$. □

Proposition 3.2.8 *Each neighborhood* $\mathcal{U}(\pi_0; J)$ *of the point* $\pi_0 = \pi(z^0_0, z^1_0, \dots, z^j_0; 0, \theta^1_0, \dots, \theta^j_0)$ *in the* J*-topology (Jacobson topology) with* $z^1_0\theta^1_0 \gtrless 0$ *contains a set of the form* $\{\pi(z^0;\lambda)\}$, *where* (z^0,λ) *runs through a set of the form* $\mathcal{U}(z^0_0;\mathcal{Z}_0)\times\{\lambda\in\mathbb{R}:\lambda\gtrless N\}$, *whereas* $\mathcal{U}(z^0_0;\mathcal{Z}_0)$ *is a small neighborhood, and* N *is a sufficiently large number (see formulas* (3.2.2), (3.2.3) *and the accompanying text).*

Proof It can be assumed that $\mathcal{U}(\pi_0; J) = U_A = \{\pi : \|\pi A\| > 1\}$, where $A = \sum_l\prod_k A_{lk}$, $A_{lk} = a_{lk}P_{lk}$, and P_{lk} denotes a ψdo in $\bar{\Psi}^0(\mathcal{M})$ with principal symbol $\Phi_{lk}(z^0_0;\cdot)$. According to part 1 of Lemma 3.2.3, the inequality $\|\pi_0 A\| > 1$ implies the estimate

$$\|\pi(z^0_0, z^1_0; 0, \theta^1_0/|\theta^1_0|)A\| > 1.$$

Recalling the definition of $\pi(z^0_0, z^1_0; 0, \theta^1_0/|\theta^1_0|)$ (see (3.1.6) and (3.1.4)), we now deduce from Lemma 3.2.7 that

$$\sum_l\prod_k a_{lk}(z^0_0, z^1_0;\cdot)\mathfrak{D}_{lk}(\pm 1); \mathcal{B}L_2(\mathbb{R}^{m-d_0-1})\| > 1$$

in the case $z^1_0\theta^1_0 \gtrless 0$. [We assume that z^1_0 coincides with the north pole. A point θ^1 with positive (negative) last coordinate appears as θ_+ (θ_-) in Lemma 3.2.7. The symbols Ψ_{lk} of the operators $\mathfrak{D}_{lk}(\lambda)$ are determined by the equalities $\Psi_{lk}(\eta,\lambda) = \Phi_{lk}(z^0_0;\eta,\lambda,\theta^0)|_{\theta^0=0}$.] The proof is concluded by using Lemma 3.2.6. □

Let us proceed to the case $z^1_0\theta^1_0 = 0$. Instead of Lemma 3.2.6, we use:

Lemma 3.2.9 *In the notation of Lemma 3.2.6, the inequality*

$$\|\sum_l\prod_k a_{lk}\mathfrak{A}_{lk}(\lambda); \mathcal{B}L_2(S^{n-1})\| \geq \|\sum_l\prod_k a_{lk}(z^1_0;\cdot)\mathfrak{D}_{lk}(0); \mathcal{B}L_2(\mathbb{R}^{n-1})\|$$

holds for all $\lambda\in\mathbb{R}$.

Proof Passing in (2.2.41) to the limit as $t \to +\infty$, we obtain

$$\|\mathfrak{A}(\lambda); \mathcal{B}L_2(S^{n-1})\| \geq \|\sum_l \prod_k a_{lk}(z_0^1; \cdot)\mathfrak{D}_{lk}(0); \mathcal{B}L_2(\mathbb{R}^{n-1})\| - \varepsilon. \tag{3.2.12}$$

The purpose now is to eliminate the cut-off functions ζ_{lk} and η_{lk} in the operator $\mathfrak{A}(\lambda)$. We choose a sequence of functions $\chi_q \in C_c^\infty(S_+^{n-1})$, $0 \leq \chi_q \leq 1$, equal to 1 near the pole $(0', 1)$ and with supports shrinking to $(0', 1)$. The inequality (3.2.12) holds for each operator $\mathfrak{A}(\lambda)\chi_q$. Suppose that $\{u_q\} \subset L_2(S^{n-1})$, $\|u_q\| = 1$, and $\|\mathfrak{A}(\lambda)\chi_q u_q\| \geq \|\mathfrak{A}(\lambda)\chi_q; \mathcal{B}L_2(S^{n-1}\| - \varepsilon$. This and (3.2.11) imply that

$$\begin{aligned} \|\mathfrak{B}(\lambda)\chi_q u_q\| &\geq \|\mathfrak{A}(\lambda)\chi_q u_q\| - \|T(\lambda)\chi_q u_q\| \geq \\ &\geq \|\mathfrak{A}(\lambda)\chi_q; \mathcal{B}L_2 S^{n-1})\| - \|T(\lambda)\chi_q u_q\| - \varepsilon. \end{aligned} \tag{3.2.13}$$

It is clear that the sequence $\{\chi_q u_q\}$ converges weakly to zero. Since $T(\lambda)$ is a compact operator on $L_2(S^{n-1})$, we have $\|T(\lambda)\chi_q u_q\| \to 0$ as $q \to \infty$. Therefore, by (3.2.12) and (3.2.13),

$$\|\mathfrak{B}(\lambda); \mathcal{B}L_2(S^{n-1})\| \geq \|\mathfrak{B}(\lambda)\chi_q u_q\| \geq \|\sum \prod a_{lk}(z_0^1; \cdot)\mathfrak{D}_{lk}(0); \mathcal{B}L_2(\mathbb{R}^{n-1})\| - 2\varepsilon.$$

□

Proposition 3.2.10 *Let* $z_0^1\theta_0^1 = 0$. *Then, every neighborhood* $\mathcal{U}(\pi_0; J)$ *of* $\pi_0 = \pi(z_0^0, z_0^1, \ldots, z_0^j; 0, \theta_0^1, \ldots, \theta_0^j)$ *in the* J*-topology contains some set* $\{\pi(z^0, \lambda)\}$, *where* z^0 *runs through a small enough neighborhood* $\mathcal{U}(z_0^0; \mathcal{Z}_0)$ *and* $\lambda \in \mathbb{R}$.

Proof As in the proof of Proposition 3.2.8, suppose that $\mathcal{U}(\pi_0; J) = U_a$ and A has the previous meaning. Applying Theorem 3.1.4 to the algebra generated on $L_2(\mathbb{R}^{m-d_0-1})$ by the operators of the form $a(z_0^0, z_0^1; \cdot)\mathfrak{D}(0)$, where $\mathfrak{D}(0) = F_{\eta\to x}^{-1}\Psi(\eta, 0)F_{y\to\eta}$, we get that π_0 is a representation of this algebra. (Formally speaking, we use not Theorem 3.1.4 itself, but a result obtained in its proof.) Therefore, the inequality $\|\pi_0 A\| > 1$ implies

$$\|\sum_l \prod_k a_{lk}(z_0^0, z_0^1; \cdot)\mathfrak{D}_{lk}(0); \mathcal{B}L_2(\mathbb{R}^{m-d_0-1})\| > 1.$$

This and Lemma 3.2.9 imply that $\|\pi(z_0^0; \lambda)A; \mathcal{B}L_2(S^{m-d_0-1})\| > 1$. □

Thus, in Propositions 3.2.8 and 3.2.10, we have determined which representations $\pi(z^0; \lambda)$ fall in the neighborhood $\mathcal{U}(\pi_0; J)$ of $\pi_0 = \pi(z_0^0, z_0^1, \ldots, z_0^j; 0, \theta_0^1, \ldots, \theta_0^j)$ with $\theta_0^1 \neq 0$. We now turn to the representations $\pi(z^0, \theta^0)$.

Proposition 3.2.11 *Each neighborhood $\mathcal{U}(\pi_0; J)$ contains some set $\{\pi(z^0; \theta^0)\}$, where z^0 varies in a small enough neighborhood $\mathcal{U}(z_0^0; \mathcal{Z}_0)$ and θ^0 does in the whole sphere $S^*(\mathcal{Z}_0)_{z^0}$.*

Proof As before, assume that $\mathcal{U}(\pi_0; J) = U_A$, where A is a finite combination of generators of $\mathcal{A}$. If $\|\pi_0 A\| > 1$, then there exist $\lambda \in \mathbb{R}$ such that $\|\pi(z_0^0; \lambda)A\| > 1$ (Propositions 3.2.8 and 3.2.10). According to part 3 of Lemma 3.2.3, the inequality $\|\pi(z_0^0; \lambda)A\| > 1$ implies that $\|\pi(z_0^0; \theta^0)A\| > 1$ for each $\theta^0 \in S^*(\mathcal{Z}_0)_{z_0^0}$. It remains to use the continuity of the function $S^*(\mathcal{Z}_0) \ni (z^0; \theta^0) \mapsto \|\pi(z_0^0; \theta^0)A\|$. □

We turn to a comparison of the J- and Σ-topologies.

Proposition 3.2.12 *For each neighborhood $\mathcal{U}(\pi_0; \Sigma)$, there exists a neighborhood $\mathcal{U}(\pi_0; J)$ belonging to it.*

Proof Suppose that the functions Φ, Ψ and the operator A are the same as in part a) (i.e., in the case $\theta_0^0 \neq 0$), but now the support of Ψ does not intersect the set of points for which $\theta^1 = 0$. Then, $\|\pi(z^0, \dots, z_0^q; \Theta_0^q)A\| > 1$ for $q = 0, \dots, j-1$. As before, the neighborhood $U_A = \{\pi : \|\pi A\| > 1\}$ does not contain representations $\pi(z_0, \dots, z^q; \theta^0, \dots, \theta^q)$, $q \geq 1$, that are not in $\mathcal{U}(\pi_0; \Sigma)$. It is also clear that U_A does not contain representations of the type $\pi(z^0, \dots, z^q; \lambda)$, $1 \leq q \leq k$; these representations annihilate A, since $\Psi = 0$ for $\theta^1 = 0$. The one-dimensional representations are excluded for the same reason as in part a) (i.e., U_A is replaced by the intersection $U = U_A \cap U_B$, which now does not contain one-dimensional representations).

According to Proposition 3.2.11, the neighborhood U contains the set $\{\pi(z^0; \theta^0) : z^0 \in \mathcal{U}(z_0^0, \mathcal{Z}_0), \theta^0 \in S^*(\mathcal{Z}_0)_{z^0}\}$; the same set also appears in $\mathcal{U}(\pi_0; \Sigma)$. If $z_0^1\theta_0^1 = 0$, then $\{\pi(z^0; \lambda) : z^0 \in \mathcal{U}(z_0^0; \mathcal{Z}_0), \lambda \in \mathbb{R}\}$ is a subset of both $\mathcal{U}(\pi_0; \Sigma)$ and U.

Suppose finally that $z_0^1\theta_0^1 \gtrless 0$ and $\mathcal{U}(\pi_0; \Sigma) \supset \{\pi(z^0, \lambda) : z^0 \in \mathcal{U}(z_0^0; \mathcal{Z}_0), \lambda \gtrless N\}$. Below, we determine an operator $P \in \mathcal{A}$ such that $\pi_0 \in U_P = \{\pi \in \widehat{\mathcal{A}} : \|\pi P\| > 1\}$ and

$$U_P \cap \{\pi(z^0; \lambda) : z^0 \in \mathcal{U}(z_0^0; \mathcal{Z}_0), \lambda \in \mathbb{R}\} \subset \{\pi(z^0, \lambda) : z^0 \in \mathcal{U}(z_0^0; \mathcal{Z}_0), \lambda \gtrless N\}.$$

The set $U \cap U_P$ is open in the J-topology and contains only those $\pi(z^0, \lambda)$ belonging to $\mathcal{U}(\pi_0; \Sigma)$. By the choice of A, $U \cap U_P$ does not contain representations not in $\mathcal{U}(\pi_0; \Sigma)$. Thus, $\pi_0 \in U \cap U_P \subset \mathcal{U}(\pi_0; \Sigma)$. It remains to produce the operator P.

Let Φ and let Ψ be defined just as in part 3 of Lemma 3.2.3 (these are new functions, not connected with those used above for the operator A). Choose an arbitrary ξ_0^j and denote by $\xi_0 \in S^{m-1}$ a point for which $\Phi(z_0^0; \xi_0) = \Psi(z_0^0, z_0^1, \dots, z_0^j; \xi_0^j, 0, \theta_0^1, \dots, \theta_0^j)$. We assume that $\Phi(z_0^0; \xi_0) > 1$ and $\Phi(z^0; \xi) = 0$ for ξ outside a small neighborhood of ξ_0 on S^{m-1} or for z^0 outside a small neighborhood of z_0^0 on $\mathcal{M}$. For $m > n$, the inclusion $\xi_0 \in S^{m-d_0-1} \subset S^{m-1}$ is valid, where $S^{m-d_0-1} = \{\eta \in \mathbb{R}^m : |\eta| = 1, \eta_{m-d_0+1} = \cdots =$

$\eta_m = 0\}$. Furthermore, let $a \in C(S^{m-d_0-1})$ be such that $a(z_0^1) \geq 1$ and $a = 0$ outside a small neighborhood of z_0^1 on S^{m-d_0-1}. We introduce the operator $P = aQ$, where Q is a ψdo in $\bar{\Psi}^0(\mathcal{M})$ with principal symbol Φ. Let us ensure that P has the necessary properties. Since

$$\pi(z_0^0, z_0^1, \dots, z_0^j; \varphi_0^j, \xi_0^j, \theta_0^1, \dots, \theta_0^j)P = a(z_0^1)\Psi(z_0^0, z_0^1, \dots, z_0^j; \xi_0^j, 0, \theta_0^1, \dots, \theta_0^j) > 1,$$

the inequality $\|\pi_0 P\| > 1$ follows from (3.2.4), i.e., $\pi_0 \in U_P$. By the inclusion $S^{m-d_0-1} \subset S^{m-1}$, we can regard z_0^1 as a point in $\mathbb{R}^m$. Recalling the definition of ξ_0, we obtain the equality $z_0^1\xi_0 = z_0^1\theta_0^1$. Suppose for definiteness that$z_0^1\theta_0^1 > 0$. Since the supports of a and Φ are small, the function $(\varphi, \omega) \mapsto a(\varphi)\Phi(z_0^0; \omega, 0)$ is equal to zero on the set

$$\{(\varphi, \omega) \in S^{n-1} \times S^{n-1} : \varphi\omega \leq 0\}.$$

Therefore, Lemma 5.4.8 in [21] leads to

$$\lim_{\lambda \to -\infty} \|\pi(z_0^0; \lambda)P; \mathcal{B}L_2(S^{n-1})\| = 0. \tag{3.2.14}$$

(An analogous relation was used in the proof of Theorem 2.2.13, where in particular, the mentioned lemma from [21] was formulated.) If the inclusion $\{\lambda : \|\pi(z_0^0; \lambda)P\| > 1\} \subset \{\lambda \in \mathbb{R} : \lambda > N\}$ holds, then P possesses all the needed properties. However, if the latter inclusion does not hold, then P must be replaced by the new operator P' described below.

Since the function $(\varphi, \omega) \mapsto a(\varphi)\Phi(z_0^0; \omega, 0)$ is equal to zero for $\varphi\omega = 0$, it follows that $\pi(z_0^0; \lambda)P \in \mathcal{K}L_2(S^{n-1})$ for all $\lambda \in \mathbb{R}$ (Theorem 2.2.12). Let $\chi \in C(\mathbb{R})$ satisfy the conditions $0 \leq \chi \leq 1$, $\chi(\lambda) = 1$ for $\lambda < N$, and $\chi(\lambda) = 0$ for $\lambda > N + 1$. Then, it follows from (3.2.14) that $\chi(\cdot)\pi(z_0^0; \cdot)P \in C_0(\mathbb{R}) \otimes \mathcal{K}L_2(S^{n-1})$. Assume that $\mathcal{L}_{cont}$ is an algebra generated on $L_2(\mathbb{R}^m)$ by the operators of the form $a(\varphi)F^{-1}_{\xi\mapsto x}\Phi(\xi)F_{y\mapsto\xi}$, where $\varphi = x^{(1)}/|x^{(1)}|$, $x = (x^{(1)}, x^{(2)})$, $x^{(1)} = (x_1, \dots, x_n)$, $x^{(2)} = (x_{n+1}, \dots, x_m)$, and $n = m - d_0 - 1$; the functions a and Φ are continuous on spheres S^{n-1} and S^{m-1}, respectively. Let also $\mathfrak{S}_{cont}$ be an algebra spanned by the operator-valued functions $\mathbb{R} \ni \lambda \mapsto \mathfrak{B}(\lambda) = a(\varphi)E_{\omega\to\varphi}(\lambda)^{-1}\Psi(\omega)E_{\psi\to\omega}(\lambda)$ with the norm $\sup\{\|\mathfrak{B}(\lambda); \mathcal{B}L_2(S^{n-1})\|, \lambda \in \mathbb{R}\}$. We introduce a mapping p on the generators of $\mathcal{L}_{cont}$ by

$$p(aF^{-1}\Phi(\xi)F) = a(\varphi)E_{\omega\to\varphi}(\cdot)^{-1}\Phi(\omega, 0)E_{\psi\to\omega}(\cdot) \in \mathfrak{S}_{cont},$$

where $\omega = \xi^{(1)}/|\xi^{(1)}|$, $\xi^{(1)} = (\xi_1, \dots, \xi_n)$. It can be shown that p extends to an epimorphism $p : \mathcal{L}_{cont} \to \mathfrak{S}_{cont}$ (see [25], Lemma 2.2, and Proposition 2.3). Therefore, since $P \in \mathcal{L}_{cont}$, there exists an element $P' \in \mathcal{L}_{cont}$ satisfying $p(P') = (1 - \chi(\cdot)\pi(\cdot)P$. We verify that P' is the desired operator. The inclusion $\{\lambda : \|\pi(\lambda)P'\| > 1\} \subset \{\lambda \in \mathbb{R} : \lambda > N\}$ obviously follows from the definition of P'. The restriction of the representation $\pi(z_0^0, \ldots, z_0^j; \varphi_0^j, \xi_0^j, 0, \theta_0^1, \ldots, \theta_0^j)$ to $\mathcal{L}_{cont}$ can be written in the form $\pi(\varphi_0, \xi_0)$, $\varphi_0 = z_0^1$, $(\varphi_0, \xi_0) \in S^{n-1} \times S^{n-1}$. This one-dimensional representation of $\mathcal{L}_{cont}$ is also a representation of the quotient algebra $\mathfrak{S}_{cont}/C_0(\mathbb{R}) \otimes \mathcal{K}L_2(S^(n-1))$. Therefore,

$$\pi(z_0^0, \ldots, z_0^j; \varphi_0^j, \xi_0^j, 0, \theta_0^1, \ldots, \theta_0^j)P' = \pi(\varphi_0, \xi_0)P' = \pi(\varphi_0, \xi_0)P$$
$$= \pi(z_0^0, \ldots, z_0^j; \varphi_0^j, \xi_0^j, 0, \theta_0^1, \ldots, \theta_0^j)P,$$

i.e., the operator P' inherits the necessary properties of P. □

Proposition 3.2.13 *Every neighborhood* $\mathcal{U}(\pi_0; J)$ *contains some neighborhood* $\mathcal{U}(\pi_0; \Sigma)$.

Proof According to Propositions 3.2.8 and 3.2.10, every neighborhood $\mathcal{U}(\pi_0; J)$ contains some set of the form $\mathcal{U}(z_0^0; \mathcal{Z}_0) \times \{\lambda \in \mathbb{R} : \lambda \gtreqless N\}$ if $z_0^1\theta_0^1 > 0$ or of the form $\mathcal{U}(z_0^0; \mathcal{Z}_0) \times \mathbb{R}$ if $z_0^1\theta_0^1 = 0$. Proposition 3.2.11 ensures that $\mathcal{U}(\pi_0; J)$ is a subset of $\{(z^0; \theta^0) : z^0 \in \mathcal{U}(z_0^0; \mathcal{Z}_0), \theta^0 \in S^*(\mathcal{Z}_0)_{z^0}\}$. The sets of the form (3.2.1) fall in $\mathcal{U}(\pi_0; J)$ in view of assertion 1 in Lemma 3.2.3. It remains to recall the definition of $\mathcal{U}(\pi_0; \Sigma)$ (see the list of neighborhoods before Theorem 3.2.1). □

(c) The case $\theta_0^0 = 0, \ldots, \theta_0^i = 0$, $\theta^{i+1} \neq 0$, $i \geq 1$. Here, we confine ourselves to a verification that every neighborhood $\mathcal{U}(\pi_0; J)$ contains the sets of all those types that occur in $\mathcal{U}(\pi_0; \Sigma)$ (according to the list of neighborhoods). The formal proofs of the analogues of Propositions 3.2.12 and 3.2.13 are left to the reader.

Now, let $\pi_0 = \pi(z_0^0, z_0^1, \ldots, z_0^j; 0, \ldots, 0, \theta^{i+1}, \ldots, \theta_0^j)$, $\theta^{i+1} \neq 0$. It can be assumed that $\mathcal{U}(\pi_0; J) = U_A = \{\pi \in \hat{\mathcal{A}} : \|\pi A\| > 1\}$, where $A = \sum_l \prod_k A_{lk}$ and A_{lk} are generators of $\mathcal{A}$.

Proposition 3.2.14 *The set* U_A *contains products of the form (3.2.1), where* $(z^0, \ldots, z^q) = (z_0^0, \ldots, z_0^q)$, $\Theta^q = \Theta_0^q$, $q = i+1, \ldots, j$, *and neighborhoods* $\mathcal{U}((z_0^0, \ldots, z_0^q); \mathcal{Z}_q)$ *and* $\mathcal{U}(\Theta_0^q; S^{d_q-1})$ *(the factors in these products) are sufficiently small.*

Proof Assertion 1 of Lemma 3.2.3 implies the inequalities

$$\|\pi_0 A\| \leq \|\pi(z_0^0, \ldots, z_0^{j-1}; \Theta_0^{j-1})A\| \leq \cdots \leq \|\pi(z_0^0, \ldots, z_0^{i+1}; \Theta_0^{i+1})A\|. \quad (3.2.15)$$

Therefore, the estimate $\|\pi_0 A\| > 1$ and the continuity of the functions $\mathcal{Z}_q \times S^{d_q-1} \ni (z^0, \ldots, z^q; \Theta^q) \mapsto \|\pi(z^0, \ldots, z^q; \Theta^q)A\|$ lead to the relations $\|\pi(z^0, \ldots, z^q; \Theta^q)A\| > 1$ for all

$$(z^0, \ldots, z^q; \Theta^q) \in \mathcal{U}(z_0^0, \ldots, z_0^q; \mathcal{Z}_q) \times \mathcal{U}(\Theta_0^q; S^{d_q-1})$$

with $q = i+1, \ldots, j$. □

Proposition 3.2.15 *The set U_A contains products of the form $\mathcal{U}(z_0^0, \dots, z_0^p; \mathcal{Z}_p) \times S^{d_p - 1}$, where $p = 1, \dots, i$ and $\mathcal{U}(z_0^0, \dots, z_0^p; \mathcal{Z}_p)$ is a sufficiently small neighborhood.*

Proof Since $\pi(z_0^0, \dots, z_0^{i+1}; \Theta_0^{i+1}) = \pi(z_0^0, \dots, z^{i+1}; 0, \dots, 0, \theta_0^{i+1}/|\theta^{i+1}|)$, it follows from (3.2.15) and part 2 of Lemma 3.2.3 that $\|\pi(z_0^0, \dots, z_0^i; \lambda_i^*)A\| > 1$ for some $\lambda_i^* \in \mathbb{R}$. Then in view of part 3 of the same lemma

$$\|\pi(z_0^0, \dots, z_0^i; \theta^0, \dots, \theta^i)A\| > 1 \tag{3.2.16}$$

for all $\theta^0, \dots, \theta^i$, $|\theta^0|^2 + \dots + |\theta^i|^2 = 1$. This means that U_A contains a set of the form $\mathcal{U}(z_0^0, \dots, z_0^i; \mathcal{Z}_i) \times S^{d_i - 1}$. Formula (3.2.16) gives, in particular, the estimate $\|\pi(z_0^0, \dots, z_0^i; 0, \dots, 0, \theta^i)A\| > 1$ for $|\theta^i| = 1$ and thereby enables us to repeat the argument just used with i replaced by $i - 1$. This procedure continues to the end of the proof. □

Proposition 3.2.16

(1) *U_A contains the products $\mathcal{U}(z_0^0, \dots, z_0^r; \mathcal{Z}_r) \times \mathbb{R}$ (i.e., the representations $\pi(z^0, \dots, z^r; \lambda_r)$, where $(z^0, \dots, z^r) \in \mathcal{U}(z_0^0, \dots, z_0^r; \mathcal{Z}_r)$ and $\lambda_r \in \mathbb{R}$), whereas $r = 1, \dots, i - 1$.*
(2) *U_A contains the set $\mathcal{U}(z_0^0, \dots, z_0^i; \mathcal{Z}_i) \times \{\lambda_i \in \mathbb{R} : \lambda_i \gtrless N\}$ if $z_0^{i+1}\theta_0^{i+1} \gtrless 0$ or the set $\mathcal{U}(z_0^0, \dots, z_0^i; \mathcal{Z}_i) \times \mathbb{R}\}$ if $z_0^{i+1}\theta_0^{i+1} = 0$.*

Proof (1) We consider the algebra $\mathfrak{S}(z_0^0, \dots, z_0^{i-1})$ by the operator-valued functions $\mathbb{R} \ni \lambda_{i-1} \mapsto \pi(z_0^0, \dots, z_0^{i-1}; \lambda_{i-1})B : L_2(S^{m-d_{i-1}-1}) \to L_2(S^{m-d_{i-1}-1})$, where $B \in \mathcal{A}$. The norm in $\mathfrak{S}(z_0^0, \dots, z_0^{i-1})$ is defined by

$$\|\pi(z_0^0, \dots, z_0^{i-1}; \cdot)B\| = \sup\{\|\pi(z_0^0, \dots, z_0^{i-1}; \lambda)B\|, \lambda \in \mathbb{R}\}.$$

This algebra is generated by the functions of the form

$$\lambda_{i-1} \mapsto a(z_0^0, \dots, z_0^{i-1}; \cdot)E(\lambda_{i-1})^{-1}\Psi(\omega)E(\lambda_{i-1})$$

with $\Psi \in C(S^{m-d_{i-1}-1})$. We assume that the point z_0^i coincides with the north pole of the sphere $S^{m-d_{i-1}-1}$. Lemma 3.2.9 gives the inequality

$$\|\sum_l \prod_k a_{lk}(z_0^0, \dots, z_0^{i-1}; \cdot)\mathfrak{A}_{lk}(\lambda_{i-1}); \mathcal{B}L_2(S^{m-d_{i-1}-1})\| \geq \tag{3.2.17}$$

$$\geq \|\sum_l \prod_k a_{lk}(z_0^0, \dots, z_0^{i-1}, z_0^i; \cdot)\mathfrak{D}_{lk}(0); \mathcal{B}L_2(\mathbb{R}^{m-d_{i-1}-1})\|,$$

for all $\lambda_{i-1} \in \mathbb{R}$, where this time $\mathfrak{A}_{lk}(\lambda) = E(\lambda)^{-1}\Psi_{lk}(\omega)E(\lambda)$, $\mathfrak{D}_{lk}(0) = F^{-1}_{\eta\to x}\Psi_{lk}(\eta,0)F_{y\to\eta}$, and Ψ_{lk} is a homogeneous function of zero degree on $\mathbb{R}^{m-d_{i-1}-1}\setminus 0$. The algebra generated by the operators of the form

$$a(z_0^0,\ldots,z_0^i;\cdot)F^{-1}\Psi(\eta,0)F : L_2(\mathbb{R}^{m-di-1-1}) \to L_2(\mathbb{R}^{m-di-1-1}),$$

is isomorphic to the algebra $\mathfrak{S}(z_0^0,\ldots,z_0^i)$ of operator-valued functions $\mathbb{R} \ni \lambda_i \mapsto \pi(z_0^0,\ldots,z_0^i;\lambda_i)B$. This and (3.2.17) give us that

$$\|\pi(z_0^0,\ldots,z_0^{i-1};\lambda_{i-1})A\| \geq \sup\{\|\pi(z_0^0,\ldots,z_0^i;\lambda_i)A\|, \lambda_i \in \mathbb{R}\}. \tag{3.2.18}$$

In verifying Proposition 3.2.15, from the estimate $\|\pi_0 A\| > 1$, we derived the inequality $\|\pi(z_0^0,\ldots,z_0^i;\lambda_i^*)A\| > 1$ for some $\lambda_i^* \in \mathbb{R}$. Using (3.2.18), we obtain $\|\pi(z_0^0,\ldots,z_0^{i-1};\lambda_{i-1})A\| > 1$ for all $\lambda_{i-1} \in \mathbb{R}$. This means that U_A contains the set $\mathcal{U}(z_0^0,\ldots,z_0^{i-1};\mathcal{Z}_{i-1}) \times \mathbb{R}$. The argument can be repeated with i replaced by $i-1$ and with an arbitrary real number chosen as λ_{i-1}^*. As a result, we obtain the inclusion $\mathcal{U}(z_0^0,\ldots,z_0^{i-2};\mathcal{Z}_{i-2}) \times \mathbb{R} \subset U_A$ and so on.

(2) In the first part of the proof, Lemma 3.2.9 was applied to the elements of the algebra $\mathfrak{S}(z_0^0,\ldots,z_0^{i-1})$ to derive (3.2.17). We replace $\mathfrak{S}(z_0^0,\ldots,z_0^{i-1})$ by the algebra $\mathfrak{S}(z_0^0,\ldots,z_0^{i-1},z_0^i)$ and use Lemma 3.2.6 instead of Lemma 3.2.9. Then, we obtain

$$\underline{\lim}_{\lambda_i\to\pm\infty}\|\sum_l\prod_k a_{lk}(z_0^0,\ldots,z_0^i;\cdot)\mathfrak{A}_{lk}(\lambda_i);\mathcal{B}L_2(S^{m-d_i-1})\| \geq$$
$$\geq \|\sum_l\prod_k a_{lk}(z_0^0,\ldots,z_0^i,z_0^{i+1};\cdot)\mathfrak{D}_{lk}(\pm 1);\mathcal{B}L_2(\mathbb{R}^{m-d_i-1})\|.$$

Applying Lemma 3.2.9 to the elements of $\mathfrak{S}(z_0^0,\ldots,z_0^{i-1},z_0^i)$, we arrive at the inequality

$$\|\sum_l\prod_k a_{lk}(z_0^0,\ldots,z_0^{i-1},z_0^i;\cdot)\mathfrak{A}_{lk}(\lambda_i);\mathcal{B}L_2(S^{m-d_i-1})\| \geq$$
$$\geq \|\sum_l\prod_k a_{lk}(z_0^0,\ldots,z_0^i,z_0^{i+1};\cdot)\mathfrak{D}_{lk}(0);\mathcal{B}L_2(\mathbb{R}^{m-d_i-1})\|$$

for all $\lambda_i \in \mathbb{R}$.

Now it remains to repeat the proof of Proposition 3.2.8 with the obvious changes in the case $z_0^{i+1}\theta_0^{i+1} \gtrless 0$ and to repeat the proof of Proposition 3.2.10 in the case $z_0^{i+1}\theta_0^{i+1} = 0$. □

We remark that together with the representations considered in Propositions 3.2.14–3.2.16 the neighborhood U_A includes close representations of the corresponding types obtained by freezing along subchains of strata.

3^0. **Neighborhoods of points in** $\mathcal{Z}_j \times \mathbb{R}$. We proceed to a discussion of neighborhoods of points in $\pi_0 = \pi(z_0^0, \dots z_0^j; \lambda_j^0)$. Let $A = \sum_l \prod_k A_{lk}$, A_{lk} being generators of the algebra $\mathcal{A}$, and suppose that $\|\pi(z_0^0, \dots z_0^j; \lambda_j^0)A\| > 1$. Using inequality (3.2.18) successively for $i = j, j-1, \dots, 0$, we obtain the inclusions $\mathcal{U}(z_0^0, \dots, z_0^i; \mathcal{Z}_i) \times \mathbb{R} \subset U_A$ for $i = 0, \dots, j-1$ and sufficiently small neighborhoods $\mathcal{U}(z_0^0, \dots, z_0^i; \mathcal{Z}_i)$; here, as before, $U_A = \{\pi \in \widehat{\mathcal{A}} : \|\pi A\| > 1\}$. According to part 3 of Lemmă 3.2.3, the estimate $\|\pi(z_0^0, \dots, z_0^j; \theta^0, \dots, \theta^j)A\| > 1$ holds for all $\theta^0, \dots, \theta^j$ such that $|\theta^0|^2 + \dots + |\theta^j|^2 = 1$. 1. This, together with part 1 of the same lemma and the continuity of the functions $\mathcal{Z}_i \times S^{d_i - 1} \ni (z^0, \dots, z^i; \Theta^i) \mapsto \|\pi(z^0, \dots, z^i; \Theta^i)A\|$, gives us that $\mathcal{U}(z_0^0, \dots, z_0^i; \mathcal{Z}_i) \times S^{d_i-1} \subset U_A, i = 0, \dots, j$. Continuity of the function

$$\mathcal{Z}_j \times \mathbb{R} \ni (z^0, \dots, z^j; \lambda_j) \mapsto \|\pi(z^0, \dots, z^j; \lambda^j)A\|$$

supplies the inclusion $\mathcal{U}(z_0^0, \dots, z_0^j; \mathcal{Z}_j) \times \mathcal{U}(\lambda_j^0; \mathbb{R}) \subset U_A$. The whole of the foregoing amounts to:

Proposition 3.2.17 *Every neighborhood* $\mathcal{U}(\pi_0; J)$ *with* $\pi_0 = \pi(z_0^0, \dots, z_0^j; \lambda_j^0)$ *contains some neighborhood* $\mathcal{U}(\pi_0; \Sigma)$.

This part of the proof concludes with a verification of the following assertion.

Proposition 3.2.18 *Every neighborhood* $\mathcal{U}(\pi_0; \Sigma)$ *of the point* $\pi_0 = \pi(z_0^0, \dots, z_0^j; \lambda_j^0)$ *contains some neighborhood* $\mathcal{U}(\pi_0; J)$.

Proof The product $C_0(\mathbb{R}) \otimes \mathcal{K}L_2(S^{m-d_j-1})$ is contained in the algebra of operator-valued functions $\mathbb{R} \ni \lambda_j \ni \pi(z_0^0, \dots, z_0^j; \lambda_j)B$, where $B \in \operatorname{com} \mathcal{A}$. We choose an element B satisfying the conditions

$$\pi(z_0^0, \dots, z_0^j; \cdot)B \in C_0(\mathbb{R}) \otimes \mathcal{K}L_2(S^{m-d_j-1}), \quad \|\pi(z_0^0, \dots, z_0^j; \lambda_j^0)B\| > 1\|$$

and $\pi(z_0^0, \dots, z_0^j; \lambda_j)B = 0$ for $\lambda_j \notin [\lambda_j^0 - \varepsilon, \lambda_j^0 + \varepsilon]$, where ε is a given positive number. Let $a \in \mathfrak{M}(\mathcal{T})$ be a function in Lemma 3.2.5 such that

$$\|a(z_0^0, \dots, z_0^j; \cdot)\pi(z_0^0, \dots, z_0^j; \lambda_j^0)B\| = \|\pi(z_0^0, \dots, z_0^j; \lambda_j^0)aB\| > 1.$$

Let $A = aB$ and $\mathcal{U}(\pi_0; J) = U_A = \{\pi \in \widehat{\mathcal{A}} : \|\pi A\| > 1\}$. We check that $U_A \subset \mathcal{U}(\pi_0; \Sigma)$.

The representations obtained by "deeper" localization (i.e., the representations of the form $\pi(z_0^0, \dots, z_0^j, z_0^{j+1}; \cdot)$, $\pi(z_0^0, \dots, z_0^{j+1}, z_0^{j+2}; \cdot)$, etc.) annihilate the operator A to property 3 of the function a in Lemma 3.2.5. All the one-dimensional representations also vanish on A, since $A \in \operatorname{com} \mathcal{A}$. Every representation $\pi(z^0, \dots, z^i; \cdot)$, $0 \le i \le j$, for points $(z^0, \dots, z^i)$ not belonging to a small neighborhood $\mathcal{U}(z_0^0, \dots, z_0^i; \mathcal{Z}_i)$ is equal to zero on A in view of part 1 in Lemma 3.2.5. Hence, all these representations do not fall in U_A. It remains to recall the definition of $\mathcal{U}(\pi_0; \Sigma)$.

4^0. **Neighborhoods of points in** $S^*(\mathfrak{C})$. Finally, to conclude the proof of Theorem 3.2.1, we must still compare neighborhoods of the one-dimensional representations of $\mathcal{A}$ in the J- and Σ-topologies. Typical neighborhoods in the space Σ of inner points of the manifold $S^*(\mathfrak{C})$ are the same as in the cospherical bundle $S^*(\mathcal{M} \setminus |\mathcal{T}|)$. Let us consider points $(z^0, \dots, z^j; \varphi^j, \xi) \in \mathcal{Z}_j \times (S^{m-d_j-1} \setminus |\mathcal{T}(s_0, \dots, s_j)|) \times S^{m-1}$ in the boundary of the manifold $S^*(\mathfrak{C})$. We write the point $(z_0^0, \dots, z_0^j; \varphi_0^j, \xi_0)$ in the form $\pi_0 = \pi(z_0^0, \dots, z_0^j; \varphi_0^j, \xi_0^j, \theta_0^0, \dots, \theta_0^j)$ (as in Sect. 3.2.1). □

Proposition 3.2.19 *For every neighborhood $\mathcal{U}(\pi_0; J)$, there exists a neighborhood $\mathcal{U}(\pi_0; \Sigma)$ such that $\mathcal{U}(\pi_0; \Sigma) \subset \mathcal{U}(\pi_0; J)$.*

Proof Suppose that $A \in \mathcal{A}$, $\|\pi_0 A\| > 1, \mathcal{U}(\pi_0; J) = U_A = \{\pi \in \widehat{\mathcal{A}} : \|\pi A\| > 1\}$.

(a) $\theta_0^0 \neq 0$. By (3.2.4), $\|\pi(z_0^0, \dots, z_0^j; \Theta_0^j)A\| > 1$. Then, part 1 in Lemma 3.2.3 leads to the inequalities $\|\pi(z_0^0, \dots, z_0^i; \Theta_0^i)A\| > 1$, $i = 0, \dots, j$. Since the functions $(z^0, \dots, z^i; \Theta^i) \mapsto \|\pi(z^0, \dots, z^i; \Theta^i)A\|$ are continuous, we have $\mathcal{U}(\pi_0; \Sigma) \subset U_A$ for "small" neighborhoods $\mathcal{U}(\pi_0; \Sigma)$.

(b) $\theta_0^0 = 0$, $\theta_0^1 \neq 0$. Again, using (3.2.4) and part 1 of Lemma 3.2.3, we obtain the inequalities $\|\pi(z_0^0, \dots, z_0^i; \Theta_0^i)A\| > 1$ for $i = 1, \dots, j$ (that is, U_A contains the sets $\mathcal{U}(z_0^0, z_0^1, \dots, z_0^i; \mathcal{Z}_i) \times \mathcal{U}(\Theta_0^i; S^{d_i-1})$). By part 2 of Lemma 3.2.3, the estimate $\|\pi(z_0^0, z_0^1; \Theta_0^1)A\| = \|\pi(z_0^0, z_0^1; 0, \theta_0^1/|\theta_0^1|)A\| > 1$ implies that $\|\pi(z_0^0; \lambda^*)A\| > 1$ for some $\lambda^* \in \mathbb{R}$. Then, part 3 of the same lemma means that $\|\pi(z^0; \theta^0)A\| > 1$ for all $\{(z^0, \theta^0) : z^0 \in \mathcal{U}(z_0^0; \mathcal{Z}_0), \theta^0 \in S^*(\mathcal{Z}_0)_{z^0}\}$. Since $\|\pi(z_0^0, \dots, z_0^j; \Theta_0^j)A\| > 1$, it follows from Proposition 3.2.8 (in the case $z_0^1\theta_0^1 \gtrless 0$) that $\mathcal{U}(z_0^0; \mathcal{Z}_0) \times \{\lambda \in \mathbb{R} : \lambda \gtrless N\} \subset U_A$, and we obtain from Proposition 3.2.10 (in the case $z_0^1\theta_0^1 = 0$) the inclusion $\mathcal{U}(z_0^0; \mathcal{Z}_0) \times \mathbb{R} \subset U_A$. Therefore, the small neighborhood $\mathcal{U}(\pi_0; \Sigma)$ belongs to U_A.

(c) $\theta_0^0, \dots, \theta_0^i = 0$, θ_0^{i+1}. As in the preceding case, $\|\pi(z_0^0, \dots, z_0^j; \Theta_0^j)A\| > 1$. Using now Propositions 3.2.14–3.2.16 instead of Propositions 3.2.8 and 3.2.10, we again obtain the inclusion $\mathcal{U}(\pi_0; \Sigma) \subset U_A$.

(d) $\theta_0^0 = 0, \dots, \theta_0^j = 0$. It follows from the inequality $\|\pi_0 A\| > 1$ and (3.2.5) that $\|\pi(z_0^0, \dots, z_0^j; \lambda_j^*)A\| > 1$ for some $\lambda_j^* \in \mathbb{R}$. Then, part 3 of Lemma 3.2.3 gives us that $\|\pi(z_0^0, \dots, z_0^j; \Theta^j)A\| > 1$ for all $\Theta^j = (\theta, \dots, \theta^j)$ with $|\Theta^j| = 1$. Setting $\theta_0 = 0, \dots, \theta^{j-1} = 0$ and using Propositions 3.2.14–3.2.16, we see that U_A contains the sets (3.2.2) and (3.2.3) for $i = j$, $(z^0, \dots, z^p) = (z_0^0, \dots, z_0^p)$, and small $\mathcal{U}(z_0^0, \dots, z_0^p; \mathcal{Z}_p)$, $p = 0, \dots, j$.

We proceed to the representations $\pi(z^0, \dots, z^j; \lambda_j)$. If $\varphi_0^j\xi_0^j = 0$, then (3.2.6) ensures the inclusion $\pi(z_0^0, \dots, z_0^j; \lambda_j) \subset U_A$ for all $\lambda_j \in \mathbb{R}$. Suppose that $\varphi_0^j\xi_0^j \gtrless 0$ and let $\chi \in$

$C(S^{m-d_j-1})$ be a function such that $0 \le \chi \le 1$, $\chi(\varphi_0^j) = 1$, and $\operatorname{supp}\chi \cap |\mathcal{T}(s_0, \dots, s_j)| = \emptyset$, i.e., it vanishes on the discontinuities of the coefficients $a(z_0^0, \dots, z_0^j; \cdot)$, $a \in \mathfrak{M}(\mathcal{T})$. It can be assumed that $A = \sum_l \prod_k A_{lk}$, where A_{lk} are generators of the algebra $\mathcal{A}$. We write the operators $\pi(z_0^0, \dots, z_0^j; \lambda_j)A_{lk}$ in the form

$$\pi(z_0^0, \dots, z_0^j; \lambda_j)A_{lk} = \mathfrak{A}_{lk}(\lambda_j) = E_{\omega\to\varphi}(\lambda_j)^{-1}\Psi_{lk}(\varphi, \omega)E_{\psi\to\omega}(\lambda_j) \in \mathcal{B}L_2(S^{m-d_j-1}).$$

Let Q be a sufficiently large natural number. We have

$$\|\sum_l \prod_k \mathfrak{A}_{lk}(\lambda_j)\| \ge \|\chi^Q \sum_l \prod_k \mathfrak{A}_{lk}(\lambda_j)\| \ge \|\sum_l \chi^{Q-Q_l} \prod_k \chi\mathfrak{A}_{lk}(\lambda_j) + K(\lambda_j)\|, \tag{3.2.19}$$

where $\|\cdot\| = \|\cdot; \mathcal{B}L_2(S^{m-d_j-1})\|$ and $K(\cdot) \in C_0(\mathbb{R}) \otimes \mathcal{K}L_2(S^{m-d_j-1})$ (we used Proposition 1.2.11). The functions $(\varphi, \omega) \mapsto \chi(\varphi)\Psi_{lk}(\varphi, \omega)$ are continuous on $S^{m-d_j-1} \times S^{m-d_j-1}$. Since $\pi_0 A = \sum_l \prod_k \Psi_{lk}(\varphi_0^j, \xi_0^j)$ and $|\pi_0 A| > 1$, it follows that

$$\|\sum_l \chi^{Q-Q_l} \prod_k \chi\mathfrak{A}_{lk}(\lambda_j)\| > 1$$

for $\lambda_j \gtrless N$ with some $N \in \mathbb{R}$ (see Theorem 2.2.13). This and (3.2.19) give us that $\|\pi(z_0^0, \dots, z_0^j; \lambda_j)A\| > 1$, if $\lambda_j \gtrless N'$. Hence,

$$\mathcal{U}(z_0^0, \dots, z_0^j; \mathcal{Z}_j) \times \{\lambda_j \in \mathbb{R} : \lambda_j \gtrless N'\} \subset U_A.$$

Summarizing all the foregoing, we arrive at the relation $\mathcal{U}(\pi_0; \Sigma) \subset U_A$ some neighborhood $\mathcal{U}(\pi_0; \Sigma)$. □

Proposition 3.2.20 *For every neighborhood $\mathcal{U}(\pi_0; \Sigma)$ with $\pi_0 = \pi(z_0^0, \dots, z_0^j; \varphi_0^j, \xi_0^j, \theta_0^0, \dots, \theta_0^j)$, there exists a neighborhood $\mathcal{U}(\pi_0; J)$ such that $\mathcal{U}(\pi_0; J) \subset \mathcal{U}(\pi_0; \Sigma)$.*

The proof will only be sketched. We look for an element $A \in \mathcal{A}$ such that $U_A = \{\pi \in \widehat{\mathcal{A}} : \|\pi A\| > 1\}$ can be taken as $\mathcal{U}(\pi_0; J)$. In cases a) and b) (see Proposition 3.2.19), the difficulty lies in the fact that U_A need not contain "large" subsets of the spheres S^{d_i-1} appearing in the products $\mathcal{U}(z_0^0, z_0^1, \dots, z_0^i; \mathcal{Z}_i) \times \mathcal{U}(\Theta_0^i; S^{d_i-1})$. We show, for example, how this difficulty is overcome for $j = 0$, i.e., for the points $\pi(z_0^0; \varphi_0^0, \xi_0^0, \theta_0^0)$, where $\theta_0^0 \ne 0$ and $\varphi_0^0 \in S^{m-d_0-1} \setminus |\mathcal{T}(s_0)|$. Let $\Phi \in C(\mathbb{R}^m \setminus 0)$ be a homogeneous function of zero degree such that $\Phi(\xi_1, \dots, \xi_{d_0}, 0) \equiv 0$. For every $\chi \in C(S^{d_0-1})$, the homogeneous function Ψ of zero degree defined for $|\theta^0| = |\xi^0| = 1$ by the equality $\Psi(\xi^0, \theta^0) = \chi(\theta^0)\Phi(\xi^0, \theta^0)$ belongs to the class $C(\mathbb{R}^m \setminus 0)$. Therefore, there exist an $a \in \mathfrak{M}(\mathcal{T})$ and a χ satisfying $a(z^0; \varphi^0)\Psi(\xi^0, \theta^0) = 0$ outside a small neighborhood of the point

$(z_0^0; \varphi_0^0, \xi_0^0, \theta_0^0)$ and $a(z_0^0; \varphi_0^0)\Psi(\xi_0^0, \theta_0^0) > 1$. It is not hard to verify that the operator in $\mathcal{A}$ defined in a coordinate neighborhood of $z_0^0 \in s_0 \subset \mathcal{M}$ by $A = a(x)F_{\eta \to x}\Psi(\eta)F_{y \to \eta}$: $L_2(\mathbb{R}^m) \to L_2(\mathbb{R}^m)$, is the desired one.

In cases (c) or (d), no new difficulties arise in general under the condition $z_0^{i+1}\theta_0^{i+1} = 0$ or $\varphi_0^j\xi_0^j = 0$ because $\mathcal{U}(\pi_0; \Sigma)$ contains the set $\mathcal{U}(z_0^0, \dots, z_0^i; \mathcal{Z}_i) \times \mathbb{R}$ or $\mathcal{U}(z_0^0, \dots, z_0^j; \mathcal{Z}_j) \times \mathbb{R}$, respectively. If, for example, $z_0^{i+1}\theta_0^{i+1} > 0$ in case c), then we have to produce a neighborhood $\mathcal{U}(\pi_0; J)$ containing the set $\mathcal{U}(z_0^0, \dots, z_0^i; \mathcal{Z}_i) \times \{\lambda \in \mathbb{R} : \lambda > N\}$ only for sufficiently large N. Such a neighborhood can be constructed by using the considerations connected with the operator P in the proof of Proposition 3.2.12.

3.3 Solving Series

3.3.1 Construction of a Solving Series. Formulation of the Theorem

Let $\mathcal{T}$ be an admissible partition of manifold $\mathcal{M}$. The union of all strata with dimension no greater than j is called the j-dimensional skeleton of $\mathcal{T}$ and denoted by $\mathcal{T}_j$. Generally, the sequence $\emptyset = \mathcal{T}_{-1}, \mathcal{T}_0, \dots, \mathcal{T}_{m-1}$ can contain coinciding elements. We form a subsequence $\mathcal{T}_{-1}, \mathcal{T}_{m_1}, \dots, \mathcal{T}_{m_k}$, where $-1 \le m_k \le m-1$ and $\mathcal{T}_{m_p} = \mathcal{T}_{m_q}$ only for $p = q$. To this goal, moving from left to right, we include to the subsequence every element not encountered before. In what follows, we, instead of $\mathcal{T}_{m_j}$, will write simply $\mathcal{T}_j$, where $\dim \mathcal{T}_j = m_j$. Introduce all kinds of skeleton collections

$$(\mathcal{T}_{j_1}, \dots, \mathcal{T}_{j_q}), \quad 1 \le j_1 < \dots < j_q,$$

of the partition $\mathcal{T}$. To simplify notation once again, we write $(\mathcal{T}_{j_1}, \dots, \mathcal{T}_{j_q})$ instead of $(j_1, \dots, j_q)$. Thus, in a row $(j_1, \dots, j_q)$ the element j_k stands for the skeleton of dimension m_{j_k}.

We put in order the set of skeleton collections according to the rule: $(j_1, \dots, j_q) \prec (l_1, \dots, l_p)$ if and only if either $j_1 = l_1, \dots, j_h = l_h$ and $j_{h+1} > l_{h+1}$ or $p > q$ and $j_1 = l_1, \dots, j_q = l_q$. It is clear that one of any two skeleton collections is subject to the other. Introduce the empty collection $T_\emptyset$ and assume that $T_\emptyset$ is subject to the minimal collection $T_{\min} = (k)$. Collections T and T' are called neighboring if $T \prec T'$, and there is no such a collection T'' that $T \prec T'' \prec T'$. Thus, all collections form an increasing sequence of neighbors

$$T_\emptyset \prec T_1 \prec \dots \prec T_Q, \tag{3.3.1}$$

where $T_1 = T_{\min} = (k)$ and $T_Q = T_{\max} = (1, \dots k)$. The passage from $T = (j_1, \dots, j_q)$ to the neighboring $T' \succ T$ can be performed by one of the following operations:

(1) If $j_q < k$, then $T' = (j_1, \dots, j_q, k)$ (the number of components increases).

(2) If $j_q = k > 1$ and either $j_{q-1} < k-1$ or $q = 1$, then $T' = (j_1, \dots, j_{q-1}, k-1)$ (the number of components does not vary).
(3) If $j_q = k$, $j_{q-1} = k-1, \dots, j_{q-p} = k-p > 1$ and either $j_{q-p-1} < k-p-1$ or $q-p = 1$, then $T' = (j_1, \dots, j_{q-p-1}, j_{q-p} - 1)$ (the number of components decreases).

If none of the above operations is applicable, then T coincides with the maximal collection $(1, \dots, k)$. Sequence (3.3.1) can be written explicitly:

$$T_\emptyset \prec (k) \prec (k-1) \prec (k-1, k) \prec (k-2) \prec (k-2, k) \prec (k-2, k-1) \prec \quad (3.3.2)$$
$$\prec (k-2, k-1, k) \prec (k-3) \prec \cdots \prec (1, 2, \dots, k).$$

With every collection T_ν in (3.3.1), we associate an ideal $\mathcal{A}(T_\nu)$ in $\mathcal{A}$. To do that, we need some notation. Let I be the intersection of the kernels of all one-dimensional representations of the algebra $\mathcal{A}$,

$$I = \bigcap \ker \pi_1(x, \xi), \quad (3.3.3)$$

where (x, ξ) runs over the cospherical bundle $S^*(\mathfrak{C})$, $x \in \mathfrak{C}$, and $\xi \in S^*(\mathfrak{C})_x$, see (3.1.8). Denote the intersection of the kernels of all representations of the form (3.1.6) by $\Theta(T_\emptyset)$,

$$\Theta(T_\emptyset) = \bigcap_{\setminus T_\emptyset} \ker \pi(z^0, \dots, z^j; \theta^0, \dots, \theta^j). \quad (3.3.4)$$

Moreover, let $\Lambda(T_\emptyset)$ stand for the intersection of the kernels of all representations of the form (3.1.7),

$$\Lambda(T_\emptyset) = \bigcap_{\setminus T_\emptyset} \ker \pi(z^0, \dots, z^j; \lambda). \quad (3.3.5)$$

We set

$$\mathcal{A}(T_\emptyset) = I \bigcap \Theta(T_\emptyset) \bigcap \Lambda(T_\emptyset). \quad (3.3.6)$$

Thus, $\mathcal{A}(T_\emptyset)$ is the intersection of the kernels of all irreducible representations of $\mathcal{A}$, except for the identity representation.

We introduce the ideal $\Theta(T_1)$ for $m_k > 0$ as the intersection of the kernels of all representations (3.1.6) except for representations of the form $\pi(z^0; \theta^0)$, where a point z^0 runs over all the strata s of dimension m_k, while $(z^0; \theta^0)$ runs over the cospherical bundle

$S^*(s)$ over each stratum s. Let us write $\Theta(T_1)$ in the form

$$\Theta(T_1) = \bigcap_{\setminus T_1} \ker \pi(z^0, \dots, z^j; \theta^0, \dots, \theta^j) \tag{3.3.7}$$

(the $\setminus T_1$ indicates the representations that are absent in the intersection considered). The representations $\pi(z^0; \theta^0)$, where $z^0 \in s$, are not defined for the zero-dimensional strata s (however, generally, the representations $\pi(z^0, z^1; \theta^1)$ and similar ones make sense). Therefore, in the case of $m_k = 0$, we obtain

$$\Theta(T_1) = \Theta(T_\emptyset). \tag{3.3.8}$$

In a similar manner, we set

$$\Lambda(T_1) = \bigcap_{\setminus T_1} \ker \pi(z^0, \dots, z^j; \lambda), \tag{3.3.9}$$

assuming that the right-hand side of (3.3.9) is the intersection of the kernels of all representations (3.1.7) except for the representations $\pi(z^0; \lambda)$, where z^0 runs over all m_k-dimensional strata s and λ runs the real axis.

Suppose that the ideals $\Theta(T)$, $\Lambda(T)$ have already been defined for all T in the sequence (3.3.1) such that $T \prec T_\nu$, while $\Theta(T)$ and $\Lambda(T)$ are given as the intersections of some primitive ideals, i.e., the kernels of irreducible representations. We introduce ideals $\Theta(T_\nu)$, $\Lambda(T_\nu)$. Let $T_\nu = (j_1, \dots, j_q)$ and consider all chains of strata $s_1, \dots, s_q$ satisfying $s_\alpha \in \operatorname{st}(s_\beta)$ for $\alpha > \beta$ and $\dim s_\alpha = m_{j_\alpha}$, $\alpha = 1, \dots, q$. If there is no such a chain, then we set

$$\Theta(T_\nu) = \Theta(T_{\nu-1}), \quad \Lambda(T_\nu) = \Lambda(T_{\nu-1}). \tag{3.3.10}$$

Assume now that some chains with the mentioned property exist and $\pi(z^1, \dots, z^q; \theta^1, \dots, \theta^q)$ are the representations obtained by localization along such a chain. From (3.3.2), it follows that the kernels of these representations participate in the intersection, which defines the ideal $\Theta(T_{\nu-1})$. Removing the kernels $\ker \pi(z^1, \dots, z^q; \theta^1, \dots, \theta^q)$ of all such representations from the mentioned intersection, we obtain the ideal $\Theta(T_\nu)$,

$$\Theta(T_\nu) = \bigcap_{\setminus T_\nu} \ker \pi(z^1, \dots, z^q; \theta^1, \dots, \theta^q). \tag{3.3.11}$$

Analogously, removing the $\ker \pi(z^0, \dots, z^j; \lambda)$ from the intersection of ideals equal to $\Lambda(T_{\nu-1})$, we obtain the ideal

$$\Lambda(T_\nu) = \bigcap_{\setminus T_\nu} \ker \pi(z^0, \dots, z^j; \lambda). \tag{3.3.12}$$

Furthermore, if $T_{\nu+1} = (j_1)$ provided $m_{j_1} = 0$ (which means that the collection $T_{\nu+1}$ is represented by the zero-dimensional skeleton), then the same reason as in the case of (3.3.8) leads to

$$\Theta(T_{\nu+1}) = \Theta(T_\nu). \tag{3.3.13}$$

Continuing the process, we define the ideals $\Theta(T)$ and $\Lambda(T)$ for all collections T in the sequence (3.3.1). In particular, $\Theta(T_Q) = \mathcal{A}$ and $\Lambda(T_Q) = \mathcal{A}$. Introduce the ideals

$$\mathcal{A}(T) := I \cap \Theta(T) \cap \Lambda(T), \tag{3.3.14}$$

where I is the same ideal as in (3.3.3). It is evident that if $T \prec T'$, then $\Theta(T) \subset \Theta(T')$, $\Lambda(T) \subset)\Lambda(T')$ and therefore $\mathcal{A}(T) \subset \mathcal{A}(T')$. Among $\mathcal{A}(T)$, identical ideals can occur; see (3.3.10). Let $\mathbf{T}_0 := T_\emptyset$, $\mathbf{T}_1 := T_{i_1}, \ldots, \mathbf{T}_{p-1} := T_{i_{p-1}}$, $\mathbf{T}_p := T_Q$ be a subsequence of sequence (3.3.1) such that the composition series

$$\mathcal{A}(\mathbf{T}_0) \subset \mathcal{A}(\mathbf{T}_1) \subset \cdots \subset \mathcal{A}(\mathbf{T}_p) \subset \mathcal{A} \tag{3.3.15}$$

consists of all different ideals of the form $\mathcal{A}(T)$ (and of only those ones); note that always $i_1 = 1$, i.e., $\mathbf{T}_1 = T_1 = T_{\min} = (k)$, see (3.3.1).

Generally, the series (3.3.15) is not solving. To obtain a solving composition series, we introduce intermediate ideals $J(\mathbf{T}_j)$. For this purpose, we consider neighboring ideals $\mathcal{A}(\mathbf{T}_j)$ and $\mathcal{A}(\mathbf{T}_{j+1})$ in (3.3.15). According to (3.3.14), we have $\mathcal{A}(\mathbf{T}_j) = I \cap \Theta(\mathbf{T}_j) \cap \Lambda(\mathbf{T}_j)$. Set

$$\mathcal{J}(\mathbf{T}_j) = I \cap \Theta(\mathbf{T}_{j+1}) \cap \Lambda(\mathbf{T}_j) \tag{3.3.16}$$

for $j = 0, \ldots, p-1$. If the zero-dimensional skeleton $\mathcal{T}_0$ of the partition $\mathcal{T}$ of $\mathcal{M}$ is not empty, then for a certain ν, the equality

$$\mathcal{J}(\mathbf{T}_\nu) = \mathcal{A}(\mathbf{T}_\nu) \tag{3.3.17}$$

holds (see (3.3.13)), while for $j \neq \nu$, we have the strict inclusions

$$\mathcal{A}(\mathbf{T}_j) \subset \mathcal{J}(\mathbf{T}_j) \subset \mathcal{A}(\mathbf{T}_{j+1}). \tag{3.3.18}$$

Thus, if the skeleton $\mathcal{T}_0$ is empty, then (3.3.18) holds for all $j = 0, \ldots, p-1$, and we obtain the composition series

$$0 \subset \mathcal{A}(\mathbf{T}_0) \subset \mathcal{J}(\mathbf{T}_0) \subset \mathcal{A}(\mathbf{T}_1) \subset \mathcal{J}(\mathbf{T}_1) \subset \cdots \subset \mathcal{A}(\mathbf{T}_p) \subset \mathcal{A}. \tag{3.3.19}$$

If $\mathcal{T}_0 \neq \emptyset$, then, for the number ν in (3.3.17), the series (3.3.19) contains the link $\mathcal{A}(\mathbf{T}_\nu) \subset \mathcal{A}(\mathbf{T}_{\nu+1})$ instead of $\mathcal{A}(\mathbf{T}_\nu) \subset \mathcal{J}(\mathbf{T}_\nu) \subset \mathcal{A}(\mathbf{T}_{\nu+1})$.

Let $Z_0 \times \cdots \times Z_q$ be the submanifold defined for a chain of strata $s_0, \ldots, s_q$ as at the beginning of Sect. 3.2.1. We say that a chain $s_0, \ldots, s_q$ is connected with a skeleton collection $\mathbf{T}_\nu = (j_0, \ldots, j_q)$ if $\dim s_\alpha = m_{j_\alpha}$, $\alpha = 0, \ldots, q$ (compared with the description of formulas (3.3.10) and (3.3.11)). Denote by $\boldsymbol{\Upsilon}_\nu$ the union of manifolds of the form $Z_0 \times \cdots \times Z_q$ corresponding to all chains connected with the collection $\mathbf{T}_\nu$. Finally, we set $\delta_\nu = \dim m_{j_q}$.

Theorem 3.3.1 *The composition series (3.3.19) is solving. Moreover,*

$$\mathcal{A}(\boldsymbol{T}_0) \simeq \mathcal{K}L_2(\mathcal{M}), \tag{3.3.20}$$

$$\mathcal{A}/\mathcal{A}(\boldsymbol{T}_p) \simeq C(S^*(\mathfrak{C})), \tag{3.3.21}$$

where $C(S^*(\mathfrak{C}))$ is the cospherical bundle over the compact $\mathfrak{C}$ (see the description of $\mathfrak{C}$ after (3.1.5)) and $\mathcal{A}(\boldsymbol{T}_p) = \mathrm{com}\mathcal{A}$. If the zero-dimensional skeleton $\mathcal{T}_0$ of the partition $\mathcal{T}$ of $\mathcal{M}$ is empty, then

$$\mathcal{J}(\boldsymbol{T}_j)/\mathcal{A}(\boldsymbol{T}_j) \simeq C_0(S^*(\boldsymbol{\Upsilon}_j)) \otimes \mathcal{K}L_2(\mathbb{R}^{m-\delta_j}), \tag{3.3.22}$$

where $j = 0, \ldots, p-1$, and

$$\mathcal{A}(\boldsymbol{T}_j)/\mathcal{J}(\boldsymbol{T}_{j-1}) \simeq C_0(\boldsymbol{\Upsilon}_j \times \mathbb{R}) \otimes \mathcal{K}L_2(S^{m-\delta_j-1}), \tag{3.3.23}$$

where $j = 1, \ldots, p$. If $\mathcal{T}_0 \neq \emptyset$, then for the number ν in (3.3.13), instead of the link $\mathcal{A}(\boldsymbol{T}_\nu) \subset \mathcal{J}(\boldsymbol{T}_\nu) \subset \mathcal{A}(\boldsymbol{T}_{\nu+1})$, the series (3.3.19) contains $\mathcal{A}(\boldsymbol{T}_\nu) \subset \mathcal{A}(\boldsymbol{T}_{\nu+1})$ and the two formulas (3.3.22) for $j = \nu$ and (3.3.23) for $j = \nu + 1$ are replaced by the only relation

$$\mathcal{A}(\boldsymbol{T}_{\nu+1})/\mathcal{A}(\boldsymbol{T}_\nu) \simeq C(|\mathcal{T}_0| \times \mathbb{R}) \otimes \mathcal{K}L_2(S^{m-1}). \tag{3.3.24}$$

3.3.2 Proof of Theorem 3.3.1

1^0. **Verification of formulas (3.3.20) and (3.3.21).** We first consider (3.3.20). From Theorem 3.1.4, it follows that all irreducible representations of the algebra $\mathcal{A}$, except for the identity one, vanish on the ideal $\mathcal{A}(\mathbf{T}_0)$. Since the algebra $\mathcal{A}$ is irreducible, the ideal $\mathcal{A}(\mathbf{T}_0)$ is irreducible as well. Moreover, the ideal contains some compact operators, for example, the commutators $[a, P]$, where $a \in C^\infty(\mathcal{M})$, $a|\mathcal{T}_{m-1} = 0$, and P is a ψDO in $\bar{\Psi}^0(\mathcal{M})$. Therefore, $\mathcal{K}L_2(\mathcal{M}) \subset \mathcal{A}(\mathbf{T}_0)$. The identity representation is the only irreducible representation of $\mathcal{A}(\mathbf{T}_0)$, so $\mathcal{A}(\mathbf{T}_0) \simeq \mathcal{K}L_2(\mathcal{M})$.

Let us turn to (3.3.21). It was noticed before (3.3.14) that $\Theta(\mathbf{T}_p) = \Lambda(\mathbf{T}_p) = \mathcal{A}$. Recall that $\mathbf{T}_p = T_Q$ and therefore, by virtue of (3.3.14), $\mathcal{A}(\mathbf{T}_p) = I$, where I is the intersection of the kernels of all one-dimensional representations of $\mathcal{A}$. Hence, $\mathcal{A}(\mathbf{T}_p) = \mathrm{com}\mathcal{A}$. It remains to take into account that, according to Theorem 3.1.4, the list (3.1.8) contains any one-dimensional representation of $\mathcal{A}$.

2^0. **Verification of formulas (3.3.22).** We first consider the case $j = 0$ and the equality

$$\mathcal{J}(\mathbf{T}_0)/\mathcal{A}(\mathbf{T}_0) \simeq C_0(S^*(\mathbf{\Upsilon}_0)) \otimes \mathcal{K}L_2(\mathbb{R}^{m-\delta_0});$$

here, $\mathcal{J}(\mathbf{T}_0) = I \cap \Theta(\mathbf{T}_1) \cap \Lambda(\mathbf{T}_0)$, the ideal $\Theta(\mathbf{T}_1) = \Theta(T_1)$ is defined by (3.3.7) (for $m_k > 0$, which is supposed to be fulfilled for the time being), and $\Lambda(\mathbf{T}_0)$ is given by (3.3.5). Thus, the spectrum $\mathcal{J}(\mathbf{T}_0)^\wedge$ consists of representations of the form $\pi(z^0; \theta^0)$, where z^0 runs over the union $\mathbf{\Upsilon}_0$ of all strata of (maximal) dimension m_k and $(z^0; \theta^0)$ runs over the cospherical bundle $S^*(\mathbf{\Upsilon}_0)$.

Let us consider the algebra $\pi(z^0; \theta^0)\mathcal{A}$ for a fixed point $(z^0; \theta^0)$. Note that the representations $\pi(z^0; \lambda)$ of $\mathcal{A}$ arose in the localization procedure as representations of the algebra $\pi(z^0; \theta^0)\mathcal{A}$. Therefore, inclusion $A \in \ker \pi(z^0; \lambda)$ implies $\pi(z^0; \lambda)\pi(z^0; \theta^0)A = 0$. It is clear that all one-dimensional representations of $\pi(z^0; \theta^0)\mathcal{A}$ are also representations of $\mathcal{A}$. Hence, all irreducible representations of $\pi(z^0; \theta^0)\mathcal{A}$ are described by Theorem 2.3.12. It now follows from this Theorem that $\pi(z^0; \theta^0)A \in \mathcal{K}L_2(\mathbb{R}^{m-m_k}_{z^0})$ for all operators A in $\mathcal{J}(\mathbf{T}_0)$. (We use the same notation for the space $\mathbb{R}^{m-m_k}_{z^0}$ and its lifting to the fiber $S^*(\mathbf{\Upsilon}_0)_{z^0}$; the same is true for $S^{m-m_k-1}_{z^0}$.)

We introduce a continuous field of elementary algebras on $S^*(\mathbf{\Upsilon}_0)$. To this goal with every point $(z^0; \theta^0) \in S^*(\mathbf{\Upsilon}_0)$, we associate the algebra $\mathcal{K}L_2(\mathbb{R}^{m-m_k}_{z^0})$. Any element A of the ideal $\mathcal{J}(\mathbf{T}_0)$ gives rise to the vector field

$$F_A : (z^0; \theta^0) \mapsto \pi(z^0; \theta^0)A \in \mathcal{K}L_2(\mathbb{R}^{m-m_k}_{z^0}). \tag{3.3.25}$$

All representations of $\pi(z^0; \theta^0)$ vanish on the ideal $\mathcal{A}(\mathbf{T}_0)$, so the field F_A depends only on the residue class $[A]$ of an operator A in $\mathcal{J}(\mathbf{T}_0)/\mathcal{A}(\mathbf{T}_0)$. On the set of vector fields of the form (3.3.25), we introduce the norm

$$\|F_A; \mathcal{F}_0\| = \sup\{\|\pi(z^0; \theta^0)A; \mathcal{K}L_2(\mathbb{R}^{m-m_k}_{z^0})\|; (z^0; \theta^0) \in S^*(\mathbf{\Upsilon}_0)\};$$

in what follows, $\mathcal{F}_0$ stands for the algebra of fields (3.3.25) with that norm. Assume in addition that Γ_0 is the set of vector fields that are limits of the fields in $\mathcal{F}_0$ with respect to the local uniform convergence. Taking the elements in Γ_0 as continuous vector fields, we introduce the continuous field $(\{\mathcal{K}L_2(\mathbb{R}^{m-m_k}_{z^0})\}, \Gamma_0)$ of elementary algebras on $S^*(\mathbf{\Upsilon}_0)$.

According to Theorem 3.1.4, the spectrum of $\mathcal{J}(\mathbf{T}_0)/\mathcal{A}(\mathbf{T}_0)$ can be parametrized by the points of the set $S^*(\boldsymbol{\Upsilon}_0)$. Therefore, for the residue class $[A] \in \mathcal{J}(\mathbf{T}_0)/\mathcal{A}(\mathbf{T}_0)$ of the operator $A \in \mathcal{J}(\mathbf{T}_0)$, there holds the equality

$$\|[A]\| = \sup\{\|\pi(z^0;\theta^0)A; \mathcal{K}L_2(\mathbb{R}^{m-m_k}_{z^0})\|; (z^0;\theta^0) \in S^*(\boldsymbol{\Upsilon}_0)\}.$$

Hence, the mapping $[A] \mapsto F_A$ is an isomorphism of C^*-algebras

$$\mathcal{J}(\mathbf{T}_0)/\mathcal{A}(\mathbf{T}_0) \simeq \mathcal{F}_0. \tag{3.3.26}$$

From Theorem 3.2.1, it follows that the Jacobson topology on the spectrum $(\mathcal{J}(\mathbf{T}_0)/\mathcal{A}(\mathbf{T}_0))^\wedge$ coincides with the topology of cospherical bundle $S^*(\boldsymbol{\Upsilon}_0)$, i.e., the space $(\mathcal{J}(\mathbf{T}_0)/\mathcal{A}(\mathbf{T}_0))^\wedge$ is finite-dimensional, Hausdorff, and locally compact. The set $\{\pi \in (\mathcal{J}(\mathbf{T}_0)/\mathcal{A}(\mathbf{T}_0))^\wedge : \|\pi A\| \geq \alpha\}$ is compact for any $\alpha > 0$ and every operator $A \in \mathcal{J}(\mathbf{T}_0)$ ([3], 3.3.7). This means that every vector field F_A, $S^*(\boldsymbol{\Upsilon}_0) \ni \pi \mapsto \pi(A)$, tends to zero at infinity. The algebra defined by the continuous field $(\{\mathcal{K}L_2(\mathbb{R}^{m-m_k}_{z^0})\}, \Gamma_0)$ is isomorphic to $\mathcal{F}_0$ ([3], 10.5.4).

The field $(\{\mathcal{K}L_2(\mathbb{R}^{m-m_k}_{z^0})\}, \Gamma_0)$ is locally trivial; as local trivializations, one can use those of the normal bundle over $\boldsymbol{\Upsilon}_0$. (In connection with the normal bundle over $\boldsymbol{\Upsilon}_j$, see the definition of the fiber $\mathbb{R}^{m-d_q}(z^0,\ldots,z^q)$ over a point $(z^0,\ldots,z^q) \in \mathcal{Z}_q = Z_0 \times \cdots \times Z_q$ given before Theorem 3.1.4.) Proposition 1.3.18 provides the triviality of this field. Therefore, $\mathcal{F}_0 \simeq C_0(S^*(\boldsymbol{\Upsilon}_0)) \otimes \mathcal{K}L_2(\mathbb{R}^{m-m_k})$. Taking into account (3.3.26), we obtain (3.3.22) for $j=0$.

Assume now that $m_k = 0$; then, $k = 1$, and the series (3.3.19) is of the form $0 \subset \mathcal{A}(\mathbf{T}_0) \subset \mathcal{A}(\mathbf{T}_1) \subset \mathcal{A}$. Thus, instead of (3.3.22) for $j=0$ and (3.3.23) for $j=1$, we must prove the isomorphism

$$\mathcal{A}(\mathbf{T}_1)/\mathcal{A}(\mathbf{T}_0) \simeq C_0(|\mathcal{T}_0| \times \mathbb{R}) \otimes \mathcal{K}L_2(S^{m-1}); \tag{3.3.27}$$

here, $\mathcal{A}(\mathbf{T}_1) = I = \operatorname{com}\mathcal{A}$, $\mathcal{A}(\mathbf{T}_0) = I \cap \Lambda(\mathbf{T}_0)$, and $\Lambda(\mathbf{T}_0) = \bigcap_{(z,\lambda)} \ker \pi(z;\lambda)$, while λ runs over the real axis and z runs over the finite set $|\mathcal{T}_0|$. According to (3.3.20), $\mathcal{A}(\mathbf{T}_0) = \mathcal{K}L_2(\mathcal{M})$.

We consider functions of the form $\mathbb{R} \ni \lambda \mapsto \pi(z;\lambda)A \in \mathcal{B}L_2(S^{m-1})$, where $A \in \mathcal{A}(\mathbf{T}_1)$ and z is a fixed point in $|\mathcal{T}_0|$. The representations $\pi(z;\lambda)$ vanish on the ideal $\mathcal{A}(\mathbf{T}_0)$, so the function $\pi(z;\cdot)A$ depends only on the residue class $[A]$ in $\mathcal{A}(\mathbf{T}_1)/\mathcal{A}(\mathbf{T}_0)$. Such a function is an element of the algebra $\mathfrak{S}(z)$. From the definition of the ideal $\mathcal{A}(\mathbf{T}_1)$, it follows that the ideal spanned by the functions in $\mathfrak{S}(z)$ belongs to the kernel of every one-dimensional representation of $\mathfrak{S}(z)$ and, consequently, coincides with the ideal $\operatorname{com}\mathfrak{S} \simeq C_0(\mathbb{R}) \otimes \mathcal{K}L_2(S^{m-1})$. This leads to (3.3.27).

We now turn to (3.3.22) for $j > 0$. Assume that $\mathbf{T}_j = (j_1,\ldots,j_q)$, the ideals $\Theta(\mathbf{T}_j)$ and $\Lambda(\mathbf{T}_j)$ are defined in (3.3.11) and (3.3.12), $\mathcal{A}(\mathbf{T}_j) = I \cap \Theta(\mathbf{T}_j) \cap \Lambda(\mathbf{T}_j)$, and $\mathcal{J}(\mathbf{T}_j) = I \cap \Theta(\mathbf{T}_{j+1}) \cap \Lambda(\mathbf{T}_j)$. From Theorem 3.1.4 and the definition of the ideals $\mathcal{A}(\mathbf{T}_j)$ and $\mathcal{J}(\mathbf{T}_j)$, it follows that the spectrum $(\mathcal{J}(\mathbf{T}_j)/\mathcal{A}(\mathbf{T}_j))^\wedge$ consists of the representations $\pi(z^1, \dots, z^q; \theta^1, \dots, \theta^q)$, where $z := (z^1, \dots, z^q) \in \boldsymbol{\Upsilon}_j$ and $\theta := (\theta^1, \dots, \theta^q) \in S^*(\boldsymbol{\Upsilon}_j)_z$. For a fixed point (z, θ), we consider the algebra $\pi(z; \theta)\mathcal{J}(\mathbf{T}_j)$. From Theorem 3.1.4 (and the localization procedure in its proof), it follows that the element $\pi(z; \theta)A$ for $A \in \mathcal{J}(\mathbf{T}_j)$ depends only on the residue class $[A] \in \mathcal{J}(\mathbf{T}_j)/\mathcal{A}(\mathbf{T}_j)$ and the only irreducible representation (up to equivalence) of the quotient algebra $\mathcal{J}(\mathbf{T}_j)/\mathcal{A}(\mathbf{T}_j)$ is the identity one. Moreover, as in the case $j = 0$, there hold the inclusion $\mathcal{K}L_2(\mathbb{R}^{m-\delta_j}) \subset \pi(z; \theta)\mathcal{J}(\mathbf{T}_j)$ and the equality $\pi(z; \theta)(\mathcal{J}(\mathbf{T}_j)/\mathcal{A}(\mathbf{T}_j)) = \mathcal{K}L_2(\mathbb{R}^{m-\delta_j})$. To complete the verification of (3.3.22) for $j > 0$, it remains to repeat with evident modifications the proof of the formula for $j = 0$.

Verification of Formulas (3.3.23)

The spectrum $(\mathcal{A}(\mathbf{T}_j)/\mathcal{J}(\mathbf{T}_{j-1}))^\wedge$ can be parametrized by the points $(z; \lambda)$, where $z = (z^1, \dots, z^q) \in \boldsymbol{\Upsilon}_j$ and $\lambda \in \mathbb{R}$. The operator $\pi(z; \lambda)A$ with $A \in \mathcal{A}(\mathbf{T}_j)$ depends only on the residue class $[A] \in \mathcal{A}(\mathbf{T}_j)/\mathcal{J}(\mathbf{T}_{j-1})$. For a fixed z, the function $\lambda \mapsto \pi(z; \lambda)A$ is an element of the algebra $\mathfrak{S}(z)$. The ideal in $\mathfrak{S}(z)$ spanned by such functions with $A \in \mathcal{A}(\mathbf{T}_j)$ coincides with the ideal com $\mathfrak{S}(z) \simeq C_0(\mathbb{R}) \otimes \mathcal{K}L_2(S_z^{m-\delta_j-1})$.

Define a field of elementary algebras on $\boldsymbol{\Upsilon}_j \times \mathbb{R}$. With every point $(z; , \lambda) \in \boldsymbol{\Upsilon}_j \times \mathbb{R}$, we associate the algebra $\mathcal{K}L_2(S_z^{m-\delta_j-1})$. Denote by $\mathcal{G}_j$ the algebra of vector fields of the form

$$G_A : (z; \lambda) \mapsto \pi(z; \lambda)A \in \mathcal{K}L_2(S_z^{m-\delta_j-1}) \tag{3.3.28}$$

endowed with the norm

$$\|G_a; \mathcal{G}_j\| = \sup\{\|\pi(z; \lambda)A; \mathcal{B}L_2(S_z^{m-\delta_j-1})\|, (z; \lambda) \in \boldsymbol{\Upsilon}_j \times \mathbb{R}\}.$$

Let Δ_j stand for the set of limits of the fields in $\mathcal{G}_j$ with respect to the local uniform convergence. We introduce the continuous field of elementary algebras $(\{\mathcal{K}L_2(S_z^{m-\delta_j-1})\}, \Delta_j)$, where Δ_j plays the role of a set of continuous vector fields. The mapping $[A] \mapsto G_A$ performs an isomorphism $\mathcal{A}(\mathbf{T}_j)/\mathcal{J}(\mathbf{T}_{j-1}) \simeq \mathcal{G}_j$. The algebra defined by the field $(\{\mathcal{K}L_2(S_z^{m-\delta_j-1})\}, \Delta_j)$ is isomorphic to $\mathcal{G}_j$. The triviality of the field is provided by Proposition 1.3.18, as by verification of formulas (3.3.22). This leads to the relation (3.3.23). □

4 Pseudodifferential Operators on Manifolds with Smooth Closed Edges

In this chapter, we define pseudodifferential operators of arbitrary order on "manifolds with edges" and discuss the general properties of these operators. The presented results will be used in Chap. 5 devoted to C^*-algebras generated by pseudodifferential operators of zero order. In Chaps. 4 and 5, we limit ourselves to manifolds with smooth closed non-intersecting edges.

To clarify the definition of the operators, let us consider the subspace $x^1 = 0$ in $\mathbb{R}^m = \{x = (x^1, x^2) : x^1 = (x_1, \dots, x_n), x^2 = (x_{n+1}, \dots, x_m)\}$ as a "wedge." Our operator class representative is, for example, a μ-order Ψdo of the form

$$(2\pi)^{-m/2} \int e^{ix\xi} a(x, |x^1|\xi)\hat{u}(\xi)\, d\xi, \tag{4.0.1}$$

where a is a function subjected to the estimates

$$|x^1|^{|\alpha|} |\partial_x^\alpha \partial_\eta^\beta a(x, \eta)| \leq C(\alpha, \beta)\langle \eta \rangle^{\mu - |\beta|}$$

for all multi-indices α, β; as usual, $\hat{u}$ is the Fourier transform of u and $\langle \eta \rangle = (1 + |\eta|^2)^{1/2}$. Note that a differential operator of the form $a(x, |x^1|D_x) = \sum_{|\alpha| \leq \mu} a_\alpha(x)(|x^1|D_x)^\alpha$ is a natural and traditional object in the theory of boundary value problems in domains with edges.

Manifolds with edges that we will consider can be locally represented as a surface of the form $K \times \mathbb{R}^{m-n}$, where K is an n-dimensional conical surface smooth outside its origin. Rectifying a neighborhood of the base of the cone K, we obtain the set $\mathcal{K} \times \mathbb{R}^{m-n}$, where $\mathcal{K}$ is an open cone in $\mathbb{R}^n$. Therefore, studying Ψdo on a manifold with edges begins with consideration of operators of the form (4.0.1) in $\mathbb{R}^m_n := \{x = (x^1, x^2) \in \mathbb{R}^m : x^1 \neq 0\}$.

© The Author(s), under exclusive license to Springer Nature Switzerland AG 2023

B. Plamenevskii, O. Sarafanov, *Solvable Algebras of Pseudodifferential Operators*, Pseudo-Differential Operators 15, https://doi.org/10.1007/978-3-031-28398-7_4

4.1 Pseudodifferential Operators in $\mathbb{R}^m_n$

4.1.1 Amplitudes

Let m and n be integers, $1 \leq n \leq m$. A point $x \in \mathbb{R}^m$ will be written in the form (x^1, x^2), where $x^1 = (x_1, \ldots, x_n)$ and $x^2 = (x_{n+1}, \ldots, x_m)$. We set $\mathbb{R}^m_n = \{x = (x^1, x^2) : x^1 \neq 0\}$.

Definition 4.1.1 A function $\tilde{a} \in C^\infty(\mathbb{R}^m_n \times \mathbb{R}^m_n \times \mathbb{R}^m)$ is a *pre-amplitude* of order μ ($\mu \in \mathbb{R}$), if for any multi-indices α, β, γ there exists a constant $C_{\alpha\beta\gamma}$ such that

$$|\partial_x^\alpha \partial_y^\beta \partial_\xi^\gamma \tilde{a}(x, y, \xi)| \leq C_{\alpha\beta\gamma} |x^1|^{-|\alpha|} |y^1|^{-|\beta|} \langle\xi\rangle^{\mu-|\gamma|}$$

for all $(x, y, \xi) \in \mathbb{R}^m_n \times \mathbb{R}^m_n \times \mathbb{R}^m$. For each pre-amplitude $\tilde{a}$, we assign *an amplitude* of order μ,

$$a(x, y, \xi) = \tilde{a}(x, y, |x^1|\,\xi). \tag{4.1.1}$$

The class of all pre-amplitudes (resp., amplitudes) of order μ will be denoted by $\tilde{\Pi}^\mu(\mathbb{R}^m_n)$ ($\Pi^\mu(\mathbb{R}^m_n)$).

Definition 4.1.2 A pre-amplitude $\tilde{a}$ and the corresponding amplitude a are called *proper*, if there exist numbers $\delta \in (0, 1)$ and $\varepsilon > 0$ such that $\tilde{a}(x, y, \xi) = 0$ for $|x^1|/|y^1| \notin (\delta, \delta^{-1})$ or $|x^2 - y^2| > \varepsilon(1 + |x^1|)$. The class of proper pre-amplitudes (resp., amplitudes) of order μ will be denoted by $\tilde{\Pi}^\mu_0(\mathbb{R}^m_n)$ ($\Pi^\mu_0(\mathbb{R}^m_n)$).

The following assertions are obvious:

1. For any $\mu \in \mathbb{R}$, the classes $\tilde{\Pi}^\mu$ and Π^μ are complex vector spaces.
2. If $\mu \leq \nu$, then $\tilde{\Pi}^\mu \subset \tilde{\Pi}^\nu$ and $\Pi^\mu \subset \Pi^\nu$.
3. $\tilde{\Pi}^\mu \tilde{\Pi}^\nu \subset \tilde{\Pi}^{\mu+\nu}$, $\Pi^\mu \Pi^\nu \subset \Pi^{\mu+\nu}$ for all $\mu, \nu \in \mathbb{R}$.
4. Let $\tilde{a} \in \tilde{\Pi}^\mu$ and $\alpha, \beta, \gamma \in \mathbb{Z}^m_+$. Then, the functions

$$(x, y, \xi) \mapsto |x^1|^{|\alpha|} |y^1|^{|\beta|} \partial_x^\alpha \partial_y^\beta \partial_\xi^\gamma \tilde{a}(x, y, \xi),$$
$$(x, y, \xi) \mapsto |x^1|^{|\alpha|-|\gamma|} |y^1|^{|\beta|} \partial_x^\alpha \partial_y^\beta \partial_\xi^\gamma a(x, y, \xi)$$

belong to the classes $\tilde{\Pi}^{\mu-|\gamma|}$ and $\Pi^{\mu-|\gamma|}$, respectively.

For singular pre-amplitudes and amplitudes, Properties 1–4 are also valid, and Property 4 can be strengthened.

5. Let $p, q \in \mathbb{R}$ and let the multi-indices α, β be such that $p + q = |\alpha| + |\beta|$. If $\tilde{a} \in \tilde{\Pi}^{\mu}_0$, then the functions

$$(x, y, \xi) \mapsto |x^1|^p |y^1|^q \partial^\alpha_x \partial^\beta_y \partial^\gamma_\xi \tilde{a}(x, y, \xi),$$

$$(x, y, \xi) \mapsto |x^1|^{p-|\gamma|} |y^1|^q \partial^\alpha_x \partial^\beta_y \partial^\gamma_\xi a(x, y, \xi)$$

belong to the classes $\tilde{\Pi}^{\mu-|\gamma|}_0$ and $\Pi^{\mu-|\gamma|}_0$, respectively.
6. Let $a \in \Pi^\mu_0$ and $b(x, y, \xi) = a(y, x, \xi)$. Then, $b \in \Pi^\mu_0$. (If $a \in \Pi^\mu$, then the analogous assertion is in general false.)

In what follows, we use several times the proper amplitude constructed in the next example.

Example 4.1.3 Let the numbers δ, δ_1, and ε satisfy $0 < \delta < \delta_1 < 1$, $\varepsilon > 0$, and let the functions $\chi_1, \chi_2 \in C^\infty(\mathbb{R}^m_n \times \mathbb{R}^m_n)$ possess the following properties:

(1) χ_1 is independent of x^2, y^2 and homogeneous of zero degree in the variables x^1, y^1.
(2) $\chi_1(x, y) = 1$ if $|x^1|/|y^1| \in [\delta_1, \delta_1^{-1}]$ and $\chi_1(x, y) = 0$ if $|x^1|/|y^1| \notin [\delta, \delta^{-1}]$.
(3) $\chi_2(x, y) = \eta(|x^2 - y^2|/(1 + |x^1|))$, where $\eta \in C^\infty(\mathbb{R}_+)$, $\eta(t) = 1$ for $t < \varepsilon/2$, and $\eta(t) = 0$ for $t \geq \varepsilon$.

We set $\chi(x, y, \xi) = \chi_1(x, y)\chi_2(x, y)$. Then, $\chi \in \Pi^0_0$.

4.1.2 Pseudodifferential Operators

For each amplitude a and a function u in $C^\infty_c(\mathbb{R}^m_n)$, we introduce the expression

$$Au(x) = (2\pi)^{-m} \iint e^{i(x-y)\xi} a(x, y, \xi) u(y)\, dy d\xi. \tag{4.1.2}$$

The integral on the right-hand side of (4.1.2) exists as an iterated integral since, for a fixed $x \in \mathbb{R}^m_n$, the function

$$\xi \mapsto \int e^{i(x-y)\xi} a(x, y, \xi) u(y)\, dy$$

is in the Schwartz class $\mathcal{S}(\mathbb{R}^m)$. (This follows from the relation

$$\int e^{i(x-y)\xi} a(x, y, \xi) u(y)\, dy = \int e^{i(x-y)\xi} \langle D_y \rangle^{2N} (a(x, y, \xi) u(y))\, dy\, \langle \xi \rangle^{-2N}, \tag{4.1.3}$$

which holds for all $N \in \mathbb{Z}_+$.)

Definition 4.1.4 The operator A defined on function u in $C_c^\infty(\mathbb{R}_n^m)$ by (4.1.2), where $a \in \Pi^\mu$, is called Ψdo of order μ in $\mathbb{R}_n^m$. The class of all such Ψdo is denoted by Ψ^μ. An operator A with amplitude a will sometimes be written in the form Op a.

Let us represent A as an integral operator and estimate its kernel. If the order μ of a satisfies $\mu < -m$, then (4.1.2) can be rewritten in the form

$$Au(x) = \int G(x, y)u(y)\, dy,$$

where

$$\begin{aligned} G(x, y) &= (2\pi)^{-m} \int e^{i(x-y)\xi} a(x, y, \xi)\, d\xi \\ &= (2\pi)^{-m} |x^1|^{-m} \int e^{i(x-y)\xi/|x^1|} \tilde{a}(x, y, \xi)\, d\xi. \end{aligned}$$

Lemma 4.1.5 *Let $A \in \Psi^\mu$ and α, β be some multi-indices. If $\mu + |\alpha + \beta| < -m$, then, for any $N \in \mathbb{Z}_+$, the kernel G of A admits the estimate*

$$|\partial_x^\alpha \partial_y^\beta G(x, y)| \le C_{\alpha\beta N} |x^1|^{N-m-|\alpha+\beta|} (1+|x^1|/|y^1|)^{|\beta|} (|x^1|^2+|x-y|^2)^{-N/2}. \tag{4.1.4}$$

Proof Integrating by parts, we obtain

$$|G(x, y)| \le (2\pi)^{-m} |x^1|^{-m} \langle (x-y)/|x^1| \rangle^{-2N} \int |\langle D_\xi \rangle^{2N} \tilde{a}(x, y, \xi)|\, d\xi$$

for all $N \in \mathbb{Z}_+$. It follows from the properties of pre-amplitudes that the integral on the right-hand side of this last inequality is uniformly bounded with respect to $(x, y) \in \mathbb{R}_n^m \times \mathbb{R}_n^m$; therefore, for such (x, y),

$$|G(x, y)| \le C_N |x^1|^{2N-m} (|x^1|^2 + |x-y|^2)^{-N}.$$

If $\mu + |\alpha + \beta| < -m$, then the derivative $\partial_x^\alpha \partial_y^\beta G$ is a linear combination of functions $G_{\alpha_1\beta_1}$ of the form

$$G_{\alpha_1\beta_1}(x, y) = |x^1|^{-|\alpha+\beta_1|} |y^1|^{-|\beta_2|} \int e^{i(x-y)\xi} a_{\alpha_1\beta_1}(x, y, \xi)\, d\xi,$$

where $a_{\alpha_1\beta_1}(x, y, \xi) = (|x^1|\xi)^{\alpha_1+\beta_1}|x^1|^{|\alpha_2|}|y^1|^{|\beta_2|}\partial_x^{\alpha_2}\partial_y^{\beta_2}a(x, y, \xi)$, while $\alpha_1 \le \alpha$, $\beta_1 \le \beta$, $\alpha_2 = \alpha - \alpha_1$, and $\beta_2 = \beta - \beta_1$. Since $a_{\alpha_1\beta_1} \in \Pi^{\mu+|\alpha_1+\beta_1|}$ and $\mu + |\alpha_1 + \beta_1| < -m$, it follows that

$$|G_{\alpha_1\beta_1}(x, y)| \le C_N|x^1|^{2N-m-|\alpha+\beta|}(|x^1|/|y^1|)^{|\beta_2|}(|x^1|^2 + |x - y|^2)^{-N}.$$

Due to

$$(|x^1|/|y^1|)^{|\beta_2|} < (1 + |x^1|/|y^1|)^{|\beta|},$$

we have

$$|G_{\alpha_1\beta_1}(x, y)| \le C_N|x^1|^{2N-m-|\alpha+\beta|}(1 + |x^1|/|y^1|)^{|\beta|}(|x^1|^2 + |x - y|^2)^{-N}.$$

Thus, inequality (4.1.4) is proved for even N. Since $|x^1|(|x^1|^2+|x-y|^2)^{-1/2} \le 1$, formula (4.1.4) is valid for any $N \in \mathbb{Z}_+$. □

Definition 4.1.6 A Ψdo $A = \operatorname{Op} a$ is called *proper* if a is a proper amplitude. The class of all proper Ψdo of order μ is denoted by Ψ_0^μ. We also set $\Psi_0 = \bigcup_\mu \Psi_0^\mu$.

Example 4.1.7 Let the function $a \in C^\infty(\mathbb{R}^m_n)$ satisfy the estimate $|\partial^\alpha a(x)| \le C_\alpha |x^1|^{-|\alpha|}$ $(x \in \mathbb{R}^m_n)$ for any $\alpha \in \mathbb{Z}^m_+$. Then, the operator $C_c^\infty(\mathbb{R}^m_n) \ni u \mapsto au$ is in the class Ψ_0^0 since it has the form $\operatorname{Op}(\chi a)$, where χ is the function in Example 4.1.3. In particular, the identity operator $I : C_c^\infty(\mathbb{R}^m_n) \to C_c^\infty(\mathbb{R}^m_n)$ is a proper Ψdo of order 0.

Example 4.1.8 For every $\alpha \in \mathbb{Z}^m_+$, we set $Q_\alpha u(x) = |x^1|^{|\alpha|}D^\alpha u(x)$ for $u \in C_c^\infty(\mathbb{R}^m_n)$. Then, $Q_\alpha \in \Psi_0^{|\alpha|}$ due to the equality

$$Q_\alpha u(x) = (2\pi)^{-m}\int\int e^{i(x-y)\xi}\chi(x, y)(|x^1|\xi)^\alpha u(y)\,dyd\xi,$$

where χ is the same as above.

Remark 4.1.9 To make inequality (4.1.3) valid, it is sufficient that (together with the condition $a \in \Pi^\mu$) the function $y \mapsto a(x, y, \xi)u(y)$ has a compact support for fixed $x \in \mathbb{R}^m_n$ and $\xi \in \mathbb{R}^m$. Therefore, if $a \in \Pi_0^\mu$, then the integral (4.1.2) exists as an iterated one for all $u \in C^\infty(\mathbb{R}^m_n)$.

We also note that the mapping $a \mapsto \operatorname{Op} a$ of the set of amplitudes onto the set of Ψdo is not one-to-one. In particular, a proper Ψdo A can be defined by (4.1.2) with a non-proper amplitude a.

4.1.3 Symbols

Definition 4.1.10 A function $\tilde{a} \in C^\infty(\mathbb{R}^m_n \times \mathbb{R}^m)$ is called a *presymbol* of order μ, $\mu \in \mathbb{R}$, if for any multi-indices α, γ there exists a constant $C_{\alpha\gamma}$ such that

$$|\partial_x^\alpha \partial_\xi^\gamma \tilde{a}(x,\xi)| \le C_{\alpha\gamma} |x^1|^{-|\alpha|} \langle \xi \rangle^{\mu - |\gamma|}, \qquad (x,\xi) \in \mathbb{R}^m_n \times \mathbb{R}^m.$$

For each presymbol $\tilde{a}$, we introduce a function a,

$$a(x,\xi) = \tilde{a}(x, |x^1|\xi), \tag{4.1.5}$$

which is called a *symbol* (of order μ). The class of all presymbols (resp., symbols) of order μ is denoted by $\tilde{S}^\mu$ (resp., S^μ).

If $\tilde{a} \in \tilde{S}^\mu$, then the function $(x, y, \xi) \mapsto \tilde{a}(x,\xi)$ is a pre-amplitude of order μ. Thus, presymbols (symbols) can be regarded as pre-amplitudes (amplitudes) independent of y. Therefore, the properties of the classes $\tilde{\Pi}^\mu$ and Π^μ listed in Sect. 4.1.1 remain valid also for the classes $\tilde{S}^\mu$ and S^μ. In what follows, we often use the evident fact that the restriction of an amplitude of order μ to the set $\{(x, y, \xi) \in \mathbb{R}^m_n \times \mathbb{R}^m_n \times \mathbb{R}^m : x = y\}$ is a symbol of the same order.

Definition 4.1.11 Elements of the sets $\Pi^{-\infty} = \bigcap_\mu \Pi^\mu$, $\Pi_0^{-\infty} = \bigcap_\mu \Pi_0^\mu$, and $S^{-\infty} = \bigcap_\mu S^\mu$ $(\mu \in \mathbb{R})$ are called *amplitudes, proper amplitudes, and symbols of order* $-\infty$, respectively. An operator A is called a Ψdo (respectively, proper Ψdo) of order $-\infty$ if $A = \operatorname{Op} a$, where $a \in \Pi^{-\infty}$ $(\Pi_0^{-\infty})$. The class of all such Ψdo will be denoted by $\Psi^{-\infty}$ $(\Psi_0^{-\infty})$.

Some sufficient conditions for Ψdo to belong to the classes $\Psi^{-\infty}$ and $\Psi_0^{-\infty}$ will be given later.

Definition 4.1.12 Let $\mu \in \mathbb{R}$, $a_j \in S^{\mu-j}$, $j \in \mathbb{Z}_+$, and $a \in C^\infty(\mathbb{R}^m_n \times \mathbb{R}^m)$. We write

$$a(x,\xi) \sim \sum_{j=0}^{\infty} a_j(x,\xi), \tag{4.1.6}$$

if $a - \sum_{j=0}^{N-1} a_j \in S^{\mu-N}$ for any $N \in \mathbb{N}$.

The relation (4.1.6) obviously implies $a \in S^\mu$.

Theorem 4.1.13 *If $a_j \in S^{\mu-j}$, $j \in \mathbb{Z}_+$, then there exists a function a such that (4.1.6) is valid. If another function a' satisfies the same relation, then $a - a' \in S^{-\infty}$.*

The proof is left to the reader (see the proof of Theorem 1.1.18).

Let A be a proper Ψdo. We set $e_\xi(x) = e^{ix\xi}$ and define *a symbol σ_A of the operator A* by

$$\sigma_A(x, \xi) = e_{-\xi} A e_\xi(x). \tag{4.1.7}$$

If $A = \operatorname{Op} a$ with a proper amplitude a, then

$$\begin{aligned} \sigma_A(x, \xi) = (2\pi)^{-m} \int\int e^{i(x-y)(\eta-\xi)} a(x, y, \eta)\, dy d\eta = \\ = (2\pi)^{-m} \int\int e^{i(x-y)\theta} a(x, y, \xi + \theta)\, dy d\theta\,. \end{aligned} \tag{4.1.8}$$

(According to Remark 4.1.9, formula (4.1.7) or, equivalently, (4.1.8) makes sense.)

Example 4.1.14 The symbol of Ψdo $u \mapsto au$ in Example 4.1.7 coincides with a.

Example 4.1.15 If $Q = Q_\alpha$ is the operator in Example 4.1.8, then $\sigma_Q(x, \xi) = |x^1|^{|\alpha|} \xi^\alpha$.

Theorem 4.1.16 *Let $A \in \Psi^\mu_0$ and let $\sigma = \sigma_A$ be the symbol of A. Then:*

(1) $\sigma \in S^\mu$.

(2) *The formula*

$$Au(x) = (2\pi)^{-m/2} \int e^{ix\xi} \sigma(x, \xi) \hat{u}(\xi)\, d\xi, \quad u \in C^\infty_c(\mathbb{R}^m_n), \tag{4.1.9}$$

is valid, where $\hat{u} = Fu$ is the Fourier transform of u.

(3) *If $A = \operatorname{Op} a$, $a \in \Pi^\mu_0$, then the following asymptotic expansion holds:*

$$\sigma(x, \xi) \sim \sum_\alpha \frac{1}{\alpha!} \partial^\alpha_\xi D^\alpha_y a(x, y, \xi)\big|_{y=x}. \tag{4.1.10}$$

The function $(x, y, \xi) \mapsto \partial^\alpha_\xi D^\alpha_y a(x, y, \xi)$ is an amplitude of order $\mu - |\alpha|$. This implies that the function $(x, \xi) \mapsto \partial^\alpha_\xi D^\alpha_y a(x, y, \xi)|_{y=x}$ is in the class $S^{\mu-|\alpha|}$. Thus, the terms of the asymptotic series (4.1.10) are symbols of decreasing orders. Before proving the theorem, we state the following assertion.

Lemma 4.1.17 *Let $a \in \Pi_0^\mu$ and $h \in C([0,1])$. For any $t \in [0,1]$, we set*

$$\sigma_t(x,\xi) = (2\pi)^{-m} \int\int e^{i(x-y)\theta} a(x,y,\xi+t\theta)u(y)\,dyd\theta$$

and define the function σ by

$$\sigma(x,\xi) = \int_0^1 \sigma_t(x,\xi)h(t)\,dt.$$

Then, $\sigma \in S^\mu$.

Proof of the Lemma Let $\tilde{\sigma}_t$, $\tilde{\sigma}$, and $\tilde{a}$ be the functions corresponding to σ_t, σ, and a due to formulas (4.1.1) and (4.1.5). We have

$$\tilde{\sigma}_t(x,\xi) = (2\pi)^{-m} \int\int e^{i(x-y)\theta} \tilde{a}(x,y,\xi+t|x^1|\theta)u(y)\,dyd\theta, \tag{4.1.11}$$

$$\tilde{\sigma}(x,\xi) = \int_0^1 \tilde{\sigma}_t(x,\xi)h(t)\,dt.$$

It suffices to verify that

$$|\partial_x^\alpha \partial_\xi^\gamma \tilde{\sigma}_t(x,\xi)| \le C_{\alpha\gamma}|x^1|^{-|\alpha|}\langle\xi\rangle^{\mu-|\gamma|}, \tag{4.1.12}$$

where $C_{\alpha\gamma}$ are constants independent of $t \in [0,1]$. In (1.13), we substitute $y \to |x^1|y$, $\theta \to \theta/|x^1|$ and integrate by parts. As a result, we obtain

$$\tilde{\sigma}_t(x,\xi) = (2\pi)^{-m} \int\int e^{i(x-y)\theta} \langle D_\theta\rangle^{2N}[\langle D_y\rangle^{2N} b_t(x,y,\xi,\theta)\langle\theta\rangle^{-2N}] \times \tag{4.1.13}$$
$$\times \langle X-y\rangle^{-2N}\,dyd\theta,$$

where $X = x/|x^1|$, $b_t(x,y,\xi,\theta) = \tilde{a}(x,|x^1|y,\xi+t\theta)$. Since $\tilde{a}$ is a proper pre-amplitude, there exists a number $\delta \in (0,1)$ such that $b_t(x,y,\xi,\theta) = 0$ for $|y^1| \notin [\delta,\delta^{-1}]$. In what follows, we therefore assume that $|y^1| \ge \delta$. Since

$$|\partial_y^{\beta_1}\partial_\theta^{\beta_2} b_t(x,y,\xi,\theta)| \le C'_{\beta_1\beta_2}|y^1|^{-|\beta_1|}\langle\xi+t\theta\rangle^{\mu-|\beta_2|}, \quad \beta_1,\ \beta_2 \in \mathbb{Z}_+^m,$$

we have

$$|\partial_y^{\beta_1}\partial_\theta^{\beta_2} b_t(x,y,\xi,\theta)| \le C''_{\beta_1\beta_2}\langle\xi\rangle^\mu\langle\theta\rangle^{|\mu|}.$$

This and (4.1.13), in view of the relation $|\partial^\beta(\langle\theta\rangle^{-2N})| \le C_{\beta N}\langle\theta\rangle^{-2N}$, imply the estimate

$$|\tilde{\sigma}_t(x,\xi)| \le C_N \langle\xi\rangle^\mu \iint \langle X-y\rangle^{-2N}\langle\theta\rangle^{\mu-2N}\,dyd\theta,$$

which for $N > (m+|\mu|)/2$ can be rewritten in the form $|\tilde{\sigma}_t| \le C\langle\xi\rangle^\mu$. Thus, the inequality (4.1.12) is proven for $\alpha = \gamma = 0$.

For arbitrary multi-indices α and γ, this inequality follows from the foregoing argument, since $|x^1|^{|\alpha|}\partial_x^\alpha\partial_\xi^\gamma\tilde{\sigma}_t(x,\xi)$ is a sum of integrals of the form (4.1.11) with $\tilde{a}$ replaced by pre-amplitudes of order $\mu - |\gamma|$. Note that it suffices to consider the case $|\alpha+\gamma| = 1$, since one can then reason by induction. □

Proof of Theorem 4.1.16 Using formula (4.1.8) and the Taylor expansion

$$a(x,y,\xi) = \sum_{|\alpha|\le N-1}\frac{1}{\alpha!}\partial_\xi^\alpha a(x,y,\xi)\theta^\alpha +$$

$$+\sum_{|\alpha|=N}\frac{N}{\alpha!}\int_0^1(1-t)^{N-1}\partial_\xi^\alpha a(x,y,\xi+t\theta)\,dt\,\theta^\alpha,$$

we obtain

$$\sigma(x,\xi) = \sum_{|\alpha|\le N-1}\frac{1}{\alpha!}\sigma^{(\alpha)}(x,\xi) + \sum_{|\alpha|=N}\frac{N}{\alpha!}\sigma^{(\alpha)}(x,\xi)\,, \tag{4.1.14}$$

where

$$\sigma^{(\alpha)}(x,\xi) = (2\pi)^{-m}\int\int e^{i(x-y)\theta}\partial_\xi^\alpha a(x,y,\xi)\theta^\alpha\,dyd\theta = \partial_\xi^\alpha D_y^\alpha a(x,y,\xi)\big|_{y=x}$$

for $|\alpha| \le N-1$ and

$$\sigma^{(\alpha)}(x,\xi) = \int_0^1 (1-t)^{N-1}\sigma_t^{(\alpha)}(x,\xi)\,dt,$$

$$\sigma_t^{(\alpha)}(x,\xi) = (2\pi)^{-m}\int\int e^{i(x-y)\theta}\partial_\xi^\alpha D_y^\alpha a(x,y,\xi+t\theta)\,dyd\theta$$

for $|\alpha| = N$. Setting $r_N(x,\xi) = \sum_{|\alpha|=N}(N/\alpha!)\sigma^{(\alpha)}(x,\xi)$, we rewrite equation (4.1.14) as

$$\sigma(x,\xi) = \sum_{|\alpha|\le N-1}\frac{1}{\alpha!}\sigma^{(\alpha)}(x,\xi) + r_N(x,\xi).$$

The function $(x, y, \xi) \mapsto \partial_\xi^\alpha D_y^\alpha a(x, y, \xi)$ is an amplitude of order $\mu - |\alpha|$. Lemma 4.1.17 now implies that $r_N \in S^{\mu-N}$. This proves assertions (3) and (1) of the theorem.

It remains to prove formula (4.1.9). Let $u \in C_c^\infty(\mathbb{R}_n^m)$. Formula (4.1.8) implies the following inequality:

$$\int e^{i(x-y)\xi}\sigma(x,\xi)\hat{u}(\xi)\,d\xi = (2\pi)^{-m}\int e^{ix\xi}\hat{u}(\xi)\,d\xi \times \\ \times \int\int e^{i(x-y)(\eta-\xi)}a(x,y,\eta)\,dyd\eta. \tag{4.1.15}$$

Let us show that for a fixed $x \in \mathbb{R}_n^m$ the function

$$(\xi, \eta) \mapsto \hat{u}(\xi)\int e^{i(x-y)(\eta-\xi)}a(x,y,\eta)\,dy \tag{4.1.16}$$

is rapidly decreasing as $|\xi| + |\eta| \to \infty$. Since $\hat{u} \in \mathcal{S}(\mathbb{R}^m)$, it is sufficient to verify the inequality

$$\left|\int e^{i(x-y)(\eta-\xi)}a(x,y,\eta)\,dy\right| \le C_N(x)\langle\eta\rangle^{\mu-2N}\langle\xi\rangle^{2N}$$

for any $N \in \mathbb{Z}_+$. Using the relations $a(x, y, \eta) = \tilde{a}(x, y, |x^1|\eta)$ and $|\partial_y^\beta \tilde{a}(x, y, \eta)| \le C_\beta |x^1|^{-|\beta|}\langle\eta\rangle^\mu$, we obtain

$$\left|\int e^{i(x-y)(\eta-\xi)}a(x,y,\eta)\,dy\right| \le \int |\langle D_y\rangle^{2N}a(x,y,\eta)|\,dy\,\langle\eta-\xi\rangle^{-2N} \le \\ \le C_N'(x)(1+|x^1|^{-2N})\langle|x^1|\eta\rangle^\mu\langle\eta\rangle^{-2N}\langle\xi\rangle^{2N} \le C_N(x)\langle\eta\rangle^{\mu-2N}\langle\xi\rangle^{2N},$$

as needed. Due to the rapid decrease of the function (4.1.16), one can change the order of integration on the right-hand side of (4.1.15):

$$\int e^{i(x-y)\xi}\sigma(x,\xi)\hat{u}(\xi)\,d\xi = (2\pi)^{-m}\int\int e^{i(x-y)\eta}a(x,y,\eta)\,dyd\eta \times \\ \times \int e^{iy\xi}\hat{u}(\xi)\,d\xi = (2\pi)^{-m/2}\int\int e^{i(x-y)\eta}a(x,y,\eta)u(y)\,dyd\eta = (2\pi)^{m/2}Au(x).$$

□

Theorem 4.1.18

(1) *A proper* Ψ*do* A *is in the class* Ψ_0^μ, $\mu \in [-\infty, +\infty)$, *if and only if* $\sigma_A \in S^\mu$.

(2) *Formula (4.1.10) for the symbol of a proper* Ψ*do* A $A = \operatorname{Op} a$ *remains valid when the amplitude* a *is not proper.*
(3) $\Psi^\mu \cap \Psi_0 = \Psi_0^\mu \quad (\forall \mu \in [-\infty, +\infty))$.
(4) $\bigcap_\mu \Psi_0^\mu = \Psi_0^{-\infty}$.

Proof Assertions (1)–(3) can be proved with the same technique using a function in Example 4.1.3. We prove only the first assertion. If $A \in \Psi_0^\mu$, then $\sigma_A \in S^\mu$ according to Theorem 4.1.16. Conversely, let $\sigma_A \in S^\mu$ and $A = \operatorname{Op} a$, where a is some proper amplitude. For any function $\chi \in C^\infty(\mathbb{R}^m_n \times \mathbb{R}^m_n)$ and any $x_0 \in \mathbb{R}^m_n$, we have

$$\operatorname{Op} \sigma_A(\chi(x_0, \cdot)u)(x) = \operatorname{Op} a(\chi(x_0, \cdot)u)(x), \quad x \in \mathbb{R}^m_n.$$

Setting $x_0 = x$, we obtain $\operatorname{Op}(\sigma_A \chi) = \operatorname{Op}(a\chi)$. Now let χ be a function from Example 4.1.3 such that the equality $a\chi = a$ holds. Then, $\operatorname{Op}(\sigma_A \chi) = \operatorname{Op} a = A$. Since $\sigma_A \chi \in \Pi_0^\mu$, we have $A \in \Psi_0^\mu$, as needed. Assertion (4) directly follows from (1). □

The following theorem shows that any function $a \in S^\mu$ is the symbol of some proper Ψdo modulo elements of $S^{-\infty}$.

Theorem 4.1.19 *Let* μ *be an arbitrary real number, and let* $a \in S^\mu$. *There exists a* Ψ*do* $A \in \Psi_0^\mu$ *such that* $a - \sigma_A \in S^{-\infty}$.

Proof Let χ denote a function constructed in Example 4.1.3 (the numbers δ, δ_1, ε in the construction of χ are chosen arbitrarily), and define an operator A by $A = \operatorname{Op}(\chi a)$. Since $\chi a \in \Pi_0^\mu$, we have $A \in \Psi_0^\mu$ and, by Theorem 4.1.16,

$$\sigma_A(x, \xi) \sim \sum_\alpha \frac{1}{\alpha!} \partial_\xi^\alpha D_y^\alpha (\chi(x, y) a(x, \xi))\big|_{y=x} = a(x, \xi)$$

(remind that $\chi = 1$ in a neighborhood of the diagonal $x = y$). According to the definition of an asymptotic expansion, the obtained relation means that $a - \sigma_A \in S^{\mu - N}$ for all $N \in \mathbb{N}$. □

By Theorems 4.1.16 and 4.1.19, the mapping $\Psi^\mu \ni A \mapsto \sigma_A \in S^\mu \quad (\forall \mu \in \mathbb{R})$ induces an isomorphism $\Psi^\mu / \Psi^{-\infty} \cong S^\mu / S^{-\infty}$ of linear spaces.

4.1.4 Composition of Ψdo: Adjoint Operator

A proper Ψdo maps the set $C_c^\infty(\mathbb{R}^m_n)$ into itself. Therefore, the composition of such Ψdo makes sense.

Theorem 4.1.20 *Let $A_j \in \Psi_0^{\mu_j}$, let σ_j be the symbol of A_j, $j = 1, 2$, and let $A = A_1 A_2$. Then:*

(1) $A \in \Psi_0^{\mu_1+\mu_2}$.
(2) *The symbol σ of A admits the asymptotic expansion*

$$\sigma(x, \xi) \sim \sum_{\alpha} \frac{1}{\alpha!} \partial_\xi^\alpha \sigma_1(x, \xi) D_x^\alpha \sigma_2(x, \xi). \tag{4.1.17}$$

Proof

(1) Let $a_j \in \Pi_0^{\mu_j}$ and $A_j = \mathrm{Op}\, a_j$, $j = 1, 2$. We define the transposed Ψdo ${}^tA_2 \in \Psi_0^{\mu_2}$,

$${}^tA_2u(x) = (2\pi)^{-m} \int\int e^{i(x-y)\xi} a_2(y, x, -\xi)u(y)\, dyd\xi.$$

Let ${}^t\sigma_2$ denote the symbol of ${}^tA_2$ and define the dual symbol σ_2' by $\sigma_2'(x, \xi) = {}^t\sigma_2(x, -\xi)$. It is easily seen that

$$A_2u(x) = (2\pi)^{-m} \int\int e^{i(x-y)\xi} \sigma_2'(y, \xi)u(y)\, dyd\xi. \tag{4.1.18}$$

Setting $v = A_2u$ in

$$A_1v(x) = (2\pi)^{-m/2} \int e^{ix\xi} \sigma_1(x, \xi)\hat{v}(\xi)\, d\xi,$$

we obtain

$$Au(x) = (2\pi)^{-m} \int\int e^{i(x-y)\xi} \sigma_1(x, \xi)\sigma_2'(y, \xi)u(y)\, dyd\xi \tag{4.1.19}$$

(note that the function $(x, y, \xi) \mapsto \sigma_1(x, \xi)\sigma_2'(y, \xi)$ is not an amplitude). Rewrite now the operator A_j, $j = 1, 2$, using the amplitude a_j:

$$A_ju(x) = (2\pi)^{-m} \int\int e^{i(x-y)\xi} a_j(x, y, \xi)u(y)\, dyd\xi.$$

Simple calculations reveal that

$$Au(x) = (2\pi)^{-m} \int\int e^{i(x-y)\xi} a(x, y, \xi)u(y)\, dyd\xi, \tag{4.1.20}$$

where

$$a(x, y, \xi) = (2\pi)^{-m} \int\int e^{i(x-y)\theta} a_1(x, z, \xi + \theta) a_2(z, y, \xi)\, dz d\theta.$$

Since a_j, $j = 1, 2$, is a proper amplitude, there exist numbers $\delta_j \in (0, 1)$ and $\varepsilon_j > 0$ such that $a_j(x, y, \xi) = 0$ for $|x^1|/|y^1| \notin (\delta_j, \delta_j^{-1})$ or $|x^2 - y^2| > \varepsilon_j(1 + |x^1|)$. We set $\delta = \delta_1\delta_2$ and $\varepsilon = \varepsilon_1 + \varepsilon_2$. Then, $a(x, y, \xi) = 0$ for $|x^1|/|y^1| \notin [\delta, \delta^{-1}]$ or $|x^2 - y^2| \geq \varepsilon(1 + |x^1|)$. Let χ be a function in Example 4.1.3, chosen such that $\chi a = a$. Reasoning as in the proof of Theorem 4.1.18, we deduce from (4.1.19) and (4.1.20) that

$$Au(x) = (2\pi)^{-m} \int\int e^{i(x-y)\xi} \sigma_1(x, \xi)\sigma_2'(y, \xi)\chi(x, y)u(y)\, dy d\xi. \tag{4.1.21}$$

Since the function $(x, y, \xi) \mapsto \sigma_1(x, \xi)\sigma_2'(y, \xi)\chi(x, y)$ is a proper amplitude of order $\mu_1 + \mu_2$, assertion (1) is now proven.

(2) Since $\chi = 1$ in a neighborhood of the diagonal $\{(x, y) \in \mathbb{R}_n^m \times \mathbb{R}_n^m : x = y\}$, equality (4.1.21) and Theorem 4.1.16 yield the asymptotic expansion

$$\sigma(x, \xi) \sim \sum_\alpha \frac{1}{\alpha!} \partial_\xi^\alpha [\sigma_1(x, \xi) D_x^\alpha \sigma_2'(x, \xi)]. \tag{4.1.22}$$

Then, formula (4.1.18) can be written in the form

$$A_2u(x) = (2\pi)^{-m} \int\int e^{i(x-y)\xi} ({}^t\sigma_2(y, -\xi))u(y)\, dy d\xi.$$

Since $A_2 = {}^t({}^tA_2)$, the operator ${}^tA_2$ can be similarly expressed via σ_2:

$${}^tA_2u(x) = (2\pi)^{-m} \int\int e^{i(x-y)\xi} \sigma_2(y, -\xi)u(y)\, dy d\xi.$$

Introduce an appropriate cut-off function χ and obtain

$${}^tA_2u(x) = (2\pi)^{-m} \iint e^{i(x-y)\xi} \sigma_2(y, -\xi)\chi(x, y)u(y)\, dy d\xi.$$

Again, applying Theorem 4.1.16, we arrive at

$${}^t\sigma_2(x, \xi) \sim \sum_\alpha \frac{1}{\alpha!} \partial_\xi^\alpha D_x^\alpha \sigma_2(x, -\xi),$$

or, equivalently, at

$$\sigma_2'(x,\xi) \sim \sum_\alpha \frac{(-1)^{|\alpha|}}{\alpha!} \partial_\xi^\alpha D_x^\alpha \sigma_2(x,\xi). \tag{4.1.23}$$

The asymptotic expansion (4.1.17) can be deduced from (4.1.22) and (4.1.23) with the same combinatorial reasoning as in the usual theory of Ψdo.

□

We now turn to the description of an adjoint Ψdo. Let $A = \mathrm{Op}\, a$ with $a \in \Pi_0^\mu$. The properties of amplitudes listed in Sect. 4.1.1 imply that, for any $\tau \in \mathbb{R}$, the function ${}_\tau a^* : (x, y, \xi) \mapsto (|y^1|/|x^1|)^{2\tau}\, \overline{a(y, x, \xi)}$ is also in the class Π_0^μ. Therefore, the operator ${}_\tau A^* := \mathrm{Op}\, {}_\tau a^*$ is a proper Ψdo of order μ. The Ψdo ${}_\tau A^*$ is called adjoint to A with respect to the inner product

$$(\cdot\,,\cdot)_\tau = \int |x^1|^{2\tau} u(x)\overline{v(x)}\, dx,$$

since the relation $(Au\,, v)_\tau = (u\,,\, {}_\tau A^* v)_\tau$ u holds for all $v \in C_c^\infty(\mathbb{R}_n^m)$. The asymptotic expansion for the symbol ${}_\tau\sigma^*$ of ${}_\tau A^*$ has the form

$${}_\tau\sigma^*(x,\xi) \sim |x^1|^{-2\tau} \sum_\alpha \frac{1}{\alpha!} \partial_\xi^\alpha D_y^\alpha \left(|y^1|^{2\tau}\, \overline{a(y, x, \xi)}\right)\Big|_{y=x}. \tag{4.1.24}$$

4.1.5 Conditions for Ψdo to Belong to Classes $\Psi_0^{-\infty}$ and $\Psi^{-\infty}$

Remind (see Definition 4.1.11) that $\Psi_0^{-\infty} = \mathrm{Op}\, \Pi_0^{-\infty}$, $\Psi^{-\infty} = \mathrm{Op}\, \Pi^{-\infty}$.

Proposition 4.1.21 *Let $A = \mathrm{Op}\, a$ be a proper Ψdo, whereas the amplitude a has a zero of infinite order on the set $\{(x, y, \xi) \in \mathbb{R}_n^m \times \mathbb{R}_n^m \times \mathbb{R}^m : x = y\}$. Then, $A \in \Psi_0^{-\infty}$.*

Proof According to Theorem 4.1.18, it is sufficient to prove that $\sigma_A \in S^{-\infty}$. This inclusion immediately follows from the condition of the Proposition and from formula (4.1.10). □

To expand Proposition 4.1.21 to the case of non-proper Ψdo, we need the next lemma.

Lemma 4.1.22 *Let the function $H \in C^\infty(\mathbb{R}_n^m \times \mathbb{R}_n^m)$ satisfy the following conditions:*

(1) *There exists a number $\varepsilon_0 > 0$ such that $H(x, y) = 0$ for $|x - y| < \varepsilon_0 |x^1|$.*
(2) *For all $\alpha, \beta \in \mathbb{Z}_+^m$ and $N \in \mathbb{Z}_+$*

$$|\partial_x^\alpha \partial_y^\beta H(x, y)| \le C_{\alpha\beta N} |x^1|^{N-|\alpha|} |y^1|^{-|\beta|} |x - y|^{-N}. \tag{4.1.25}$$

Let also v be a function in the Schwartz class $\mathcal{S}(\mathbb{R}^m)$. Then, the function

$$b : (x, y, \eta) \mapsto e^{i(x-y)\eta} H(x, y) v(|x^1|\eta)$$

belongs to the class $\Pi^{-\infty}$.

Proof The first condition of the lemma implies that there exists a number $\varepsilon_1 > 0$ such that

$$|x - y| \geq \varepsilon_1(|x^1| + |y^1|) \quad \text{for} \quad (x, y) \in \operatorname{supp} H.$$

Indeed, if $|y^1| \leq 2|x^1|$, then $|x - y| \geq (\varepsilon_0/3)(|x^1| + |y^1|)$, and if $|y^1| \geq 2|x^1|$, then $|x - y| \geq |y^1|/2 \geq (|x^1| + |y^1|)/3$. We set $\tilde{b}(x, y, \eta) = b(x, y, \eta/|x^1|)$. The inclusion $b \in \Pi^{-\infty}$ is equivalent to $\tilde{b} \in \overset{\sim}{\Pi}^{-\infty}$, i.e., to the estimates

$$|\partial_x^\alpha \partial_y^\beta \partial_\eta^\gamma \tilde{b}(x, y, \eta)| \leq C|x^1|^{-|\alpha|}|y^1|^{-|\beta|}\langle\eta\rangle^{-M}$$

for all $\alpha, \beta, \gamma \in \mathbb{Z}_+^m$ and $M \in \mathbb{Z}_+$ (here and in what follows, we do not indicate dependence of the constant factors on $\alpha, \beta, \ldots, M$ and so on). Since the function $(x, y, \xi) \mapsto \partial_x^\alpha \partial_y^\beta \partial_\eta^\gamma \tilde{b}(x, y, \eta)$ is a linear combination of the functions

$$d_{\alpha_1\beta_1\gamma_1}(x, y, \eta) = (\partial_x^{\alpha_1} \partial_y^{\beta_1} \partial_\eta^{\gamma_1} e^{i(x-y)\eta/|x^1|})(\partial_x^{\alpha_2} \partial_y^{\beta_2} H(x, y))(\partial_\eta^{\gamma_2} v(\eta)), \tag{4.1.26}$$

where $\alpha_1 + \alpha_2 = \alpha$, $\beta_1 + \beta_2 = \beta$, $\gamma_1 + \gamma_2 = \gamma$, it suffices to prove that

$$|d_{\alpha_1\beta_1\gamma_1}(x, y, \eta)| \leq C|x^1|^{-|\alpha|}|y^1|^{-|\beta|}\langle\eta\rangle^{-M}$$

for any $\alpha_1 \leq \alpha$, $\beta_1 \leq \beta$, $\gamma_1 \leq \gamma$. By induction on $|\alpha_1 + \beta_1 + \gamma_1|$, one can easily establish that the first factor on the right-hand side of (4.1.26) is estimated by the sum of a finite number of terms of the form

$$e_{k\delta}(x, y, \eta) = |x^1|^{-|\alpha_1+\beta_1|-k}|x - y|^k|\eta^\delta|, \qquad k \in \mathbb{Z}_+, \ \delta \in \mathbb{Z}_+^m.$$

Since $v \in \mathcal{S}(\mathbb{R}^m)$, it follows that $|\eta^\delta \partial^{\gamma_2} v(\eta)| \leq C\langle\eta\rangle^{-M}$ for all $M \in \mathbb{Z}_+$. Taking into account the estimates (4.1.25), we conclude that

$$\begin{aligned} &\left|e_{k\delta}(x, y, \eta)\partial_x^{\alpha_2}\partial_y^{\beta_2} H(x, y)\, \partial_\eta^{\gamma_2} v(\eta)\right| \leq \qquad (4.1.27)\\ &\leq C\,(|x^1|^{-|\alpha_1+\beta_1|-k}|x - y|^k)(|x^1|^{N-|\alpha_2|}|y^1|^{-|\beta_2|}|x - y|^{-N})\langle\eta\rangle^{-M} =\\ &= C\,|x^1|^{-|\alpha|}|y^1|^{-|\beta|}(|x^1|^{N-k-|\beta_1|}|y^1|^{|\beta_1|}|x - y|^{k-N})\langle\eta\rangle^{-M} \end{aligned}$$

for any M and $N \in \mathbb{Z}_+$. Let $N \geq k + |\beta_1|$. Then,

$$|x-y|^{N-k} \geq (\varepsilon_1(|x^1|+|y^1|))^{N-k} \geq \varepsilon_1^{N-K}|x^1|^{N-k-|\beta_1|}|y^1|^{|\beta_1|}$$

on supp$H$. From here and (4.1.27), we derive

$$\left|e_{k\delta}(x,y,\eta)\partial_x^{\alpha_2}\partial_y^{\beta_2}H(x,y)\partial_\eta^{\gamma_2}v(\eta)\right| \leq C'|x^1|^{-|\alpha|}|y^1|^{-|\beta|}\langle\eta\rangle^{-M}$$

on the same set. □

Proposition 4.1.23 *Let a be an amplitude having a zero of infinite order on the set $\{(x,y,\xi) \in \mathbb{R}_n^m \times \mathbb{R}_n^m \times \mathbb{R}^m : x = y\}$. Then, the operator $A = \mathrm{Op}\, a$ is in the class $\Psi^{-\infty}$.*

Proof We represent the operator A in the form $A = \mathrm{Op}\,(\chi a) + \mathrm{Op}\,(\chi_1 a)$, where χ is a function in Example 4.1.3, $\chi_1 = 1 - \chi$. According to Proposition 4.1.21, we have $\mathrm{Op}\,(\chi a) \in \Pi_{-\infty}^0$. It then follows from the construction of the function $\chi_1(x,y) = 0$ that $|x^1|/|y^1| \in (\delta, \delta^{-1})$ and $|x^2 - y^2| < \varepsilon(1+|x^1|)$ ($\delta \in (0,1)$ and $\varepsilon > 0$ are some fixed numbers). We put $\varepsilon_0 = \min\{1-\delta, \varepsilon\}$. An elementary verification shows that if $|x-y| < \varepsilon_0|x^1|$, then $\chi_1(x,y) = 0$. Therefore, from the beginning, we can assume that

$$a(x,y,\xi) = 0 \quad \text{for} \quad |x-y| < \varepsilon_0|x^1|. \tag{4.1.28}$$

For $(x,y,\xi) \in \mathrm{supp}\, a$, this implies the relation

$$|x-y| \geq \varepsilon_1(|x^1|+|y^1|) \qquad (\varepsilon_1 > 0) \tag{4.1.29}$$

(see the proof of Lemma 4.1.22). For each $k \in \mathbb{Z}_+$, we define the function a_k by

$$a_k(x,y,\xi) = (-\Delta_\xi)^k a(x,y,\xi)\,|x-y|^{-2k}.$$

Let $a \in \Pi^\mu$. From (4.1.29), it easily follows that $a_k \in \Pi^{\mu-2k}$. If $k > (m+\mu)/2$, then

$$Au(x) = (2\pi)^{-m}\iint e^{i(x-y)\xi}a_k(x,y,\xi)u(y)\,dyd\xi = \int G(x,y)u(y)\,dy,$$

where G is the kernel of A,

$$G(x,y) = (2\pi)^{-m}\int e^{i(x-y)\xi}a_k(x,y,\xi)\,d\xi. \tag{4.1.30}$$

Let $v \in \mathcal{S}(\mathbb{R}^m)$, $(2\pi)^{-m}\int v(\eta)\,d\eta = 1$. We set

$$b(x,y,\eta) = e^{i(y-x)\eta}|x^1|^m G(x,y)v(|x^1|\eta).$$

Since

$$(2\pi)^{-m} \int\int e^{i(x-y)\eta} b(x, y, \eta) u(y)\, dy d\eta = \int G(x, y) u(y)\, dy = Au(x),$$

we have $A = \text{Op}\, b$. Thus, the proposition is proved if the inclusion $b \in \Pi^{-\infty}$ is established. To this end, it is sufficient to prove that the function $H : (x, y) \mapsto |x^1|^m G(x, y)$ satisfies the conditions of Lemma 4.1.22. Since condition (1) directly follows from (4.1.28), we need only to obtain the estimates (4.1.25).

We fix $\alpha, \beta \in \mathbb{Z}^m_+$ and take $k > (m + \mu + |\alpha + \beta|)/2$. Since $A = \text{Op}\, a_k \in \Psi^{\mu-2k}$ and $\mu - 2k + |\alpha + \beta| < -m$, it follows from Lemma 4.1.5 (with μ replaced by $\mu - 2k$ and N by $N + |\beta|$) that

$$|\partial^\alpha_x \partial^\beta_y G(x, y)| \le C|x^1|^{N-m-|\alpha|}(1 + |x^1|/|y^1|)^{|\beta|}|x - y|^{-(N+|\beta|)}$$

($N \in \mathbb{Z}_+$, $C = C(\alpha, \beta, N)$). Hence, taking into account (4.1.29), we deduce that

$$|\partial^\alpha_x \partial^\beta_y G(x, y)| \le C'|x^1|^{N-m-|\alpha|}|y^1|^{-|\beta|}|x - y|^{-N}. \tag{4.1.31}$$

Since the multi-indices α and β are arbitrarily chosen, estimates (4.1.25) follow from (4.1.31) and from the formula

$$\partial^\alpha_x \partial^\beta_y H(x, y) = \sum_{\gamma \le \alpha} \binom{\alpha}{\gamma} \partial^{\alpha-\gamma}_x |x^1|^m \cdot \partial^\gamma_x \partial^\beta_y G(x, y).$$

□

Proposition 4.1.24 *Any Ψdo A of arbitrary order μ admits the representation $A = A_0 + A_1$, where $A_0 \in \Psi^\mu_0$ and $A_1 \in \Psi^{-\infty}$.*

Proof The needed representation is provided by the equality $\text{Op}\, a = \text{Op}\, \chi a + \text{Op}\, (1-\chi)a$, where χ is a function from Example 4.1.3. □

4.1.6 Elliptic Ψdo

Definition 4.1.25 A symbol $\sigma \in S^\mu$ is called elliptic if there exist positive numbers ρ_0 and C such that, for $|\xi| \ge \rho_0$, the inequality

$$|\tilde{\sigma}(x, \xi)| \ge C|\xi|^\mu,$$

holds, where $\tilde{\sigma}$ denotes the presymbol corresponding to σ in accordance with (4.1.5). A proper Ψdo A is called elliptic if σ_A is an elliptic symbol.

If $A \in \Psi_0^\mu$ and $B \in \Psi_0^\nu$ are elliptic, then the relations $\sigma_{AB} - \sigma_A \sigma_B \in S^{\mu+\nu-1}$, $\sigma_{A^*} - \overline{\sigma}_A \in S^{\mu-1}$ imply that the operators $AB \in \Psi_0^{\mu+\nu}$, $A^* \in \Psi_0^\mu$ are elliptic.

Definition 4.1.26 A proper Ψdo B is called a parametrix of the proper Ψdo A if

$$BA = I + R_1, \quad AB = I + R_2,$$

where $R_1, R_2 \in \Psi_0^{-\infty}$ and I is the identity operator.

Theorem 4.1.27

(1) *Any elliptic operator $A \in \Psi_0^\mu$ possesses a parametrix $B \in \Psi_0^{-\mu}$.*
(2) *If B' is another proper Ψdo satisfying $B'A - I \in \Psi_0^{-\infty}$ or $AB' - I \in \Psi_0^{-\infty}$, then $B' - B \in \Psi_0^{-\infty}$.*

Proof Since σ_A is an elliptic symbol of order μ, we have

$$c|\xi|^\mu \leq |\tilde{\sigma}_A(x, \xi)| \leq C|\xi|^\mu \quad (C, c > 0, |\xi| \geq \rho_0). \tag{4.1.32}$$

Let $\tilde{\sigma}_0$ be a presymbol of order $-\mu$, satisfying $\tilde{\sigma}_0(x, \xi) = \tilde{\sigma}_A(x, \xi)^{-1}$ for $|\xi| \geq \rho_0$, and let σ_0 be the corresponding symbol. Adding to σ_0 a certain element in $S^{-\infty}$, we can assume (Theorem 4.1.19) that σ_0 is the symbol of some proper Ψdo B_0. From (4.1.32), it follows that the operator $B_0 \in \Psi_0^{-\mu}$ is elliptic. Moreover, $B_0 A = I + R_0$, $AB_0 = I + R_0'$, where $R_0, R_0' \in \Psi_0^{-1}$.

Denote the symbol of Ψdo $(-1)^j R_0^j$, $j \geq 0$, by $\sigma^{(j)}$. By Theorem 4.1.13, one can find an element $\sigma_1 \in S_0$ such that

$$\sigma_1(x, \xi) \sim \sum_{j \geq 0} \sigma^{(j)}(x, \xi).$$

Applying Theorem 4.1.19 again, we choose σ_1 so that the operator $B_1 = \mathrm{Op}\, \sigma_1$ is proper. We set $B = B_1 B_0$ and obtain

$$BA = I + R_1, \quad B \in El\Psi^{-\mu}, \quad R_1 \in \Psi_{-\infty}^0, \tag{4.1.33}$$

where the operator B is elliptic. In an analogous way, starting from $AB_0 = I + R_0'$, we find that

$$AB' = I + R_1', \quad B' \in \Psi^{-\mu}, \quad R_1' \in \Psi_0^{-\infty},$$

where B' is elliptic. Since $BAB' = B(I + R_1') = (I + R_1)B'$, we obtain $B - B' \in \Psi_0^{-\infty}$. Therefore,

$$AB = I + R_2, \quad R_2 \in \Psi_0^{-\infty}. \tag{4.1.34}$$

Formulas (4.1.33) and (4.1.34) imply the first assertion of the theorem. The second assertion follows from the fact that, in the proof of the inclusion $B' - B \in \Psi_0^{-\infty}$, we need only the relations $BA - I \in \Psi_0^{-\infty}$ and $AB' - I \in \Psi_0^{-\infty}$. □

4.2 Operators on Manifolds with Wedges

4.2.1 Admissible Diffeomorphisms of Subsets of $\mathbb{R}^m_n$

Below, a mapping $f = (f_1, \dots, f_m) : E \to \mathbb{R}^m = \mathbb{R}^n \bigoplus \mathbb{R}^{m-n}$ (E is an arbitrary set) is often written in the form $f = (f^1, f^2)$, where $f^1 = (f_1, \dots, f_n)$, $f^2 = (f_{n+1}, \dots, f_m)$. If E is an open subset of $\mathbb{R}^m$ and the mapping f is differentiable, then $\det f'$ denotes the determinant of the Jacobian matrix $f' = (\partial f_i/\partial x_j)$.

Definition 4.2.1 Let $W, W_1 \subset \mathbb{R}^m_n$ be open (not necessarily connected) sets. A diffeomorphism $f : W \to W_1$ (of class C^∞) is called admissible if:

(1) coordinate functions of the diffeomorphism f are subjected to the estimates

$$|\partial^\alpha f_i(x)| \le C_\alpha |x^1|^{1-|\alpha|}, \quad x \in W, \tag{4.2.1}$$

for all $\alpha \in \mathbb{Z}^m_+$, $|\alpha| \ge 1$; (2) there exist constants $C_0, c_0 > 0$, such that

$$c_0|x - y|/|x^1| \le |f(x) - f(y)|/|f^1(x)| \le C_0|x - y|/|x^1| \tag{4.2.2}$$

for any $x, y \in W$.

Note that condition (2) follows from the relations

$$c_0'|x^1| \le |f^1(x)| \le C_0'|x^1|, \quad x \in W, \tag{4.2.3}$$

$$c_0'|x - y| \le |f(x) - f(y)| \le C_0'|x - y|, \quad x, y \in W. \tag{4.2.4}$$

The converse is also true. To prove this, we rewrite (4.2.2) in the form

$$c_0|f^1(x)|/|x^1| \le |f(x) - f(y)|/|x - y| \le C_0|f^1(x)|/|x^1|. \tag{4.2.5}$$

Since x and y play equal roles, we have $c_0\,|f^1(x)|/|x^1| \le C_0\,|f^1(y)|/|y^1|$. Hence, the fraction $|f^1(x)|/|x^1|$ is bounded and separated from zero. Due to (4.2.5), an analogous statement holds for $|f(x) - f(y)|/|x - y|$, too. In the following lemma, we gather the properties of the admissible diffeomorphisms that will be used in the proof of Theorem 4.2.7 on the change of variables in Ψdo.

Lemma 4.2.2 *Let W, $W_1 \subset \mathbb{R}^m_n$ be open sets and $f : W \to W_1$ an admissible diffeomorphism. Then:*

(1) *The diffeomorphism $g : W_1 \to W$ inverse to f is also admissible.*
(2) *The Jacobian* $\det f'$ *is bounded away from zero on the set W.*
(3) *If $a \in \Pi^\mu$ (respectively, $a \in \Pi^\mu_0$) and* $\operatorname{supp} a \subset W \times W \times \mathbb{R}^m$, *then the function $W_1 \times W_1 \times \mathbb{R}^m \ni (x, y, \xi) \mapsto a(g(x), g(y), \xi)$ extended to be zero on $(\mathbb{R}^m_n \times \mathbb{R}^m_n \times \mathbb{R}^m) \setminus (W_1 \times W_1 \times \mathbb{R}^m)$ belongs to Π^μ (respectively, Π^μ_0).*

Proof In (4.2.2), we change $x, y \in W$ for $g(x), g(y)$ $(x, y \in W_1)$ and obtain that the diffeomorphism g satisfies condition (2) of Definition 4.2.1. Next, the inequality $|g(x) - g(y)| \le C|x - y|$ implies that the derivatives $g_{pq} = \partial g_p/\partial x_q$ $(1 \le p, q \le m)$ are bounded on W_1. Therefore, the determinant $\det g'$ is also bounded, and hence, the determinant $\det f'$ takes values separated from zero.

The estimates $|\partial^\alpha g_p(x)| \le C\,|x^1|^{1-|\alpha|}$ established previously for $|\alpha| = 1$ are equivalent to

$$|\partial^\beta g_{pq}(x)| \le C|x^1|^{-|\beta|}. \tag{4.2.6}$$

To prove inequality (4.2.6), we note that the entries $g_{pq} \circ f$ of the matrix $g' \circ f$ are rational functions of the entries f_{ij} of f'. Since the denominator $\det f'$ of these functions is bounded away from zero, we have

$$|\partial_z^\beta (g_{pq} \circ f)(z)| \le C|z^1|^{-|\beta|}, \quad \beta \in \mathbb{Z}^m_+, \ z \in W,$$

due to (4.2.1). The derivative $\partial_x^\beta g_{pq}(x)$ is the sum of functions of the form

$$\partial_z^{\beta_1} (g_{pq} \circ f)(z)\big|_{z=g(x)} \prod_k (\partial^{\gamma_k} g_{p_k}(x))^{l_k},$$

where $|\beta_1| \le |\beta|$, $\sum_k l_k|\gamma_k| = |\beta|$, $\sum_k l_k = |\beta_1|$ (the numbers p_k corresponding to distinct k may coincide; the same is true for the multi-indices γ_k). Reasoning by induction, one can assume that the estimates (4.2.6) are valid with β replaced by γ such that $|\gamma| \le |\beta| - 1$. Then, $|\partial^\gamma g_p(x)| \le C|x^1|^{1-|\gamma|}$ for $|\gamma| \le |\beta|$, and since $|\gamma_k| \le |\beta|$ for all k, we obtain

$$|\partial_z^{\beta_1}(g_{pq}\circ f)(g(x))\prod_k(\partial^{\gamma_k}g_{p_k}(x))^{l_k}|\leq C|g^1(x)|^{-|\beta_1|}|x^1|^{l_1(1-|\gamma_1|)+\dots}=$$

$$=C|g^1(x)|^{-|\beta_1|}|x^1|^{|\beta_1|-|\beta|}\leq C_1|x^1|^{-|\beta|}.$$

It remains to prove assertion (3). For short, we further write $(\mathbb{R}_n^m)^2$, W^2, etc., instead of $\mathbb{R}_n^m\times\mathbb{R}_n^m$, $W\times W$, etc. It follows from (4.2.3) and (4.2.4) that if the set $E\subset W^2\times\mathbb{R}^m$ is closed in $(\mathbb{R}_n^m)^2\times\mathbb{R}^m$, then its image under the mapping $F:(x,y,\xi)\mapsto(f(x),f(y),\xi)$ is also closed in $(\mathbb{R}_n^m)^2\times\mathbb{R}^m$. In particular, if the amplitude a satisfies the conditions of Lemma and $b(x,y,\xi)=a(g(x),g(y),\xi)$ $(x,y\in W_1)$, then the set $W_1^2\times\mathbb{R}^m$ is a neighborhood of the set supp$b=F(\text{supp}a)$ in $(\mathbb{R}_n^m)^2\times\mathbb{R}^m$. Therefore, setting $b(x,y,\xi)=0$ for $(x,y,\xi)\in((\mathbb{R}_n^m)^2\times\mathbb{R}^m)\setminus(W_1^2\times\mathbb{R}^m)$, we obtain a function of class C^∞ on $(\mathbb{R}_n^m)^2\times\mathbb{R}^m$. Let $\tilde b(x,y,\xi)=b(x,y,\xi/|x^1|)$. The inclusion $b\in\Pi^\mu$ to be established is equivalent to $\tilde b\in\overset{\sim\mu}{\Pi}$, i.e., to the estimates

$$|\partial_x^\alpha\partial_y^\beta\partial_\xi^\gamma\tilde b(x,y,\xi)|\leq C_{\alpha\beta\gamma}|x^1|^{-|\alpha|}|y^1|^{-|\beta|}\langle\xi\rangle^{\mu-|\gamma|}. \tag{4.2.7}$$

If $\tilde a$ is a pre-amplitude corresponding to a by (4.1.1), then $\tilde b(x,y,\xi)=\tilde a(x,y,\xi|g^1(x)|/|x^1|)$. In accordance with the relations $c_0|x^1|\leq|g^1(x)|\leq C_0|x^1|$, $|\partial^\alpha g(x)|\leq C_\alpha|x^1|^{1-|\alpha|}$, this implies (4.2.7).

Finally, let $a\in\Pi_0^\mu$ be a proper amplitude. According to Definition 4.1.2, there exist numbers $\delta\in(0,1)$ and $\varepsilon>0$ such that $a(x,y,\xi)=0$ for $|x^1|/|y^1|\notin(\delta,\delta^{-1})$ or $|x^2-y^2|>\varepsilon(1+|x^1|)$. It follows from (4.2.3) and (4.2.4) that the amplitude $b:(x,y,\xi)\mapsto a(g(x),g(y),\xi)$ possesses the same property. □

4.2.2 Change of Variables in Ψdo

If W is an open subset of $\mathbb{R}_n^m$, then $\rho(x,\partial W)$ denotes the distance from the point $x\in W$ to the boundary ∂W of W.

Definition 4.2.3 An open set $W\subset\mathbb{R}_n^m$ is called a privileged neighborhood of the set $E\subset W$, if there exists a number $\varepsilon>0$ such that $\rho(x,\partial W)\geq\varepsilon|x^1|$ for all $x\in E$.

Lemma 4.2.4 *Let W be a privileged neighborhood of E, $f:W\to W_1$ an admissible diffeomorphism, and $E_1=f(E)$. Then, W_1 is a privileged neighborhood of E_1.*

Proof According to Definition 4.2.3, the inequality $\rho(z,\partial W)\geq\varepsilon|z^1|$ holds for all $z\in E$ with some $\varepsilon>0$. Let $g=(g_1,\dots,g_m)$ be the diffeomorphism inverse to f. The estimate $\rho(g(x),\partial W)\leq C\rho(x,\partial W_1)$ follows from the boundedness of the derivatives

$\partial g_p/\partial x_q$; here, C is independent of $x \in W_1$. If $x \in E_1$, then $g(x) \in E$ and $\rho(x, \partial W_1) \geq C^{-1}\rho(g(x), \partial W) \geq C^{-1}\varepsilon|g_1(x)| \geq \varepsilon_1|x^1|$. □

If a is a continuous function on the product $\mathbb{R}^m_n \times \mathbb{R}^m_n \times \mathbb{R}^m$, then $\operatorname{supp}_1 a$ (respectively, $\operatorname{supp}_2 a$) denotes the projection of the support of a onto the first (respectively, second) factor of this product.

Definition 4.2.5 Let $W \subset \mathbb{R}^m_n$ be an open subset. We denote by $\Pi^\mu(W)$ the class of all amplitudes $a \in \Pi^\mu$, for which W is a privileged neighborhood of the set $\operatorname{supp}_1 a \cup \operatorname{supp}_2 a$. The class $\Pi^\mu_0(W)$ is defined by $\Pi^\mu_0(W) = \Pi^\mu_0 \cap \Pi^\mu(W)$. Set also $\Psi^\mu(W) = \operatorname{Op}(\Pi^\mu(W))$, $\Psi^\mu_0(W) = \operatorname{Op}(\Pi^\mu_0(W))$.

Our immediate objective is to prove the theorem on the change of variables in Ψdo of class $\Psi^\mu(W)$. To this end, we need the following lemma.

Lemma 4.2.6 *Let W_1 be a privileged neighborhood of the set E_1 and $g : W_1 \to W$ be an admissible diffeomorphism. There exists a number $\varepsilon_1 > 0$ such that the matrix-valued function*

$$h(x, y) = \int_0^1 g'(y + t(x - y))\, dt \tag{4.2.8}$$

is correctly defined in some neighborhood V of the set

$$\mathcal{D} = \{(x, y) : x \in E_1, |x - y| \leq \varepsilon_1 |x^1|\},$$

and, in this neighborhood, its entries h_{pq} admit the estimates

$$|\partial_x^\alpha \partial_y^\beta h_{pq}(x, y)| \leq C_{\alpha\beta} |x^1|^{-|\alpha|} |y^1|^{-|\beta|}. \tag{4.2.9}$$

The number ε_1 can be chosen so that the following inequality holds:

$$\inf\{|\det h(x, y)|;\ (x, y) \in V\} > 0. \tag{4.2.10}$$

Proof According to Definition 4.2.3, we have $\rho(x, \partial W_1) \geq \varepsilon|x^1|$ for all $x \in E_1$. We may assume that $\varepsilon < 2$. Set

$$U = \{x \in W_1 : (\exists z \in E_1)\ |x - z| < \varepsilon|z^1|/2\}.$$

Then, U is an open subset of W_1 containing E_1, and $\rho(x, \partial W_1) \geq (\varepsilon/2)(1 - \varepsilon/2)|x^1|$ if $x \in U$. Let $\varepsilon_1 = (\varepsilon/4)(1 - \varepsilon/2)$,

$$V = \{(x, y) : x \in U,\ |x - y| < 2\varepsilon_1 |x^1|\}.$$

Clearly, V is an open neighborhood of $\mathcal{D}$. If $(x, y) \in V$, then the open ball (in $\mathbb{R}^m_n$) centered at x of radius $2\varepsilon|x^1|$ contains y. Since $\rho(x, \partial W_1) \geq 2\varepsilon_1$, this ball is a subset of W_1. Thus, the function h is correctly defined for $(x, y) \in V$.

Since $2\varepsilon_1 < 1$, the condition $(x, y) \in V$ implies $xy \geq 0$. This, in turn, is followed by $|tx + (1-t)y| \geq \max\{t|x^1|, (1-t)|y^1|\}$ for any $t \in [0, 1]$. We now have

$$|\partial_x^\alpha \partial_y^\beta h_{pq}(x, y)| \leq \int_0^1 |\partial_x^\alpha \partial_y^\beta g_{pq}(y + t(x-y))|\, dt \leq$$
$$\leq Ct^{|\alpha|}(1-t)^{|\beta|}|tx + (1-t)y|^{-|\alpha+\beta|} \leq C\, |x^1|^{-|\alpha|}|y^1|^{-|\beta|}.$$

Note finally that $|h(x, y)| = |g'(x)| \geq \delta_0$, where $\delta_0 > 0$ is independent of $x \in U$. The determinant $\det h(x, y)$ and its entries satisfy (4.2.9). In particular, $|\nabla_y \det h(x, y)| \leq C|y^1|^{-1} \leq C(1-2\varepsilon)^{-1}|x^1|$ for $(x, y) \in V$, and therefore,

$$|\det h(x, y)| \geq |\det h(x, x)| - |\det h(x, x) - \det h(x, y)| \geq$$
$$\geq \delta_0 - C(1-2\varepsilon_1)^{-1}|x^1|^{-1}|x - y| \geq \delta_0 - 2C\varepsilon_1(1-2\varepsilon_1)^{-1}.$$

We subject ε_1 to the condition $2C\varepsilon_1(1-2\varepsilon_1)^{-1} \leq \delta_0/2$ and obtain $|\det h(x, y)| \geq \delta_0/2$ on V. □

Assume now that $A = \operatorname{Op} a$, $a \in \Pi^\mu(W)$, and $f : W \to W_1$, $g : W_1 \to W$ are mutually inverse diffeomorphisms. We define an operator B on functions $u \in \mathbb{R}^m_n$ by

$$Bu = [A(u \circ f)] \circ g. \tag{4.2.11}$$

Note that although the composition $u \circ f$ is defined only on W, the expression

$$A(u \circ f)(x) = (2\pi)^{-m} \int\int e^{i(x-y)\xi} a(x, y, \xi)(u \circ f)(y)\, dy d\xi$$

makes sense due to the condition $\operatorname{supp}_2 a \subset W$. Since $\operatorname{supp}_1 a \subset W$, the function $v = [A(u \circ f)] \circ g$ is in $C^\infty(W_1)$ and vanishes near ∂W_1. Setting $v(x) = 0$ for $x \in \mathbb{R}^m_n \setminus W_1$, we obtain an element of $C^\infty(\mathbb{R}^m_n)$, which is identified with Bu.

Theorem 4.2.7 *Let $f : W \to W_1$ be an admissible diffeomorphism of open subsets of $\mathbb{R}^m_n$ and $g : W_1 \to W$ the inverse diffeomorphism. If $a \in \Pi^\mu(W)$, $A = \operatorname{Op} a$, and the operator B is defined by (4.2.11), then $B \in \Psi^\mu(W_1)$.*

Proof Fix a number $\varepsilon_0 \in (0, 1)$. Let $\zeta_0 \in C^\infty(\mathbb{R}_+)$, $\zeta_0(t) = 1$ for $t < \varepsilon_0/2$, and $\zeta_0(t) = 0$ for $t > \varepsilon_0$. It is easily seen that $\zeta(x, y) = \zeta_0(|x - y|/|x^1|)$ is in class Π^0. We set $a_0 = \zeta a$, $a_1 = (1 - \zeta)a$ ($a, a_1 \in \Pi^\mu$, since $\zeta \in \Pi^0$) and represent the Ψdo A in the form $A =$

A_0+A_1, where $A_j = \operatorname{Op} a_j$, $j = 0, 1$. Let B_j denote the operator defined by (4.2.11) with A replaced by A_j. Since $B = B_0 + B_1$, to prove the inclusion $B \in \Psi^\mu(W_1)$, it suffices to prove that $B_j \in \Psi^\mu(W_1)$, $j = 0, 1$.

First, we consider the operator B_1. We have

$$B_1 u(x) = (2\pi)^{-m} \int\int e^{i(g(x)-g(y))\xi} \mathfrak{b}_1(x, y, \xi) u(y)\, dy d\xi, \quad u \in C_c^\infty(\mathbb{R}_n^m),$$

where $\mathfrak{b}_1(x, y, \xi) = a_1(g(x), g(y), \xi)|\det g'(y)|$, if $(x, y) \in W_1$, and $\mathfrak{b}_1(x, y, \xi) = 0$ otherwise. According to Lemma 4.2.2, the diffeomorphism g is admissible. Therefore, the entries g_{pq} of the determinant $\det g'$ are subject to estimates (4.2.6). This and assertion (3) of the same Lemma imply that $\mathfrak{b}_1 \in \Pi^\mu$. Moreover, since $|g(x) - g(y)|/|g^1(x)| \leq C_0|x - y|/|x^1|$, we have

$$\mathfrak{b}_1(x, y, \xi) = 0 \quad \text{for} \quad |x - y|/|x^1| \leq C_0^{-1}\varepsilon_0/2. \tag{4.2.12}$$

Let $\mathfrak{G}_1$ be the kernel of the operator A_1. In the proof of Proposition 4.1.23, we have obtained the estimate

$$|\partial_x^\alpha \partial_y^\beta \mathfrak{G}_1(x, y)| \leq C_{\alpha\beta N}|x^1|^{N-m-|\alpha|}|y^1|^{-|\beta|}|x - y|^{-N} \tag{4.2.13}$$

for all $\alpha, \beta \in \mathbb{Z}_+^m$, and $N \in \mathbb{Z}_+$. Denote by G_1 the kernel of B_1. Then,

$$G_1(x, y) = (2\pi)^{-m} \int e^{i(g(x)-g(y))\xi} \mathfrak{b}_1(x, y, \xi)\, d\xi = \mathfrak{G}_1(g(x), g(y))\, |\det g'(y)|.$$

From the properties of admissible diffeomorphisms and from (4.2.13), it easily follows that

$$\left|\partial_x^\alpha \partial_y^\beta (|x^1|^m G_1(x, y))\right| \leq C'_{\alpha\beta N}|x^1|^{N-|\alpha|}|y^1|^{-|\beta|}|x - y|^{-N}. \tag{4.2.14}$$

Let $v \in \mathcal{S}(\mathbb{R}^m)$ and $(2\pi)^{-m} \int v(\eta)\, d\eta = 1$. We set

$$b_1(x, y, \eta) = e^{i(y-x)\eta}|x^1|^m G_1(x, y) v(|x^1|\eta).$$

Due to (4.2.12) and (4.2.14), the function $H : (x, y) \mapsto |x^1|^m G_1(x, y)$ satisfies the conditions of Lemma 4.1.22. Therefore, $b_1 \in \Pi^{-\infty}$, and since $B_1 = \operatorname{Op} b_1$, we have $B_1 \in \Psi^{-\infty} \subset \Pi^\mu$. Taking into account $\operatorname{supp}_j b_1 \subset f(\operatorname{supp}_j a_1)$, $j = 1, 2,$, and Lemma 4.2.4, we conclude that $B_1 \in \Psi^\mu(W_1)$.

To prove the inclusion $B_0 \in \Psi^\mu(W_1)$, let us consider the matrix function h defined by (4.2.8). According to Lemma 4.2.6, there exists a number $\varepsilon_1 > 0$ such that the function h is correctly defined in some neighborhood V of the set

$$\mathcal{D} = \{(x, y) : x \in f(\operatorname{supp}_1 a), \ |x - y|/|x^1| \leq \varepsilon_1\}$$

and, in this neighborhood, satisfies (4.2.9) and (4.2.10). The support of the function $(x, y, \xi) \mapsto a_0(g(x), g(y), \xi)$ is a subset of

$$\mathcal{D}_1 = \{(x, y) : \ x \in f(\mathrm{supp}_1 a), \ \ |g(x) - g(y)|/|g^1(x)| \leq \varepsilon_0\}.$$

In view of $c_0|x - y|/|x^1| \leq |g(x) - g(y)|/|g^1(x)|$ for $\varepsilon_0 \leq c_0\varepsilon_1$, the inclusion $\mathcal{D}_1 \subset \mathcal{D}$ holds, and therefore, the mentioned support lies in domain V of h. Taking into account that $g(x) - g(y) = h(x, y)(x - y)$, we have

$$B_0 u(x) = (2\pi)^{-m} \int\!\!\int e^{i(g(x)-g(y))\xi} a_0(g(x), g(y), \xi)| \det g'(y)| u(y)\, dy d\xi =$$

$$= (2\pi)^{-m} \int\!\!\int e^{i(x-y)\eta} b_0(x, y, \eta) u(y)\, dy d\eta,$$

where $b_0(x, y, \eta) = a_0(g(x), g(y), {}^t h(x, y)^{-1}\eta)| \det g'(y)|| \det h(x, y)|^{-1}$. Due to Lemma 4.2.2, the function $(x, y, \xi) \mapsto a_0(g(x), g(y), \xi)$ belongs to the class Π^μ. Estimates (4.2.9) now imply that $b_0 \in \Pi^\mu$, and since W_1 is a privileged neighborhood of the set $\mathrm{supp}_1 b_0 \cup \mathrm{supp}_2 b_0$, we have $b_0 \in \Pi^\mu(W_1)$. □

If in the conditions of Theorem 4.2.7 the amplitude a is proper, then the amplitudes b_0 and b_1 constructed in the proof of the theorem will also be proper (this immediately follows from assertion (3) of Lemma 4.2.2). Thus, $B \in \Psi_0^\mu(W_1)$. At the same time, the role of a can be given to the symbol σ_A of A. Then,

$$b_0(x, y, \eta) = \zeta(g(x), g(y))\sigma_A(g(x), {}^t h(x, y)^{-1}\eta)| \det g'(y)|\, | \det h(x, y)|^{-1},$$

and therefore,

$$\sigma_B(x, \eta) \sim \sum_\alpha \frac{1}{\alpha!} \partial_\eta^\alpha D_y^\alpha (\sigma_A(g(x), {}^t h(x, y)^{-1}\eta)| \det g'(y)|\, | \det h(x, y)|^{-1}\big|_{y=x}$$

in view of Theorem 4.1.16 (we have taken into account that the function $(x, y) \mapsto \zeta(g(x), g(y))$ equals 1 in a neighborhood of the set $x = y$ and that the amplitude b_1, which vanishes in this neighborhood, does not influence σ_B). Using additional reasoning of combinatorial nature, this expansion can be transformed into

$$\sigma_B(x, \eta) \sim \sum_\alpha \frac{1}{\alpha!} \sigma_A^{(\alpha)}(z, {}^t f'(z)\eta) D_w^\alpha e^{if''(z,w)\eta}\big|_{w=z}, \tag{4.2.15}$$

where $\sigma_A^{(\alpha)}(z, \xi) = \partial_\xi^\alpha \sigma_A(z, \xi)$ and $f''(z, w) = f(w) - f(z) - f'(z)(w - z)$. Therefore, the following theorem is true.

Theorem 4.2.8 *Let the operator A in Theorem 4.2.7 be proper. Then, the Ψdo B defined by (4.2.11) is also proper, and its symbol admits the asymptotic expansion (4.2.15).* □

4.2.3 Pseudodifferential Operators on a Wedge

For any subset D for the unit sphere $S^{N-1} \subset \mathbb{R}^N$, we denote by $K(D)$ a cone in $\mathbb{R}^N \setminus 0$ with the base D,

$$K(D) = \{z \in \mathbb{R}^N; z = r\varphi, r > 0, \varphi \in D\}.$$

If D is a compact (without boundary) smooth $(n-1)$-dimensional submanifold of S^{N-1}, then the set $\mathbb{W} = K(D) \times \mathbb{R}^{m-n}$ is a smooth m-dimensional surface in $\mathbb{R}^N \bigoplus \mathbb{R}^{m-n}$. Any such a surface will be called a surface of wedge type or just a wedge. If $N = n$ and $D = S^{n-1}$, then $\mathbb{W} = \mathbb{R}^m_n$.

Let $U^{(0)}$ be an open (not necessarily connected) subset of D admitting a diffeomorphic mapping $\upsilon^{(0)}$ onto some subset of the sphere S^{n-1}, $U = K(U^{(0)}) \times \mathbb{R}^{m-n}$. We extend $\upsilon^{(0)}$ to a mapping $\upsilon = \upsilon_U : U \to \mathbb{R}^n \bigoplus \mathbb{R}^{m-n}$ by setting

$$\upsilon(r\varphi, x^2) = (r\upsilon^{(0)}(\varphi), x^2) \quad (r > 0, \varphi \in S^{N-1},\ x^2 \in \mathbb{R}^{m-n}).$$

The pair (U, υ) is called a local chart on the surface $\mathbb{W}$.

Remark 4.2.9 Replacing, if necessary, the set $U^{(0)}$ with its (open) subset, we may assume that all the derivatives of the coordinate functions of the mapping $\upsilon^{(0)}$ (in local coordinates on S^{N-1} and S^{n-1}) are bounded. In what follows, we consider only local charts for which this condition is fulfilled.

Definition 4.2.10 Let $W \subset \mathbb{R}^m_n$ be a set of the form $K \times \mathbb{R}^{m-n}$, where K is an open cone in $\mathbb{R}^n$. A function $\zeta \in C^\infty(\mathbb{R}^m_n)$ is called a homogeneous cut-off function (HCF) serving the set W, if it is independent of x^2, homogeneous of degree 0, and $\operatorname{supp}\zeta \subset W$.

Definition 4.2.11 A pseudodifferential operator (a proper Ψdo) of order μ on a wedge $\mathbb{W}$ is a linear mapping $A : C_c^\infty(\mathbb{W}) \to C^\infty(\mathbb{W})$ such that, for any local chart and any ζ_1, ζ_2 serving the set $W = \upsilon(U)$, the operator

$$C_c^\infty(\mathbb{R}^m_n) \ni u \mapsto \zeta_1(\upsilon^{-1})^* A\upsilon^*(\zeta_2 u) \equiv \zeta_1[A((\zeta_2 u) \circ \upsilon)] \circ \upsilon^{-1}$$

is in the class $\Psi^\mu(W)$ ($\Psi^\mu_0(W)$). The set of all such Ψdo is denoted by $\Psi^\mu(\mathbb{W})$ ($\Psi^\mu_0(\mathbb{W})$).

To establish the general form of a Ψdo on a wedge-type surface (Theorem 4.2.13), we need the following lemma.

Lemma 4.2.12 *Let $(U, \upsilon_U$ be (V, υ_V) local charts on the wedge $\mathbb{W}$, $U \cap V \neq \emptyset$, and let ζ and η be the HCF serving the sets $\upsilon_U(U)$ and $\upsilon_V(V)$, respectively. We set $W' = \upsilon_U(U \cap V)$, $W'' = \upsilon_V(U \cap V)$, $p = \upsilon_V \circ \upsilon_U^{-1}|W'$, $\theta = \eta \circ p$. Then:*

(1) *The function $\zeta\theta$ extended to be zero on $\mathbb{R}^m_n \setminus W'$ is an HCF serving the set W'.*
(2) *W' is a privileged neighborhood of the set* $\operatorname{supp}(\zeta\theta)$.
(3) *$p : W' \to W'$ is an admissible diffeomorphism.*

Proof The function $W' \ni (x^1, x^2) \mapsto (\zeta\theta)(x^1, x^2)$ extended to be zero on $\mathbb{R}^m_n \setminus W'$ is homogeneous of degree zero and independent of x^2; its support coincides with $\upsilon_U(\operatorname{supp}(\zeta \circ \upsilon_U)(\eta \circ \upsilon_V))$ and is contained in W'. This proves assertion (1). Assertion (2) is obvious and (3) directly follows from Remark 4.2.9. □

Now consider some finite covering $\{U_j^{(0)}\}_{j=1}^l$ of the manifold D such that for all $j, k = 1, \ldots, l$ the set $U_{jk}^{(0)} = U_j^{(0)} \cup U_k^{(0)}$ is a coordinate neighborhood. Let $\upsilon_{jk}^{(0)} : U_{jk}^{(0)} \to S^{n-1}$ be a coordinate mapping chosen in accordance with Remark 4.2.9 and $\{(U_{jk}, \upsilon_{jk})\}$ be the corresponding atlas of local charts on $\mathbb{W}$.

Theorem 4.2.13 *Let us fix a number $\mu \in \mathbb{R}$ and for each pair (j, k) choose a Ψdo $A_{jk} \in \Psi^\mu(W_{jk})$ arbitrarily. Then, the operator*

$$A = \sum_{j,k} \upsilon_{jk}^* A_{jk} (\upsilon_{jk}^{-1})^* \tag{4.2.16}$$

is a Ψdo of order μ on the surface $\mathbb{W}$. Any $A \in \Psi^\mu(\mathbb{W})$ can be written in the form (4.2.16) with suitable A_{jk}.

Proof Consider a partition of unity $\{\chi_j^{(0)}\}_{j=1}^l$ subordinate to the covering $\{U_j^{(0)}\}_{j=1}^l$ of D. For $r > 0$, $\varphi \in D$, and $x^2 \in \mathbb{R}^{m-n}$, we set $\chi_j(r\varphi, x^2) = \chi_j^{(0)}(\varphi)$ and represent A in the form $A = \sum_{j,k} \chi_j A \chi_k$. According to Definition 4.2.11, the operator

$$A_{jk} : u \mapsto (\chi_j \circ \upsilon_{jk}^{-1})(\upsilon_{jk}^{-1})^* A \upsilon_{jk}^*((\chi_k \circ \upsilon_{jk}^{-1})u), \quad u \in C_c^\infty(\mathbb{R}^m_n),$$

is in the class $\Psi^\mu(W_{jk})$ $(j, k = 1, \ldots, l)$. Since $\chi_j A \chi_k = \upsilon_{jk}^* A_{jk} (\upsilon_{jk}^{-1})^*$, the second assertion of the theorem is proved.

We now prove the first assertion. Let $a_{jk} \in \Pi^\mu(W_{jk})$, $A_{jk} = \operatorname{Op} a_{jk}$, $j, k = 1, \ldots, l$, and let the operator A be defined by (4.2.16). Since for any (j, k), the set W_{jk} is a privileged neighborhood of $\operatorname{supp}_1 a_{jk} \cup \operatorname{supp}_2 a_{jk}$ (see Definition 4.2.5), there exists an HCF ζ_{jk}, serving W_{jk} such that $\zeta_{jk}(x) a_{jk}(x, y, \xi) \zeta_{jk}(y) = a_{jk}(x, y, \xi)$. Let us consider an arbitrary chart (V, υ_V). Set $W'_{jk} = \upsilon_{jk}(U_{jk} \cap V)$ and $W''_{jk} = \upsilon_V(U_{jk} \cap V)$. If η_1 and η_2 are HCFs serving the set $W = \upsilon_V(V)$, then for $u \in C_c^\infty(\mathbb{R}^m_n)$ we have

$$\eta_1(\upsilon_V^{-1})^* A \upsilon_V^*(\eta_2 u) = \sum_{j,k}{}' [(\theta^{(1)}_{jk} \zeta_{jk} A_{jk} \zeta_{jk} \theta^{(2)}_{jk})(u \circ p_{jk})] \circ p_{jk}^{-1}, \tag{4.2.17}$$

where $p_{jk} = \upsilon_V \circ \upsilon_{jk}^{-1} | W'_{jk}$, $\theta^{(i)}_{jk} = \eta_i \circ p_{jk}$, $i = 1, 2$, and the prime on the summation sign means that the summation extends only over the pairs (j, k) for which $U_{jk} \cap V \neq \emptyset$. By Lemma 4.2.12, the function $\theta^{(i)}_{jk} \zeta_{jk}$, $i = 1, 2$, is an HCF serving W'_{jk}. This implies that $\theta^{(1)}_{jk} \zeta_{jk} A_{jk} \zeta_{jk} \theta^{(2)}_{jk} \in \Psi^\mu(W'_{jk})$. Moreover, due to the same lemma, the mapping $p_{jk} : W'_{jk} \to W'_{jk}$ is an admissible diffeomorphism. From Theorem 4.2.7, it now follows that the operator

$$C_c^\infty(\mathbb{R}^m_n) \ni u \mapsto [(\theta^{(1)}_{jk} \zeta_{jk} A_{jk} \zeta_{jk} \theta^{(2)}_{jk})(u \circ p_{jk})] \circ p_{jk}^{-1}$$

is in the class $\Psi^\mu(W'_{jk})$. Taking into account formula (4.2.17) and the inclusion $\Psi^\mu(W'_{jk}) \subset \Psi^\mu(W)$, we obtain $\eta_1(\upsilon_V^{-1})^* A \upsilon_V^* \eta_2 \in \Psi^\mu(W)$. □

Remark 4.2.14 If, on the right-hand side of (4.2.16), A_{jk} is a proper Ψdo, then the operator A defined by this formula is also proper. From the proof of Theorem 4.2.13, it now follows that the converse is also true: any $A \in \Psi_0^\mu(\mathbb{W})$ admits a representation of the form (4.2.16), where $A_{jk} \in \Psi_0^\mu(W_{jk})$.

Corollary 4.2.15 *Let $\chi_1^{(0)}, \chi_2^{(0)} \in C^\infty(D)$, where D is the base of the cone K,* $\operatorname{supp} \chi_1^{(0)} \cap \operatorname{supp} \chi_2^{(0)} = \emptyset$. *For $r > 0$, $\varphi \in D$, and $x^2 \in \mathbb{R}^{m-n}$, we set $\chi_i(r\varphi, x_2) = \chi_i^{(0)}(\varphi)$, $i = 1, 2$. Moreover, let $A \in \Psi^\mu(\mathbb{W})$, where $\mu \in \mathbb{R}$ is an arbitrary number. Then, $\chi_1 A \chi_2 \in \Psi^{-\infty}(\mathbb{W})$.*

Proof Using formula (4.2.16), we obtain $\chi_1 A \chi_2 = \sum \upsilon_{jk}^* A'_{jk} (\upsilon_{jk}^{-1})^*$, where $A'_{jk} = (\chi_1 \circ \upsilon_{jk}^{-1}) A_{jk} (\chi_2 \circ \upsilon_{jk}^{-1})$. From Definition 4.2.11, it follows that $A'_{jk} \in \Psi^\mu(W_{jk})$. In fact, $A'_{jk} \in \Psi^{-\infty}(W_{jk})$ in view of Proposition 4.1.23. □

Let $\{U_j^{(0)}\}_{j=1}^l$ be a covering of D by coordinate neighborhoods, $\{\chi_j^{(0)}\}_{j=1}^l$ a partition of unity subordinate to this covering, and let the elements of this partition be homogeneous cut-off functions. For each $j = 1, \dots, l$, we choose a function $\eta_j^{(0)} \in C^\infty(D)$ satisfying $\operatorname{supp} \eta_j^{(0)} \subset U_j^{(0)}$ and $\chi_j^{(0)} \eta_j^{(0)} = \chi_j^{(0)}$. As before, we set $\chi_j(r\varphi, x^2) = \chi_j^{(0)}(\varphi)$, $\eta_j(r\varphi, x^2) = \eta_j^{(0)}(\varphi)$. Taking into account the equality $A - \sum \eta_j A \chi_j = \sum (1 - \eta_j) A \chi_j$ and Corollary 4.2.15, we obtain the following statement.

Corollary 4.2.16 *If $A \in \Psi^\mu(\mathbb{W})$, then $A - \sum_j \eta_j A \chi_j \in \Psi^{-\infty}(\mathbb{W})$.*

Corollary 4.2.17 *Let $\mu \in \mathbb{R}$ be an arbitrary number and $A \in \Psi^{\mu}(\mathbb{W})$. Then, A admits the representation $A = A_0 + A_1$, where $A_0 \in \Psi_0^{\mu}(\mathbb{W})$, $A_1 \in \Psi_{-\infty}(\mathbb{W})$.*

To prove Corollary 4.2.17, one must use formula (4.2.16) and apply Proposition 4.1.24 to operators A_{jk}.

Concluding the section, we introduce admissible diffeomorphisms of open subsets of a wedge.

Definition 4.2.18 Let $D \subset S^N$, $D_1 \subset S^{N_1}$ be smooth compact $(n-1)$-dimensional manifolds, and let $\mathbb{W} = K(D) \times \mathbb{R}^{m-n}$, $\mathbb{W}_1 = K(D_1) \times \mathbb{R}^{m-n}$. A diffeomorphism $f : Z \to Z_1$, where $Z \subset \mathbb{W}$ and $Z_1 \subset \mathbb{W}_1$ are open subsets, is called admissible if for any two local charts (U, υ_U) and (V, υ_V) such that $U \cap Z \neq \emptyset$, $V \cap Z_1 \neq \emptyset$, and $f(U \cap Z) \cap V \cap Z_1 \neq \emptyset$, the mapping

$$\upsilon_V \circ f \circ \upsilon_U^{-1} : \upsilon_U(U \cap Z \cap f^{-1}(V \cap Z_1)) \to \upsilon_V(f(U \cap Z) \cap V \cap Z_1)$$

is an admissible diffeomorphism in the sense of Definition 4.2.1.

Now, by analogy with Definitions 4.2.3 and 4.2.5, we can introduce classes $\Psi^{\mu}(Z)$ and $\Psi_0^{\mu}(Z)$ for any open $Z \subset \mathbb{W}$. These classes are invariant under admissible diffeomorphisms due to Theorem 4.2.7.

4.2.4 W-manifolds

A compact subset $\mathcal{M}$ of a topological manifold is called a stratified manifold if: (a) $\mathcal{M}$ is a finite union of pairwise disjoint topological manifolds (strata); (b) a closure of any stratum consists of the stratum itself and the union of some (possibly empty) set of strata of lower dimensions.

Let $\{s_i\}_i$ be a set of strata whose union is a stratified manifold $\mathcal{M}$. We set $m = \max \dim s_i$, $n_i = m - \dim s_i$, $I = \{i : \dim s_i < m\}$, and $\mathcal{M}_0 = \cup_{i \notin I} s_i$. Below, we assume that:

(I) $\bar{s}_i = s_i$ for $i \in I$, where $\bar{s}_i$ is the closure of s_i.
(II) For each point of s_i, $i \in I$, there exist a neighborhood $\mathcal{O} \subset \mathcal{M}_0 \cup s_i$ of this point, an m-dimensional wedge $\mathbb{W} = K \times \mathbb{R}^{m-n_i}$, an open set $V \subset \overline{\mathbb{W}}$, and a homeomorphism $\varkappa : \mathcal{O} \to V$ such that $\varkappa(s_i \cap \mathcal{O}) = V \cap (\{0\} \times \mathbb{R}^{m-n_i})$.

It follows from conditions (I) and (II) that the (m-dimensional) manifold $\mathcal{M}_0$ is dense in $\mathcal{M}$, and $\mathcal{M} \setminus \mathcal{M}_0$ is a disjoint union of compact manifolds (edges). The pair $(\mathcal{O}, \varkappa)$ in condition (II) is called a special local chart on $\mathcal{M}$. If $\mathcal{O}'$ is a coordinate neighborhood on

$\mathcal{M}_0$ and $\varkappa'$ is a homeomorphism of this neighborhood onto an open subset of $\mathbb{R}^m$, then the chart $(\mathcal{O}', \varkappa')$ is called standard.

An atlas on $\mathcal{M}$ is any finite collection $\mathfrak{C} = \{(\mathcal{O}_j, \varkappa_j)\}$ of local charts satisfying the following conditions:

(i) $\bigcup \mathcal{O}_j = \mathcal{M}$.
(ii) If $(\mathcal{O}_j, \varkappa_j), (\mathcal{O}_k, \varkappa_k) \in \mathfrak{C}$ are special charts, whereas $\mathcal{O}_j \cap s_{i_1} \neq \emptyset$, $\mathcal{O}_k \cap s_{i_2} \neq \emptyset$, where $i_1, i_2 \in I$, $i_1 \neq i_2$, then $\mathcal{O}_j \cap \mathcal{O}_k = \emptyset$.
(iii) $\overline{\mathcal{O}_j} \subset \mathcal{M}_0$ for any standard chart $(\mathcal{O}_j, \varkappa_j) \in \mathfrak{C}$.

Definition 4.2.19 An atlas $(\mathcal{O}_j, \varkappa_j)$ of local charts on a manifold $\mathcal{M}$ with edges defines on $\mathcal{M}$ a structure of w-manifold, if any two charts in this atlas are compatible, i.e., or $\mathcal{O}_j \cap \mathcal{O}_k = \emptyset$, or $\mathcal{O}_j \cap \mathcal{O}_k \neq \emptyset$ and the mapping $\varkappa_k \circ \varkappa_j^{-1} : \varkappa_j(\mathcal{O}_j \cap \mathcal{O}_k \cap \mathcal{M}_0) \to \varkappa_k(\mathcal{O}_j \cap \mathcal{O}_k \cap \mathcal{M}_0)$ is an admissible diffeomorphism of open subsets of the wedge when $(\mathcal{O}_j, \varkappa_j)$ and $(\mathcal{O}_k, \varkappa_k)$ are special charts and a diffeomorphism of smooth manifolds in other cases.

The equivalence of two atlases is defined in the usual way. Equivalent atlases define the same w-structure on $\mathcal{M}$. As follows from Definition 4.2.19, the open dense part $\mathcal{M}_0$ of $\mathcal{M}$ is a smooth manifold. In what follows, when considering local charts on w-manifolds, we always assume that these charts are compatible with the w-structure.

Definition 4.2.20 Let $\varkappa : \mathcal{O} \to \overline{\mathbb{W}}$ be a special local chart on $\mathcal{M}$ and let the function $\zeta \in C^\infty(\mathbb{W})$ possess the following properties: (1) The set $\operatorname{supp}\zeta$ is bounded, and its closure (in $\overline{\mathbb{W}}$) belongs to $V = \varkappa(\mathcal{O})$. (2) If (U, υ) is a local chart $\mathbb{W}$, then

$$|\partial^\alpha(\zeta \circ \upsilon^{-1})(x)| \leq C_\alpha |x^1|^{-|\alpha|}, \quad \alpha \in \mathbb{Z}_+^m.$$

Any function ζ having these properties is called a cut-off function serving the set V. An analogous terminology will be applied to the function $\chi = \zeta \circ \varkappa$ (extended to be zero on $\mathcal{M}_0 \setminus \mathcal{O}$) and to the set $\mathcal{O}$. If $\mathcal{O}'$ is a standard coordinate neighborhood, then any function $\chi \in C^\infty(\mathcal{M}_0)$ satisfying $\operatorname{supp}\chi \subset \mathcal{O}'$ is called a cut-off function serving the set $\mathcal{O}'$. The set of all cut-off functions serving the coordinate neighborhood $\mathcal{O} \subset \mathcal{M}$ is denoted by $\mathcal{CF}(\mathcal{O})$.

For a special coordinate neighborhood, the definition of the set $\mathcal{CF}(\mathcal{O})$ is correct in view of Remark 4.2.9.

Let s_i, $i \in I$, be an edge of the w-manifold $\mathcal{M}$, $\{\mathcal{O}_j\}_{j=1}^k$ a covering of this edge by special coordinate neighborhoods. Denote by W some neighborhood of s_i satisfying $\overline{W} \subset \bigcup_{j=1}^k \mathcal{O}_j$. Let $\chi_0, \chi_1, \dots, \chi_k$ be a partition of unity on $\mathcal{M}_0$ subordinate to the covering $\mathcal{M}_0 \setminus \overline{W}$, $\mathcal{O}_1 \cap \mathcal{M}_0, \dots, \mathcal{O}_k \cap \mathcal{M}_0$, while $\chi_j \in \mathcal{CF}(\mathcal{O}_j)$ for $j \geq 1$. Let also

$\varkappa_j = (\varkappa_j^1, \varkappa_j^2)$, $j \geq 1$, be a coordinate mapping of the neighborhood $\mathcal{O}_j$ onto an open subset of the closed surface $\overline{\mathbb{W}}_j = \overline{K}_j \times \mathbb{R}^{m-n_j}$ of wedge type. We set

$$\ell_i(w) = \chi_0(w) + \sum_{j=1}^{k} \chi_j(w)|\varkappa_j^1(w)|, \quad w \in \mathcal{M}_0. \tag{4.2.18}$$

Then, ℓ_i is a smooth positive function on $\mathcal{M}_0$, while $\ell_i = 1$ in the vicinity of any edge distinct from s_i, and $(\ell_i \circ \varkappa^{-1})(x) \sim r$, $x \in \mathcal{O} \cap \mathcal{M}_0$, for any special coordinate system $\varkappa : \mathcal{O} \to \overline{\mathbb{W}}$, $\mathcal{O} \cap s_i \neq \emptyset$; here, r is the distance from $x \in \mathbb{W} = K \times \mathbb{R}^{m-n_i}$ to the "edge" $\mathbb{R}^{m-n_i}$, and the relation $(\ell_i \circ \varkappa^{-1})(x) \sim r$ means that for any compact set $E \subset \varkappa(\mathcal{O})$, there exist constants $c'_{\varkappa,E}, c''_{\varkappa,E} > 0$ such that

$$c'_{\varkappa,E}\, r \leq (\ell_i \circ \varkappa^{-1})(x) \leq c''_{\varkappa,E}\, r\,, \quad x \in E \cap \mathbb{W}. \tag{4.2.19}$$

According to formula (4.2.18), any edge s_i complies with an infinite set of functions depending on the choice of local charts $(\mathcal{O}_j, \varkappa_j)$ and cut-off functions χ_j. Below, ℓ_i denotes an arbitrarily fixed function in this set.

Introduce a Riemannian metric on the manifold $\mathcal{M}_0$. Let $\mathfrak{C} = \{(\mathcal{O}_j, \varkappa_j)\}$ be an atlas on $\mathcal{M}$. If $(\mathcal{O}_j, \varkappa_j) \in \mathfrak{C}$ is a special chart, then the metric on $\mathcal{O}_j \cap \mathcal{M}_0$ will be defined as the pre-image of the Riemannian metric induced on $\varkappa_j(\mathcal{O}_j \cap \mathcal{M}_0)$ by the Euclidean metric of the surrounding space. For a standard chart $\{(\mathcal{O}_j, \varkappa_j)\}$, a (smooth) metric on $\mathcal{O}_j$ is defined arbitrarily. The metric on the whole manifold $\mathcal{M}_0$ is now pieced together using a partition of unity $\{\chi_j\}$, $\chi_j \in C\mathcal{F}(\mathcal{O}_j)$. If we choose another atlas $\mathfrak{C}$ or a partition of unity $\{\chi_j\}$, then the metric is changed to an equivalent metric. A smooth measure on $\mathcal{M}_0$ induced by the metric is further denoted by ν. Note that relations (4.2.19) are equivalent to the estimate $c'\mathrm{dist}(\cdot, s_i) \leq \ell_i(\cdot) \leq c''\mathrm{dist}(\cdot, s_i)$ for ℓ_i; here c', c'' are some constants, and $\mathrm{dist}(w, s_i)$ is the distance from $w \in \mathcal{M}_0$ to the edge s_i defined by the metric.

4.2.5 Ψdo on w-manifold

In Sect. 4.2.4, we have defined a Ψdo on a wedge as an operator that in any local coordinates (U, υ) is an element of some class $\Psi^\mu(W)$, $W \subset \mathbb{R}^m_n$. To this end, coordinate neighborhoods $U = K(U^{(0)}) \times \mathbb{R}^{m-n}$ of the same kind were used; for any two points on a wedge, there exists a (possibly not connected) coordinate neighborhood of this kind containing these points. If we define a Ψdo on a w-manifold in an analogous way, then an additional obstacle arises, since there exist neighborhoods of two types: special and standard. Moreover, the special neighborhoods may serve different strata (probably of distinct dimensions). This approach (in a more general situation) is developed in [29, 30]. Here, we choose another way, namely, let us define a Ψdo on a w-manifold $\mathcal{M}$ as a usual Ψdo A on the smooth manifold $\mathcal{M}_0$, which, in any special coordinate system, is a Ψdo on a wedge-type surface. Then, we have to postulate properties of the kernel of operator A outside the diagonal $\{(x, y) \in \mathcal{M}_0 \times \mathcal{M}_0 : x = y\}$.

To motivate the following definitions, let us consider any amplitude a in class $\Pi^\mu(\mathbb{R}^m_n)$. Assume that functions $\chi_1, \chi_2 \in C^\infty(\mathbb{R}^m_n)$ satisfy the estimates $|\partial^\alpha \chi_i(x)| \le C_\alpha |x^1|^{-|\alpha|}$, $i = 1, 2$ (i.e., $\chi_i \in \mathcal{CF}(\mathbb{R}^m_n)$) and that the closures (in $\mathbb{R}^m$) of their supports are disjoint. Set $b(x, y, \xi) = \chi_1(x)a(x, y, \xi)\chi_2(y)$, $B = \operatorname{Op} b$. Since the amplitude b and all its derivatives vanish as $x = y$, we have $B \in \Psi^{-\infty}(\mathbb{R}^m_n)$ in view of Proposition 4.1.24. Moreover,

$$\inf\{|x - y|;\ (x, y) \in \operatorname{supp}\chi_1 \times \operatorname{supp}\chi_2\} > 0,$$

and therefore, according to Lemma 4.1.5, for $|x^1| + |y^1| \le C$, the kernel of B satisfies the inequalities

$$|\partial_x^\alpha \partial_y^\beta G(x, y)| \le C_{\alpha\beta N} |x_1|^N |y_1|^{-|\beta|} \quad (\alpha, \beta \in \mathbb{Z}^m_+,\quad N \in \mathbb{Z}_+).$$

Extend these estimates to the case of a w-manifold. Let $\{(\mathcal{O}_j, \varkappa_j)\}$ be an (finite) atlas on $\mathcal{M}$, $\{\chi_j\}$, $\chi \in \mathcal{CF}(\mathcal{O}_j)$, a partition of unity on $\mathcal{M}_0$, and $\ell = \prod_{i \in I} \ell_i$, where ℓ_i are functions in (4.2.18).

Definition 4.2.21 We will say that the operator

$$Bu(z) = \int G(z, w)u(w)\, dv(w)\,, \quad u \in C_c^\infty(\mathcal{M}_0 \times \mathcal{M}_0), \tag{4.2.20}$$

is in the class $\Upsilon(\mathcal{M})$ if $G \in C^\infty(\mathcal{M} \times \mathcal{M})$ and for all $\alpha, \beta \in \mathbb{Z}^m_+$, $N \in \mathbb{Z}_+$, the following inequality holds:

$$\sup_{j,k} |\partial_z^\alpha \partial_w^\beta (\chi_j(z)G(z, w)\chi_k(w))| \le C_{\alpha\beta N} \ell(z)^N \ell(w)^{-|\beta|}. \tag{4.2.21}$$

The derivatives on the right-hand side of (4.2.21) are calculated in local coordinates on $\mathcal{O}_j$ and $\mathcal{O}_k$. At the same time, the coordinates on any special neighborhood $\mathcal{O}_j$ are defined using maps $\upsilon_{jp} \circ (\varkappa|\mathcal{O} \cap \mathcal{M}_0) : \varkappa_j^{-1}(U_p) \cap \mathcal{O}_j \to \mathbb{R}^m_{n_j}$, where $\{(U_{jp}, \upsilon_{jp})\}$ is a finite atlas on $\mathbb{W}_j$. If other atlases $\{(\mathcal{O}_j, \varkappa_j)\}$, $\{(U_{jp}, \upsilon_{jp})\}$ and a partition of unity $\{\chi_j\}$ are selected, then estimates (4.2.21) remain valid with some other constants $C_{\alpha\beta N}$.

Definition 4.2.22 The operator (4.2.20) is called proper ($B \in \Upsilon_0(\mathcal{M})$) if there exists a number $\delta \in (0, 1)$ such that $G(z, w) = 0$ whenever $\ell_i(z)/\ell_i(w) \notin (\delta, \delta^{-1})$ for some $i \in I$.

The estimate (4.2.21) for the operator kernel $B \in \Upsilon_0(\mathcal{M})$ takes the form

$$\sup_{j,k} |\partial_z^\alpha \partial_w^\beta (\chi_j(z)G(z, w)\chi_k(w))| \le C_{\alpha\beta N} \ell(z)^{N_1} \ell(w)^{N_2}. \tag{4.2.22}$$

($N_1, N_2 \in \mathbb{Z}_+$, $C = C(\alpha, \beta, N_1, N_2)$), whereas the adjoint kernel $G^*(x, y) = \overline{G(y, x)}$ is also subject to this estimate.

Definition 4.2.23 Ψdo A of order μ on a smooth manifold $\mathcal{M}_0$ (in the sense of standard definition) is called a Ψdo (proper Ψdo) of order μ on the w-manifold $\mathcal{M}$ if:

(1) For any local chart $\varkappa : \mathcal{O} \to V$, where V is an open subset of the closed wedge $\overline{\mathbb{W}}$, and for any two cut-off functions ζ_1, ζ_2 serving the set V, the operator[1]

$$C_c^\infty(\mathbb{W}) \ni u \mapsto \zeta_1(\varkappa^{-1})^* A \varkappa^*(\zeta_2 u)$$

is in the class $\Psi^\mu(V \cap \mathbb{W})$ ($\Psi_0^\mu(V \cap \mathbb{W})$).

(2) $\chi A \eta \in \Upsilon(\mathcal{M})$ ($\chi A \eta \in \Upsilon_0(\mathcal{M})$) for any pair of cut-off functions χ, η such that dist (suppχ, suppη) > 0.

Let $\{(\mathcal{O}_j, \varkappa_j)\}_{j=1}^l$ be an atlas on $\mathcal{M}$. Assume that a chart $(\mathcal{O}_j, \varkappa_j)$ is special if $j = 1, \ldots, k$, and standard if $j = k + 1, \ldots, l$. We set $Z_j = \varkappa_j(\mathcal{O}_j)$. Let $\chi_j, \eta_j \in \mathcal{CF}(\mathcal{O}_j)$ be functions such that:

(a) $\sum_{j=1}^l \chi_j = 1$.
(b) $\eta_j \chi_j = \chi_j$, $j = 1, \ldots, l$.
(c) dist(suppχ_j, supp$(1 - \eta_j)$) > 0, $j = 1, \ldots, l$.

From Definition 4.2.23, it follows that, for any Ψdo $A \in \Psi^\mu(\mathcal{M})$, the equality $A = \sum \eta_j A \chi_j + B$ holds with $B \in \Upsilon(\mathcal{M})$. Denoting $\zeta_{1j} = \eta_j \circ \varkappa_j^{-1}$, $\zeta_{2j} = \chi_j \circ \varkappa_j^{-1}$, we rewrite this equality in the form

$$Au = \sum \zeta_{1j}(\varkappa_j^{-1})^* A_j \varkappa_j^*(\zeta_{2j} u) + Bu, \quad u \in C_c^\infty(\mathcal{M}_0), \tag{4.2.23}$$

where $A_j \in \Psi^\mu(Z_j)$ for $1 \le j \le k$, and for $k + 1 \le j \le l$, the operator A_j is a "usual" Ψdo of order μ on $Z_j \subset \mathbb{R}^m$. If A is a proper Ψdo, then the operators A_j, $1 \le j \le k$, and B are also proper.

Suppose that the operators $A_1, \ldots, A_l, B$ satisfy the following conditions:

(1) $A_j \in \Psi^\mu(Z_j)$ ($A_j \in \Psi_0^\mu(Z_j)$), $1 \le j \le k$.
(2) For $k + 1 \le j \le l$, the operator A_j is a "usual" Ψdo of order μ on $Z_j \subset \mathbb{R}^m$.
(3) $B \in \Upsilon(\mathcal{M})$ ($B \in \Upsilon_0(\mathcal{M})$).

[1] For short, we write $\varkappa$ and $\varkappa^{-1}$ instead of $\varkappa|\mathcal{O} \cap \mathcal{M}_0$ and $\varkappa^{-1}|V \cap \mathbb{W}$.

Then, the operator A defined by (4.2.23) is in the class $\Psi^{\mu}(\mathcal{M})$ ($\Psi_0^{\mu}(\mathcal{M})$). To prove this assertion, one can repeat (with minor modifications) the proof of Theorem 4.2.13; therefore, we do not do it here.

To make compositions of proper Ψdo, we have to show that if $A \in \Psi_0^{\mu}(\mathcal{M})$ and $B \in \Upsilon^0(\mathcal{M})$, then the operators AB and BA belong to the class $\Upsilon_0(\mathcal{M})$. An elementary verification of this fact using the estimates (4.2.22) is given to the reader.

4.3 Pseudodifferential Operators in Weighted Spaces

In this section, we define a scale H_τ^s of weighted analogues of Sobolev spaces on a manifold $\mathcal{M}$ with edges and prove the boundedness of Ψdo at this scale.

4.3.1 Boundedness of Proper Ψdo of Non-positive Order in the Spaces $L_{2,\tau}$

For $\tau \in \mathbb{R}$, we denote by $L_{2,\tau} = L_{2,\tau}(\mathbb{R}_n^m)$ the completion of the set $C_c^\infty(\mathbb{R}_n^m)$ for the norm

$$\|u; L_{2,\tau}\| = \left(\int |x^1|^{2\tau} |u(x)|^2 \, dx\right)^{1/2}.$$

Theorem 4.3.1 *Let $A = \operatorname{Op} a$, where $a \in \Psi_0^{\mu}$ and $\mu \leq 0$. Then, for any $\tau \in \mathbb{R}$, the operator A defined on $C_c^\infty(\mathbb{R}_n^m)$ extends to a continuous mapping $L_{2,\tau} \to L_{2,\tau}$.*

Proof First, we assume that $\tau = 0$. Consider three cases:

(a) $\mu < -m$. Then,

$$Au(x) = \int G(x, y)u(y)\, dy, \quad u \in C_c^\infty(\mathbb{R}_n^m).$$

Since the operator A is proper, inequalities (4.1.4) imply that

$$|G(x, y)| \leq C_N |x^1|^{N-m}/(|x^1|^2 + |x - y|^2)^{N/2} \tag{4.3.1}$$

for any $N \in \mathbb{Z}_+$. Note that if $N > m/2$, then

$$\int_{\mathbb{R}^m} \frac{|x^1|^{N-m}}{(|x^1|^2 + |x - y|^2)^{N/2}}\, dy < +\infty, \tag{4.3.2}$$

whereas the integral is independent of x. Indeed, after the substitution $y \to |x^1|y$, this integral takes the form

$$\int \frac{|x^1|^N}{(|x^1|^2 + |x - |x^1|y|^2)^{N/2}}\,dy = \int \frac{dy}{(1 + ||x^1|^{-1}x - y|^2)^{N/2}}\,dy =$$
$$= \int \frac{dz}{(1 + |z|^2)^{N/2}}\,dz < +\infty.$$

Comparing (4.3.1) and (4.3.2), we obtain

$$C_1 := \sup_x \int |G(x,y)|\,dy < +\infty. \tag{4.3.3}$$

Since the operator A is proper, along with (4.3.1), the following inequality holds:

$$|G(x,y)| \le C_N |y^1|^{N-m}/(|y^1|^2 + |x-y|^2)^{N/2};$$

hence, the estimate

$$C_2 := \sup_y \int |G(x,y)|\,dx < +\infty \tag{4.3.4}$$

is also true. For u and v in $C_c^\infty(\mathbb{R}^m_n)$, we have

$$|(Au, v)|^2 \le \Big(\iint |G(x,y)||u(y)||v(x)|\,dxdy\Big)^2 \le$$
$$\le \iint |G(x,y)||u(y)|^2 dxdy \iint |G(x,y)||v(x)|^2 dxdy \le$$
$$\le C_2 \int |u(y)|^2 dy \cdot C_1 \int |v(x)|^2 dx = C_1 C_2 \|u\|^2 \|v\|^2.$$

Thus, $\|A; \mathcal{B}L_2(\mathbb{R}^m_n)\| < +\infty$.

(b) $\mu < 0$. Due to the equality $\|Au\|^2 = (A^*Au, u)$, the boundedness of A follows from the boundedness of A^*A, which in an analogous way follows from the boundedness of $(A^*A)^{2^k}$ for some $k \in \mathbb{Z}_+$. Choose k such that the order $\mu 2^k$ of $(A^*A)^{2^k}$ is less than $-m$. According to the first part of the proof, for such k, the operator $(A^*A)^{2^k}$ is bounded.

(c) $\mu = 0$. Denote by σ and σ^* the symbols of A and its adjoint operator $A^* = {}_\tau A^*$ (cf. Sect. 4.1.4). Formulas (4.1.17) and (4.1.24) imply

$$A^*Au(x) = (2\pi)^{-m/2} \int e^{ix\xi}\big(|\sigma(x,\xi)|^2 + \sigma_{-1}(x,\xi)\big)\hat{u}(\xi)\,d\xi, \quad \sigma_{-1} \in S^{-1}.$$

Let us take $M > \sup|\sigma(x,\xi)|$ and set $\sigma_0(x,\xi) = (M^2 - |\sigma(x,\xi)|^2)^{1/2}$. Then, for some $A_{-1} \in \Psi^{-1}$, we have

$$A^*Au(x) = (2\pi)^{-m/2} \int e^{ix\xi}\big(M^2 - \sigma_0(x,\xi)^2\big)\hat{u}(\xi)\,d\xi + A_{-1}u(x).$$

By Theorem 4.1.19, there exists an element $\sigma_{-\infty} \in S^{-\infty}$ such that the operator $B = \mathrm{Op}\,(\sigma_0 + \sigma_{-\infty})$ is proper. Comparing

$$B^*Bu(x) = (2\pi)^{-m/2} \int e^{ix\xi}\sigma_0(x,\xi)^2\hat{u}(x,\xi)\,d\xi + A'_{-1}u(x), \quad A'_{-1} \in \Psi^{-1},$$

with the formula for A^*A, we obtain

$$A^*A = M^2 - B^*B + A''_{-1}, \quad A''_{-1} \in \Psi^{-1}. \tag{4.3.5}$$

Since the operator (of multiplication by) M^2 is a proper Ψdo, it follows from (4.3.5) that the inclusion $A''_{-1} \in \Psi_0^{-1}$ holds. Then, by part (b) of the proof, the operator $A''_{-1} : L_2(\mathbb{R}^m_n) \to L_2(\mathbb{R}^m_n)$ is continuous. Therefore,

$$\begin{aligned}\|Au;\, L_2\|^2 = (A^*Au, u) = M^2\|u\|^2 - \|Bu\|^2 + (A''_{-1}u, u) \leq \\ \leq (M^2 + \|A''_{-1}\|)\|u\|^2.\end{aligned}$$

This proves the theorem if $\tau = 0$. For arbitrary τ, the theorem is now evident since the boundedness of a Ψdo A in $L_{2,\tau}(\mathbb{R}^m_n)$ is equivalent to the boundedness of $d^{-\tau}Ad^{\tau}$ in $L_2(\mathbb{R}^m_n)$, where d is the function defined by $\mathbb{R}^m_n \ni x \mapsto d(x) = |x^1|$.

□

4.3.2 Pseudodifferential Operators in the Spaces H^s_τ

This section has an overview character. For each $s \in \mathbb{R}$, we denote by Λ_s some elliptic Ψdo of order s and by R_s a parametrix of Λ_s. Then, $R_s\Lambda_s = I + T_s$, $T_s \in \Psi_0^{-\infty}$. Define the space $H^s_\tau = H^s_\tau(\mathbb{R}^m_n)$ for $\tau, s \in \mathbb{R}$ as the completion of $C^\infty_c(\mathbb{R}^m_n)$ for the norm

$$\|u\|_{s,\tau} = (\|\Lambda_s u;\, L_{2,\tau}\|^2 + \|T_s u;\, L_{2,\tau}\|^2)^{1/2}. \tag{4.3.6}$$

The function $C^\infty_c(\mathbb{R}^m_n) \ni u \mapsto \|u\|_{s,\tau}$ is indeed a norm: if $\|u\|_{s,\tau} = 0$, then $\Lambda_s u = T_s u = 0$ and hence, $u = (R_s\Lambda_s - T_s)u = 0$.

Theorem 4.3.2 *Assume that $\tau, s, t, \mu \in \mathbb{R}$, $\mu \leq s - t$, and $A \in \Psi_0^\mu$. Then, the operator A extends to a continuous mapping $H^s_\tau \to H^t_\tau$.*

Proof For short, let us write $\|\cdot\|$ instead of $\|\cdot; L_{2,\tau}\|$. From $I = R_s\Lambda_s - T_s R_s \Lambda_s + T_s^2$, it follows that

$$\|\Lambda_t A u\| \leq \|\Lambda_t A(R_s - T_s R_s)\Lambda_s u\| + \|\Lambda_t A T_s^2 u\|.$$

Since $R_s \in \Psi_0^{-s}$, the operator $\Lambda_t A(R_s - T_s R_s)$ is in the class $\Psi_0^{t-s+\mu}$, and since $t-s+\mu \leq 0$, this operator is continuous in $L_{2,\tau}$. This is also true for the operator $\Lambda_t A T_s$. Therefore,

$$\|\Lambda_t A u\| \leq C_1 \left(\|\Lambda_s u\| + \|T_s u\|\right). \tag{4.3.7}$$

Analogously, one can prove the inequality

$$\|T_t A u\| \leq C_2(\|\Lambda_s u\| + \|T_s u\|). \tag{4.3.8}$$

Comparing (4.3.7) and (4.3.8), we obtain $\|Au\|_{t,\tau} \leq C\|u\|_{s,\tau}$. □

Remark 4.3.3 Theorem 4.3.2 is valid for arbitrary elliptic operators $\Lambda_s \in \Psi_0^s$, $\Lambda_t \in \Psi_0^t$ and their parametrices R_s and R_t participating in the definition of the norms $\|\cdot\|_{s,\tau}$ and $\|\cdot\|_{t,\tau}$.

Corollary 4.3.4 *Any proper Ψdo of order $-\infty$ extends to a continuous mapping $H_\tau^s \to H_\tau^t$ for all τ, s, $t \in \mathbb{R}$.*

Assume that τ, $s \in \mathbb{R}$, Λ_s, and that $\tilde{\Lambda}_s$ are elliptic operators in Ψ_0^s, R_s, $\tilde{R}_s$ are parametrices of Λ_s, $\tilde{\Lambda}_s$ so that $R_s\Lambda_s = I + T_s$, $\tilde{R}_s\tilde{\Lambda}_s = I + \tilde{T}_s$, whereas T_s, $\tilde{T}_s \in \Psi_0^{-\infty}$. In accordance with (4.3.6), we define norms $\|\cdot\|_{s,\tau}$ and $\|\cdot\|_{s,\tau}^{\sim}$ on $C_c^\infty(\mathbb{R}_n^m)$.

Theorem 4.3.5 *The norms $\|\cdot\|_{s,\tau}$ and $\|\cdot\|_{s,\tau}^{\sim}$ are equivalent.*

Proof Since the norms $\|\cdot\|_{s,\tau}$ and $\|\cdot\|_{s,\tau}^{\sim}$ are interchangeable, it is sufficient to prove that $\|\cdot\|_{s,\tau}^{\sim} \leq C\|\cdot\|_{s,\tau}$ for $u \in C_c^\infty(\mathbb{R}_n^m)$. The orders of the operators $\tilde{\Lambda}_s$ and $\tilde{T}_s$ do not exceed s; therefore, Theorem 4.3.2 is followed by the inequality

$$(\|\tilde{\Lambda}_s u\|_{0,\tau}^2 + \|\tilde{T}_s u\|_{0,\tau}^2)^{1/2} \leq C\|u\|_{s,\tau}. \tag{4.3.9}$$

According to Remark 4.3.3, we can suppose that $\Lambda_0 = I$ and $T_0 = 0$. Then, $\|\cdot\|_{0,\tau} = \|\cdot; L_{2,\tau}\|$, and inequality (4.3.9) takes the form $\|u\|_{s,\tau}^{\sim} \leq C\|u\|_{s,\tau}$. □

Corollary 4.3.6 *If $s \in \mathbb{Z}_+$, then the norm $\|\cdot\|_{s,\tau}$ is equivalent to the norm*

$$\|u\|'_{s,\tau} = \Big(\sum_{|\alpha|=0}^{s} \int |x^1|^{2(\tau+|\alpha|)} |\partial^\alpha u(x)|^2\, dx\Big)^{1/2}. \tag{4.3.10}$$

Proof We restrict ourselves to proving this assertion for even s. For any multi-index $\alpha \in \mathbb{Z}^m_+$, let us denote by Q_α the differential operator $u \mapsto |x^1|^{|\alpha|}\partial^\alpha u$. Then,

$$\|u\|'_{s,\tau} = \Big(\sum_{|\alpha|=0}^{s} \|Q_\alpha u;\, L_{2,\tau}\|^2\Big)^{1/2}.$$

Since $Q_\alpha \in \Psi_0^{|\alpha|}$, Theorem 4.3.2 implies the estimate $\|u\|'_{s,\tau} \le C\,\|u\|_{s,\tau}$. To obtain the converse estimate, we define a norm $\|\cdot\|'_{s,\tau}$ using the elliptic operator $\Lambda_s = \sum_{|\alpha|=0}^{s} c_\alpha Q_\alpha$. The inequality $\|\Lambda_s u;\, L_{2,\tau}\| \le C\|u\|'_{s,\tau}$ is then obvious, and we have only to prove $\|T_s u;\, L_{2,\tau}\| \le C\|u\|'_{s,\tau}$. However, this is also obvious, since the order of the (proper) Ψdo T_s is non-positive and that is why $\|T_s u\,;\, L_{2,\tau}\| \le C\|u;\, L_{2,\tau}\| \le C\|u\|'_{s,\tau}$. □

Proposition 4.3.7 *For $s \ge t$, the identity operator $C_c^\infty(\mathbb{R}^m_n) \to C_c^\infty(\mathbb{R}^m_n)$ extends to an embedding $H^s_\tau \to H^t_\tau$, i.e., continuous injective mapping $I^{s,t}_\tau : H^s_\tau \to H^t_\tau$.*

Proof Since $I \in \Psi_0^0$ and $t \le s$, the operator I extends to a continuous mapping $I^{s,t}_\tau : H^s_\tau \to H^t_\tau$ (Theorem 4.3.2), and it suffices only to prove that the operator $I^{s,t}_\tau$ is injective. In other words, we have to show that if a sequence $\{u_k\}_{k=1}^\infty \subset C_c^\infty(\mathbb{R}^m_n)$ is a Cauchy sequence in the norm $\|\cdot\|_{s,\tau}$ and tends to zero in the norm $\|\cdot\|_{t,\tau}$, then $\|u_k\|_{s,\tau} \to 0$.

Suppose first that $t = 0$ and that s is an even number. According to Corollary 4.3.6, the sequence $\{u_k\}$ is a Cauchy sequence in the norm $\|\cdot\|'_{s,\tau}$. Therefore, for each $\alpha \in \mathbb{Z}^m_+$, $|\alpha| \le s$, there exists a function $v_\alpha \in L_{2,\tau}$ satisfying

$$\lim_{k\to\infty} \||x^1|^{|\alpha|}\partial^\alpha u_k - v_\alpha;\, L_{2,\tau}\| = 0. \tag{4.3.11}$$

Let $w \in C_c^\infty(\mathbb{R}^m_n)$. From (4.3.11), the inequality

$$\lim_{k\to\infty} \int \partial^\alpha u_k(x) w(x)\, dx = \int v_\alpha(x) w(x)\, dx$$

follows. At the same time, since $\|u_k;\, L_{2,\tau}\| = \|u_k\|'_{0,\tau} \to 0$, we have

$$\lim_{k\to\infty} \int \partial^\alpha u_k(x) w(x)\, dx = \lim_{k\to\infty} \int u_k(x)(-\partial)^\alpha w(x)\, dx = 0.$$

Thus, $v_\alpha = 0$ almost everywhere, and hence, $|x^1|^{|\alpha|}\partial^\alpha u_k \to 0$ in $L_{2,\tau}$ for all $\alpha \in \mathbb{Z}^m_+$, $|\alpha| \le s$. Then, $\|u_k\|_{s,\tau} \to 0$ due to Corollary 4.3.6.

Suppose now that $s, t \in \mathbb{R}$, $s - t \in \mathbb{Z}_+$, while $s - t$ is an even number. From the definitions of the norms $\|\cdot\|_{s,\tau}$ and $\|\cdot\|_{t,\tau}$, it now follows that:

(a) $\{\Lambda_s u_k\}$, $\{T_s u_k\}$ are Cauchy sequences in the norm of $L_{2,\tau}$.
(b) The sequences $\{\Lambda_t u_k\}$, $\{T_t u_k\}$ tend to zero in this norm.
We need to show that then

$$\lim \|\Lambda_s u_k; L_{2,\tau}\| = 0, \quad \lim \|T_s u_k; L_{2,\tau}\| = 0 \quad (k \to \infty). \tag{4.3.12}$$

Since $T_s u_k = T_s R_t \Lambda_t u_k - T_s T_t u_k$, the second relation in (4.3.12) follows from (b) and the inclusions $T_s R_t$, $T_s \in \Psi_0^{-\infty}$. Prove now the first relation. Set $v_k = \Lambda_t u_k$, $k \in \mathbb{N}$. According to the definition of the norm in H_τ^{s-t}, we have

$$\|v_k - v_l\|_{s-t,\tau} = (\|\Lambda_{s-t}(v_k - v_l); L_{2,\tau}\|^2 + \|T_{s-t}(v_k - v_l); L_{2,\tau}\|^2)^{1/2}.$$

One can assume that $\Lambda_s = \Lambda_{s-t}\Lambda_t$ and rewrite the preceding equality in the form

$$\|v_k - v_l\|_{s-t,\tau} = (\|\Lambda_s u_k - \Lambda_s u_l); L_{2,\tau}\|^2 + \|T_{s-t}(v_k - v_l); L_{2,\tau}\|^2)^{1/2}. \tag{4.3.13}$$

From the definition of the sequence $\{v_k\}$, taking into account of (b), we deduce that the second summand on the right-hand side of (4.3.13) tends to zero as $k, l \to \infty$. The first summand also tends to zero in view of (a). Thus, $\{v_k\}$ is a Cauchy sequence in the norm $\|\cdot\|_{s-t,\tau}$. Moreover, $\|v_k\|_{0,\tau} \sim \|v_k; L_{2,\tau}\| = \|\Lambda_t u_k; L_{2,\tau}\| \to 0$. Since $s - t \in 2\mathbb{Z}_+$, the conditions of the foregoing case are fulfilled, and we conclude that $\lim \|v_k\|_{s-t,\tau} = 0$. All the more

$$\lim \|\Lambda_{s-t} v_k; L_{2,\tau}\| = \lim \|\Lambda_s u_k; L_{2,\tau}\| = 0,$$

as needed.

Let finally $s, t \in \mathbb{R}$ be arbitrary numbers satisfying $t \leq s$. Take $r \in (-\infty, t]$ so that the inclusion $s - t \in 2\mathbb{Z}_+$ holds. The obvious relation $I_\tau^{s,r} = I_\tau^{t,r} \circ I_\tau^{s,t}$ and the injectivity of the mapping $I_\tau^{s,r}$ imply that $I_\tau^{s,t}$ is injective. □

Theorem 4.3.8 *For any τ, $s \in \mathbb{R}$, the bilinear form*

$$C_c^\infty(\mathbb{R}_n^m) \times C_c^\infty(\mathbb{R}_n^m) \ni (u, v) \mapsto \int u(x)v(x)\,dx$$

extends to a pairing between the spaces H_τ^s and $H_{-\tau}^{-s}$, i.e., to a continuous bilinear form on $H_\tau^s \times H_{-\tau}^{-s}$. The spaces H_τ^s and $H_{-\tau}^{-s}$ are mutually conjugate with respect to this pairing.

Theorem 4.3.9 *For any τ, $s \in \mathbb{R}$, the bilinear form*

$$C_c^\infty(\mathbb{R}_n^m) \times C_c^\infty(\mathbb{R}_n^m) \ni (u, v) \mapsto \int |x^1|^{2\tau} u(x) v(x)\, dx$$

extends to a pairing between the spaces H_τ^s and H_τ^{-s}. These spaces are mutually conjugate with respect to this pairing.

We conclude this section with a theorem on continuity of non-proper Ψdo in the scale of H_τ^s.

Theorem 4.3.10 *Let τ, s, t, $\mu \in \mathbb{R}$, $|\tau| < n/2$, $\mu \leq s - t$, and let $A \in \Psi^\mu$. Then, the operator A extends to a continuous mapping $H_\tau^s \to H_\tau^t$.*

4.3.3 Pseudodifferential Operators on Spaces with Weighted Norms on w-Manifolds

Let $\mathbb{W}$ be an m-dimensional wedge[2] and $\{(U_j, \upsilon_j)\}_{j=1}^l$ a partition of unity formed by homogeneous cut-off functions serving the sets U_j (cf. Definition 4.2.10). For any τ, $s \in \mathbb{R}$, the space $H_\tau^s(\mathbb{W})$ can be defined as the completion of $C_c^\infty(\mathbb{W})$ with respect to the norm

$$\|u; H_\tau^s(\mathbb{W})\| = \Big(\sum_{j=1}^l \|(\upsilon_j^{-1})^*(\chi_j u)\|_{s,\tau}^2\Big)^{1/2}$$

(the function $\chi_j u$ is extended on $\mathbb{R}^m \setminus \upsilon_j(U_j)$ as zero).

Let $\mathcal{M}$ be an m-dimensional w-manifold and $\mathcal{M}_0$ its open dense part (the union of strata of maximal dimension). Consider an atlas $\{(\mathcal{O}_j, \varkappa_j)\}_{j=1}^l$ on $\mathcal{M}$ consisting of special local charts $\{(\mathcal{O}_j, \varkappa_j)\}_{j=1}^k$ and standard charts $\{(\mathcal{O}_j, \varkappa_j)\}_{j=k+1}^l$. Remind that for $1 \leq j \leq k$, the image $\upsilon_j(\mathcal{O}_j)$ of the neighborhood $\mathcal{O}_j$ is an open subset of some closed surface $\overline{\mathbb{W}_j}$ of wedge type. Let $\{s_i\}_{i\in I}$ be the set of strata satisfying $\dim s_i < m$. To each index $j = 1, \ldots, k$, there corresponds an index $i \in I$ equal to the index of the stratum that intersects with the neighborhood $\mathcal{O}_j$. We denote this correspondence by h and write $i = h(j)$. To each stratum s_i, $i \in I$, we assign a real number τ_i in an arbitrary manner. We denote the set $\{\tau_i\}_{i\in I}$ by τ.

[2] We use notions and notations introduced in Sects. 4.2.3–4.2.5.

For any number $s \in \mathbb{R}$ and any set $\tau \in \mathbb{R}^I$, we introduce the space $H^s_\tau(\mathcal{M})$ as the completion of $C^\infty_c(\mathcal{M}_0)$ with respect to the norm

$$\|u; H^s_\tau(\mathcal{M})\|$$
$$= \Big(\sum\nolimits_{j=1}^{k} \|(\varkappa_j^{-1})^*(\chi_j u); H^s_{\tau_i}(\mathbb{W}_j)\|^2 + \sum\nolimits_{j=k+1}^{l} \|(\varkappa_j^{-1})^*(\chi_j u); H^{(s)}(\mathbb{R}^m)\|^2\Big)^{1/2};$$

here, $i = h(j)$, $\{\chi_j\}_{j=1}^{l}$ is a partition of unity on $\mathcal{M}_0$ formed by cut-off functions $\chi_j \in \mathcal{CF}(\mathcal{O}_j)$, $H^{(s)}(\mathbb{R}^m)$ is the Sobolev space in $\mathbb{R}^m$. Another choice of the atlas $\{(\mathcal{O}_j, \varkappa_j)\}$ and the functions $\{\chi_j\}$ leads to an equivalent norm in $H^s_\tau(\mathcal{M})$.

Theorem 4.3.11 *Assume that* $\tau = \{\tau_i\}_{i\in I}$, s, t, $\mu \in \mathbb{R}$, $\mu \le s - t$, *and* $A \in \Psi^\mu_0(\mathcal{M})$. *Then,* A *extends to a continuous map* $H^s_\tau(\mathcal{M}) \to H^t_\tau(\mathcal{M})$. *If* $A \in \Psi^\mu(\mathcal{M})$, *then the statement remains valid under the additional condition* $\tau_i \in (-n_i/2, n_i/2)$, $i \in I$.

Proof In view of formula (4.2.23), it is sufficient to apply Theorems 4.3.2 and 4.3.10 and use that if $B \in \Upsilon_0(\mathcal{M})$, then the mapping $B : H^s_\tau(\mathcal{M}) \to H^t_\tau(\mathcal{M})$ is continuous for all $s, t \in \mathbb{R}$, and $\tau \in \mathbb{R}^I$; in the case $B \in \Upsilon(\mathcal{M})$, the mapping $B : H^s_\tau(\mathcal{M}) \to H^t_\tau(\mathcal{M})$ is continuous for $\tau = \{\tau_i\} \in \mathbb{R}^I$ satisfying $\tau_i \in (-n_i/2, n_i/2)$ and for arbitrary $s, t \in \mathbb{R}$. □

C^*-Algebra of Pseudodifferential Operators on Manifold with Edges

5

In this chapter, we consider w-manifolds and pseudodifferential operators subject to some additional "regularity" conditions near edges. The spectrum of the C^*-algebra generated by these operators is studied.

5.1 Classes Ψ^μ

Definition 5.1.1 A function $\tilde{a} \in C^\infty(\mathbb{R}^m_n \times \mathbb{R}^m)$ is called a presymbol of orderμ, $\mu \in \mathbb{C}$, if

(1) $\|\partial_r^k \partial_{x^2}^\beta \partial_\xi^\gamma \tilde{a}(r\,\cdot, x^2, \xi); C^q(S^{n-1})\| \leq C_{kq\beta\gamma} \langle \xi \rangle^{\operatorname{Re}\mu - |\gamma|}$ for all $k, q \in \mathbb{Z}_+$, $\beta \in \mathbb{Z}_+^{m-n}$, $\gamma \in \mathbb{Z}_+^m$.
(2) There exists the limit $\tilde{a}^{(0)}(x, \xi) = \lim_{t \to +\infty} t^{-\mu} \tilde{a}(x, t\xi)$, $(x, \xi) \in \mathbb{R}^m_n \times (\mathbb{R}^m \setminus 0)$.
(3) The function $(x, \xi) \mapsto \tilde{a}(x, \xi) - \zeta(\xi)\tilde{a}^{(0)}(x, \xi)$, where $\zeta \in C^\infty(\mathbb{R}^m)$, $\zeta(\xi) = 0$ for $|\xi| < 1/2$ and $\zeta(\xi) = 1$ for $|\xi| > 1$, satisfies condition 1) with μ replaced by $\mu - 1$.

Set $\Omega = \mathbb{R} \times S^{n-1} \times \mathbb{R}^{m-n}$, $\Omega_+ = \mathbb{R}_+ \times S^{n-1} \times \mathbb{R}^{m-n}$ and denote by q the bijection $\Omega_+ \times \mathbb{R}^m \to \mathbb{R}^m_n \times \mathbb{R}^m : (r, \varphi, x^2, \xi) \mapsto (r\varphi, x^2, \xi)$. Assume that $\overline{C}^\infty(\mathbb{R}^m_n \times \mathbb{R}^m) = \{f \circ \mathsf{q}^{-1}; f \in C^\infty(\overline{\Omega}_+ \times \mathbb{R}^m)\}$. Condition 1) of Definition 5.1.1 implies that all presymbols belong to $\overline{C}^\infty(\mathbb{R}^m_n \times \mathbb{R}^m)$.

Definition 5.1.2 To each presymbol $\tilde{a}$ of arbitrary order μ, we associate a function $a(x, \xi) = \tilde{a}(x, |x^1|\xi)$, which is called a symbol (of the same order). The class of all presymbols (symbols) of order μ is denoted by $\tilde{\mathsf{S}}^\mu(\mathbb{R}^m_n)$ $(\mathsf{S}^\mu(\mathbb{R}^m_n))$.

© The Author(s), under exclusive license to Springer Nature Switzerland AG 2023

B. Plamenevskii, O. Sarafanov, *Solvable Algebras of Pseudodifferential Operators*, Pseudo-Differential Operators 15, https://doi.org/10.1007/978-3-031-28398-7_5

If $a \in \bigcup_\mu \mathsf{S}^\mu(\mathbb{R}^m_n)$, then the formula

$$(Au)(x) = (2\pi)^{-m/2} \int e^{ix\xi} a(x,\xi)\hat{u}(\xi)\, d\xi$$

defines a linear operator $A = \operatorname{Op} a$ from $C^\infty_c(\mathbb{R}^m_n)$ to $C^\infty(\mathbb{R}^m_n)$. The kernel of this operator is defined by the oscillatory integral

$$G(x,y) = (2\pi)^{-m} \int e^{i(x-y)\xi} a(x,\xi)\, d\xi, \quad (x,y) \in \mathbb{R}^m_n \times \mathbb{R}^m_n.$$

Definition 5.1.3 An operator $A = \operatorname{Op} a$ with $a \in \mathsf{S}^\mu$ is called a pseudodifferential operator of order μ on $\mathbb{R}^m_n$ of class $\mathit{\Psi}^\mu(\mathbb{R}^m_n)$. We say that an operator A is proper if for some $\delta \in (0,1)$ the support of its kernel is contained in the domain $\{(x,y) : |x^1|/|y^1| \in (\delta, \delta^{-1}),\ |x^2 - y^2| < \delta^{-1}(1 + |x^1|)\}$. We denote by $\mathit{\Psi}^\mu_0(\mathbb{R}^m_n)$ the class of all proper Ψdo of order μ.

The similarity of the notations Ψ^μ and $\mathit{\Psi}^\mu$ for old and new classes of Ψdo does not lead to confusion, since only new classes are discussed in this section. It is easy to prove that if $A \in \mathit{\Psi}^\mu_0(\mathbb{R}^m_n)$, $B \in \mathit{\Psi}^\nu_0(\mathbb{R}^m_n)$, then $A^* \in \mathit{\Psi}^{\bar{\mu}}_0(\mathbb{R}^m_n)$, $AB \in \mathit{\Psi}^{\mu+\nu}_0(\mathbb{R}^m_n)$. (The operators A^* and AB belong to old classes $\Psi^{\mathrm{Re}\mu}_0(\mathbb{R}^m_n)$ and $\Psi^{\mathrm{Re}(\mu+\nu)}_0(\mathbb{R}^m_n)$, and it remains only to prove that the symbols of these operators satisfy the requirements of Definition 5.1.1.) The definitions of the classes $\mathit{\Psi}(\mathbb{W})$ and $\mathit{\Psi}_0(\mathbb{W})$ of pseudodifferential operators on a wedge $\mathbb{W}$ are quite similar to the definitions of $\Psi(\mathbb{W})$ and $\Psi_0(\mathbb{W})$ in Sect. 4.2.3. For the classes $\mathit{\Psi}^\mu(\mathbb{W})$ and $\mathit{\Psi}^\mu_0(\mathbb{W})$, Theorem 4.2.13 remains valid.

Assume that $\mathbb{W} = K \times \mathbb{R}^{m-n}$, where K is a smooth n-dimensional conical surface (in some space $\mathbb{R}^N$) with base D. We set $\Omega(\mathbb{W}) = \mathbb{R} \times D \times \mathbb{R}^{m-n}$ and $\Omega_+(\mathbb{W}) = \mathbb{R}_+ \times D \times \mathbb{R}^{m-n}$. Denote by $\mathfrak{p}_\mathbb{W}$ the bijection $\Omega_+(\mathbb{W}) \to \mathbb{W} : (r, \varphi, x^2) \mapsto (r\varphi, x^2)$. This bijection can be extended into a continuous mapping $\overline{\mathfrak{p}}_\mathbb{W} : \overline{\Omega_+(\mathbb{W})} \to \overline{\mathbb{W}}$. If V is an open subset of $\overline{\mathbb{W}}$, then its preimage $\overline{\mathfrak{p}}^{-1}_\mathbb{W}(V)$ is a manifold with (possibly empty) boundary.

Let us consider m-dimensional surfaces $\mathbb{W}_1 = K_1 \times \mathbb{R}^{m-n}$, $\mathbb{W}_2 = K_2 \times \mathbb{R}^{m-n}$, some points $(0, x^2) \in \overline{\mathbb{W}}_1$, $(0, y^2) \in \overline{\mathbb{W}}_2$, and their neighborhoods $V_1 \subset \overline{\mathbb{W}}_1$, $V_2 \subset \overline{\mathbb{W}}_2$.

Definition 5.1.4 A mapping $f : V_1 \to V_2$ is called an admissible diffeomorphism if $f \circ (\overline{\mathfrak{p}}_1 | \overline{\mathfrak{p}}^{-1}_1(V_1)) = \overline{\mathfrak{p}}_2 \circ F$, where $F : \overline{\mathfrak{p}}^{-1}_1(V_1) \to \overline{\mathfrak{p}}^{-1}_2(V_2)$ is a diffeomorphism (of class C^∞) of manifolds with boundary.

It is clear that f induces a diffeomorphism of $V_1 \cap (0 \times \mathbb{R}^{m-n})$ onto $V_2 \cap (0 \times \mathbb{R}^{m-n})$. Note also that a diffeomorphism admissible in the sense of Definition 5.1.4 may be not admissible in the sense of Definitions 4.2.1 and 4.2.18. However, this is insignificant for what follows: the definition of Ψdo on a manifold involves multiplication of the operator by a cut-off function, and, in some neighborhood of the support of this function, a diffeomorphism in Definition 5.1.4 satisfies the requirements of the other mentioned definitions.

We will use the notations and notions introduced in Sect. 4.2.4. In accordance with the new definition of admissible diffeomorphisms, we change the definition of a w-manifold. Namely, let $\mathcal{M}$ be a stratified manifold satisfying conditions (I) and (II) in this section, and let $(\mathcal{O}_1, \varkappa_1)$, $(\mathcal{O}_2, \varkappa_2)$ be special local charts on $\mathcal{M}$ such that $\mathcal{O}_1 \cap \mathcal{O}_2 \neq \emptyset$. These charts are called compatible if the mapping $\varkappa_2 \circ \varkappa_1^{-1} : \varkappa_1(\mathcal{O}_1 \cap \mathcal{O}_2 \cap \mathcal{M}_0) \to \varkappa_2(\mathcal{O}_1 \cap \mathcal{O}_2 \cap \mathcal{M}_0)$ is an admissible diffeomorphism in the sense of Definition 5.1.4. Atlases and a w-structure on $\mathcal{M}$ are introduced in the same way as in 4.2.4. Note that a w-structure on $\mathcal{M}$ induces a smooth structure on each edge s_i, $i \in I$.

For further needs, we have to take some more narrow classes of cut-off functions. Let ζ be a smooth function on a wedge $\mathbb{W}$. We say that ζ is in the class $\overline{C}{}_c^\infty(\mathbb{W})$ if $\zeta = \eta \circ \mathfrak{p}_{\mathbb{W}}^{-1}$, where $\eta \in C_c^\infty(\overline{\Omega_+(\mathbb{W})})$. By definition, a function ζ in $\overline{C}{}_c^\infty(\mathbb{W})$ is serving an open set $V \in \overline{\mathbb{W}}$ if the closure (in $\overline{\mathbb{W}}$) of its support lies inside V. Let now $\mathcal{M}$ be a w-manifold, $\mathcal{O}$ a special coordinate neighborhood, $\varkappa : \mathcal{O} \to \mathbb{W}$ a coordinate diffeomorphism. A function $\chi \in C^\infty(\mathcal{M}_0)$ is called a cut-off function serving the set $\mathcal{O}$ if $\chi = \zeta \circ (\varkappa|\mathcal{M}_0)$, $\zeta \in \overline{C}{}_c^\infty(\mathbb{W})$, whereas the function ζ is subject to $\varkappa(\mathcal{O})$. If $\mathcal{O}$ is a standard coordinate neighborhood, then any function in $C_c^\infty(\mathcal{O})$ is called a cut-off function serving $\mathcal{O}$. The set of all cut-off functions serving $\mathcal{O}$ is denoted (as in Sect. 4.2) by $\mathcal{CF}(\mathcal{O})$.

Using the new cut-off functions, we can introduce a Riemannian metric on $\mathcal{M}_0$ (sf. Sect. 4.2.4) and define functions ℓ_i by (4.2.18). We denote by ν a smooth measure on $\mathcal{M}_0$ induced by the metric.

Turn to the description of classes Ψ^μ, which repeats the content of Sect. 4.2.5 with minor changes. Consider an arbitrary Ψdo in $\mathbb{R}_n^m$ of the form $\eta_1(\operatorname{Op} a)\eta_2$, where $a \in \bigcup_\mu \mathsf{S}^\mu$ is a symbol, and $\eta_1, \eta_2 \in \overline{C}{}_c^\infty(\mathbb{R}_n^m)$ are such that the closures (in $\mathbb{R}^m$) of their supports are disjoint. Let H be the kernel of this operator,

$$H(x, y) = (2\pi)^{-m} \int e^{i(x-y)\xi} \eta_1(x) a(x, \xi) \eta_2(y)\, d\xi.$$

Then, $H \circ (\mathfrak{p}, \mathfrak{p}) \in C_c^\infty(\Omega \times \Omega)|(\Omega_+ \times \Omega_+)$; for $\mathbb{W} = \mathbb{R}_n^m$ we write $\mathfrak{p}$ instead of $\mathfrak{p}_{\mathbb{W}}$. As simple calculations show (cf. Sect. 4.2.5), $|H(x, y)| \le C_N |x^1|^N$ for all $N \in \mathbb{N}$. Conversely, if a function H satisfies the mentioned properties, then the operator

$$Bu(x) = \int H(x, y) u(y)\, dy$$

is a smoothing Ψdo, i.e., it belongs to $\bigcap_{\mu \in \mathbb{C}} \Psi^\mu(\mathbb{R}_n^m)$. For further needs, we note that the mapping $H \to H \circ (\mathfrak{p}, \mathfrak{p})$ generates an isomorphism

$$\overline{C}{}_c^\infty(\mathbb{R}_n^m) \hat{\otimes} \overline{C}{}_c^\infty(\mathbb{R}_n^m) \cong C_c^\infty(\Omega \times \Omega)|(\Omega_+ \times \Omega_+).$$

Definition 5.1.5 We say that an operator

$$Bu(z) = \int H(z, w)u(w)\, dv(w), \quad u \in C_c^\infty(\mathcal{M}_0),$$

belongs to the class $\Upsilon(\mathcal{M})$ if $H \in \overline{C}_c^\infty(\mathcal{M}_0)\hat{\otimes}\overline{C}_c^\infty(\mathcal{M}_0)$ and the inequality $|H(z, w)| \leq C_N \ell(z)^N$ holds for any $N \in \mathbb{N}$. The operator B is called proper ($B \in \Upsilon_0(\mathcal{M})$) if there exists a number $\delta \in (0, 1)$ such that $H(z, w) = 0$ whenever $\ell_i(z)/\ell_i(w) \notin (\delta, \delta^{-1})$ for at least one $i \in I$.

Let $\mathcal{U}$ be an open subset of $\mathbb{R}^m$. Denote by $\mathsf{S}^\mu(\mathcal{U})$, $\mu \in \mathbb{C}$, the set of all C^∞-functions a in $\mathcal{U} \times \mathbb{R}^m$ admitting the representation $a = a_0 + a_1$ with $a_0, a_1 \in C^\infty(\mathcal{U} \times \mathbb{R}^m)$, whereas a_0 is a homogeneous function of degree μ for large $|\xi|$, and a_1 satisfies the estimates

$$\sup\{|\partial_x^\alpha \partial_\xi^\beta a_1(x, \xi)|; x \in E\} \leq C_{\alpha,\beta,E} \langle \xi \rangle^{\mathrm{Re}\mu - |\beta| - 1}$$

for all $\alpha, \beta \in \mathbb{Z}_+^m$ and any compact set E. An operator $A : C_c^\infty(\mathcal{M}_0) \to C^\infty(\mathcal{M}_0)$ is called a pseudodifferential operator of order μ on a smooth manifold $\mathcal{M}_0$ if, for each (probably non-connected) coordinate neighborhood $\mathcal{O} \subset \mathcal{M}_0$ and each diffeomorphism $\varkappa : \mathcal{O} \to \mathcal{U} \subset \mathbb{R}^m$, the mapping $A_\varkappa : u \mapsto [A(u \circ \varkappa)] \circ \varkappa^{-1}$, $u \in C_c^\infty(\mathcal{U})$, is a Ψdo of order μ in $\mathcal{U}$. The latter means that

$$A_\varkappa u(x) = (2\pi)^{-m/2} \int e^{ix\xi} a_\varkappa(x, \xi) \hat{u}(\xi)\, d\xi,$$

where $a_\varkappa$ is an element in $\mathsf{S}^\mu(\mathcal{U})$.

Put $a_\varkappa^{(0)}(x, \xi) = \lim_{t \to +\infty} t^{-\mu} a_\varkappa(x, t\xi)$. For a fixed x, the function $\xi \to a_\varkappa^{(0)}(x, \xi)$ is homogeneous of degree μ. It follows from the formula for change of variables in Ψdo that $a_\varkappa^{(0)}$ is an expression in local coordinates for some function $a^{(0)}$ well-defined on the complement of zero section of the cotangent bundle over $\mathcal{M}_0$. The function $a^{(0)}$ is called the principal symbol of Ψdo A. The class of all Ψdo of order μ on a manifold $\mathcal{M}_0$ is denoted by $\Psi^\mu(\mathcal{M}_0)$. According to the standard definition, an operator $A \in \Psi^\mu(\mathcal{M}_0)$ is called proper ($A \in \Psi_0^\mu(\mathcal{M}_0)$) if both projections from the support of its kernel—subset of the product $\mathcal{M}_0 \times \mathcal{M}_0$—onto the factors of this product are proper mappings.

The following definition of the classes $\Psi(\mathcal{M})$ is quite similar to Definition 4.2.23. Previously, let us agree that supports of cut-off functions χ_1 and χ_2 are called disjunctive if the closures of these supports (in $\mathcal{M}$) are disjoint.

Definition 5.1.6 The class $\Psi^\mu(\mathcal{M})$ of pseudodifferential operators of order μ on a w-manifold $\mathcal{M}$ consists of all operators $A \in \Psi^\mu(\mathcal{M}_0)$ satisfying the following properties:

(1) For any special coordinate system $\varkappa : \mathcal{O} \to \overline{\mathbb{W}}$ and any two cut-off functions χ_1, $\chi_2 \in \mathcal{CF}(\mathcal{O})$, the operator

$$u \mapsto [(\chi_1 A \chi_2)(u \circ \varkappa_0)] \circ \varkappa_0^{-1}, \quad u \in C_c^\infty(\mathbb{W}),$$

where $\varkappa_0 = \varkappa|(\mathcal{O} \cap \mathcal{M}_0)$, belongs to $\Psi^\mu(\mathbb{W})$;

(2) $\chi_1 A \chi_2 \in \Upsilon(\mathcal{M})$ for any pair of cut-off functions χ_1, χ_2 with disjunctive supports. We change $\Psi^\mu(\mathcal{M}_0)$, $\Psi^\mu(\mathbb{W})$, and $\Upsilon(\mathcal{M})$ for $\Psi_0^\mu(\mathcal{M}_0)$, $\Psi_0^\mu(\mathbb{W})$, and $\Upsilon_0(\mathcal{M})$ and obtain the definition of the class $\Psi_0^\mu(\mathcal{M})$ of proper Ψdo.

5.2 C^*-Algebra Generated by Proper Ψdo. Local Algebras

Let, as before, $\mathcal{M}$ be a w-manifold, $\mathcal{M}_0$ its open dense part (i.e., the union of all strata of maximal dimension), $m = \dim \mathcal{M}_0$. Let $\mathcal{A}$ denote the C^*-algebra generated in $L_2(\mathcal{M}) \equiv L_2(\mathcal{M}, \nu)$ by proper Ψdo of class $\Psi_0^0(\mathcal{M})$. Denote by $\overline{C}^\infty(\mathcal{M}_0)$ the set of functions f in $C^\infty(\mathcal{M}_0)$, that near an edge admit the representation $f = h \circ (\varkappa|\mathcal{M}_0)$, where $h \in \overline{C}\infty(\mathbb{W})$ and $\varkappa$ is a coordinate diffeomorphism. Define also $\overline{C}(\mathcal{M}_0)$ as the closure of $\overline{C}^\infty(\mathcal{M}_0)$ in the algebra $C_b(\mathcal{M}_0)$ of bounded continuous functions. Multiplication operators $(f\cdot)$, $f \in \overline{C}(\mathcal{M}_0)$, belong to $\mathcal{A}$, since $(f\cdot) \in \Psi_0^0(\mathcal{M})$ for $f \in \overline{C}^\infty(\mathcal{M}_0)$. We will consider the algebra $\overline{C}(\mathcal{M}_0)$ as a subalgebra of $\mathcal{A}$.

For each $z \in \mathcal{M}$, we put $\mathcal{I}_z = \{f \in C(\mathcal{M}) : f(z) = 0\}$ and denote by $\mathcal{J}_z$ the least ideal in $\mathcal{A}$ containing $\mathcal{I}_z$.

Proposition 5.2.1 *The following assertions are true:*

(1) The algebra $\mathcal{A}$ is irreducible.
(2) $\mathcal{K}L_2(\mathcal{M}) \subset \mathcal{A}$.
(3) The following equality holds:

$$\hat{\mathcal{A}} = \bigcup_{z \in \mathcal{M}} (\mathcal{A}/\mathcal{J}_z)^\wedge \cup [Id],$$

where $[Id]$ is the equivalence class containing the identity representation of $\mathcal{A}$ in $L_2(\mathcal{M})$. If $z \neq x$, then the sets $(\mathcal{A}/\mathcal{J}_z)^\wedge$ and $\mathcal{A}/\mathcal{J}_x)^\wedge$ of the spectrum $\hat{\mathcal{A}}$ are disjoint.

Proof One can prove that $\mathcal{A}$ is irreducible arguing as in the proof of Proposition 2.1.2. If $f_1, f_2 \in \overline{C}(\mathcal{M}_0)$ are functions with disjoint supports, then $f_1 A f_2 \in \overline{\Upsilon(\mathcal{M})} \subset \mathcal{K}L_2(\mathcal{M})$ for any $A \in \Psi_0^0(\mathcal{M})$. Thus, $\mathcal{K}L_2(\mathcal{M}) \subset \mathcal{A}$. Note that $C(\mathcal{M})$ is a subalgebra of both $\overline{C}(\mathcal{M})$ and $\mathcal{A}$. Let us apply Proposition 1.3.24 taking $C(\mathcal{M})$ and $\mathcal{K}L_2(\mathcal{M})$ as $\mathcal{C}$ and $\mathcal{J}$. It remains to take into account that $C(\mathcal{M})^\wedge = \mathcal{M}$ and that the spectrum of $\mathcal{K}L_2(\mathcal{M}) * \wedge$ consists of the only element $[Id]$. $\square$

We now turn to the description of local algebras $\mathcal{A}/\mathcal{J}_z$. The following statement is in essence similar to Proposition 2.1.5.

Proposition 5.2.2 *Assume that* $A \in \Psi_0^0(\mathcal{M}_0)$ *and that* $a^{(0)}$ *is the principal symbol of* A. *For each* $z \in \mathcal{M}_0$, *the mapping* $\mathcal{A} \mapsto a^{(0)}|S_z^*$ *generates an isomorphism* $\mathcal{A}/\mathcal{J}_z \cong C(S_z^*)$.

Turn to the local algebras $\mathcal{A}/\mathcal{J}_z$ for $z \in \mathcal{M} \setminus \mathcal{M}_0$. Let $\mathbb{W} = K \times \mathbb{R}^{m-n}$ be a wedge, where K denotes an n-dimensional conical surface in $\mathbb{R}^N$. Let also $L_2(\mathbb{W})$ be the space of functions square integrable in a Euclidean surface measure and $\mathcal{A}(\mathbb{W})$ be a subalgebra in $\mathcal{B}L_2(\mathbb{W})$ generated by operators of the form $\chi A \chi$, where $A \in \Psi_0^0(\mathbb{W})$, $\chi \in \overline{C}_c^\infty(\mathbb{W})$. For any $t > 0$ we define a unitary operator $U_t : L_2(\mathbb{W}) \to L_2(\mathbb{W})$ by $U_t u = t^{m/2} u(t\,\cdot)$. If $A \in \Psi_0^0(\mathbb{W})$ and

$$A_\upsilon u(x) = (2\pi)^{-m/2} \int e^{ix\xi} a_\upsilon(x, \xi)\hat{u}(\xi)\, d\xi\,, \quad u \in C_c^\infty(\upsilon(U)),$$

is a representation of A in local coordinates $\upsilon : U \to \mathbb{R}^m$, then

$$(U_t A U_t^{-1})_\upsilon u(x) = (2\pi)^{-m/2} \int e^{ix\xi} a_\upsilon(tx, \xi/t)\hat{u}(\xi)\, d\xi =$$

$$= (2\pi)^{-m/2} \int e^{ix\xi} \tilde{a}_\upsilon(tx, |x^1|\xi)\hat{u}(\xi)\, d\xi$$

(see definition of local coordinates on a wedge in Sect. 4.2.3). This implies that the formula

$$A_0 u = \lim_{t\to 0} (U_t A U_t^{-1})u\,, \quad u \in C_c^\infty(\mathbb{W}), \tag{5.2.1}$$

defines an operator A_0 that admits an extension to a bounded operator in $L_2(\mathbb{W})$, whereas $\|A_0\| \le \|A\|$. In local coordinates,

$$(A_0)_\upsilon u(x) = (2\pi)^{-m/2} \int e^{ix\xi} (\tilde{a}_\upsilon)_0(\varphi, |x^1|\xi)\hat{u}(\xi)\, d\xi, \tag{5.2.2}$$

where $(\tilde{a}_\upsilon)_0(\varphi, \xi) = \lim_{t\to 0} \tilde{a}_\upsilon(tx, \xi)$ and $\varphi = x^1/|x^1|$.

Denote by $\mathcal{A}_0(\mathbb{W})$ the least subalgebra in $\mathcal{B}L_2(\mathbb{W})$ containing all operators A_0 of the form (5.2.2)and by $\mathcal{J}_0(\mathbb{W})$ the ideal in $\mathcal{A}(\mathbb{W})$ generated by operators of multiplication by functions $\chi \in C_c(\overline{\mathbb{W}})$ such that $\chi(0) = 0$. According to what was said above, the mapping $\Psi_0^0(\mathbb{W}) \ni A \mapsto A_0$ extends to an epimorphism $\mathcal{A}(\mathbb{W}) \to \mathcal{A}_0(\mathbb{W})$ whose kernel contains the ideal $\mathcal{J}_0(\mathbb{W})$. Therefore, the epimorphism

$$\mathcal{A}(\mathbb{W})/\mathcal{J}_0(\mathbb{W}) \ni [A] \mapsto A_0 \in \mathcal{A}_0(\mathbb{W}) \tag{5.2.3}$$

is defined; actually, it is an isomorphism. To prove this, we note that if A_0 is an operator corresponding by formula (5.2.1) to a proper Ψdo A, then for any $\chi \in \overline{C_c^\infty}(\mathbb{W})$ the operator $\chi A_0 \chi$ is proper (this statement follows from the definition of proper Ψdo and from the equality $t^m G(tx, ty) = G_t(x, y)$, where $t > 0$ and G, G_t denote the kernels of the operators $A, U_t A U_t^{-1}$). If $\chi \in C(\overline{\mathbb{W}}) \cap \overline{C_c^\infty}(\mathbb{W})$ and $\chi(0) = 1$, then the operator A_0 is the image of the residue class $[\chi A_0 \chi]$ under mapping (5.2.3). In view of the relations $A - \chi A_0 \chi \in \mathcal{J}_0(\mathbb{W})$ and $\|\chi A_0 \chi\| \le (\sup|\chi|^2)\|A_0\|$, which implies that mapping (5.2.3) is injective.

Assume now that $\mathcal{M}$ is a w-manifold, ν is a measure on $\mathcal{M}_0$ introduced in Sect. 5.1, $z \in \mathcal{M} \setminus \mathcal{M}_0$ is an arbitrarily fixed point, $\varkappa : \mathcal{O} \to \overline{\mathbb{W}}$ is a local chart on $\mathcal{M}$, $\mathcal{O} \ni z$, and $\varkappa(z) = 0$. The isomorphism $\mathcal{A}/\mathcal{J}_z \cong \mathcal{A}(\mathbb{W})/\mathcal{J}_0(\mathbb{W})$ is implemented by the mapping

$$\mathcal{A}/\mathcal{J}_z \ni A + \mathcal{J}_z \mapsto f^{1/2}(\varkappa_0^{-1})^*(\chi A \chi)\varkappa_0^* f^{-1/2} + \mathcal{J}_0(\mathbb{W}) \in \mathcal{A}(\mathbb{W})/\mathcal{J}_0(\mathbb{W}),$$

where $\varkappa_0 = \varkappa|(\mathcal{O} \cap \mathcal{M}_0)$, $\chi \in \mathcal{CF}(\mathcal{O})$, $\chi = 1$ near z, and f is the density of the measure $(\varkappa_0)_*(\nu|\mathcal{O} \cap \mathcal{M}_0)$ with respect to the surface measure on $\mathbb{W}$. We apply to the residue classes $[(\varkappa_0^{-1})^*(\chi A \chi)\varkappa_0^*]$ and $[f]$ the mapping (5.2.3) and obtain some element of $\mathcal{A}_0(\mathbb{W})$ denoted again by A_0 and a function f_0 defined by $f_0(x) = \lim_{t\to+0} f(tx)$. Thus, there holds

Proposition 5.2.3 *For any $z \in \mathcal{M}\setminus\mathcal{M}_0$, the mapping $\mathcal{A} \ni A \mapsto f_0^{1/2} A_0 f_0^{-1/2} \in \mathcal{A}_0(\mathbb{W})$, where A_0 is the operator defined by*

$$A_0 u = \lim_{t\to 0} U_t (\varkappa_0^{-1})^*(\chi A \chi)\varkappa_0^* U_t^{-1} u\,, \quad u \in C_c^\infty(\mathbb{W}),$$

generates an isomorphism $\mathcal{A}/\mathcal{J}_z \cong \mathcal{A}_0(\mathbb{W})$. □

The operator A_0 in Proposition 5.2.3 depends on the choice of $\varkappa$. The invariance of elements of the algebra $\mathcal{A}/\mathcal{J}_z$ will be determined in Sect. 5.6.

Therefore, if $z \in \mathcal{M}_0$, then due to the isomorphism $\mathcal{A}/\mathcal{J}_z \cong C(S_z^*)$ (Proposition 5.2.2) the spectrum $(\mathcal{A}/\mathcal{J}_z)^\wedge$ is identified with S_z^*. For $z \in \mathcal{M} \setminus \mathcal{M}_0$, we have an isomorphism $\mathcal{A}/\mathcal{J}_z \cong \mathcal{A}_0(\mathbb{W})$ (Proposition 5.2.3), where $\mathbb{W} = K \times \mathbb{R}^{m-n}$ is a wedge, which is a local model for $\mathcal{M}$ in a neighborhood of z. Thus, the problem of description of the spectrum of $\mathcal{A}$ is reduced to an analogous problem for $\mathcal{A}_0(\mathbb{W})$. For further study, it is convenient to replace $\mathcal{A}_0(\mathbb{W})$ with an isomorphic algebra. To this end, we take into account that A_0 in (5.2.1) is invariant under shifts along the edge. Therefore,

$$(A_0 u)(\cdot, x^2) = F^{-1}_{\zeta\to x^2} A_0(\zeta) F_{y^2\to\zeta} u(\cdot, y^2), \tag{5.2.4}$$

where F is the Fourier transform along the edge and the function $\zeta \mapsto A_0(\zeta)$ is defined on $\mathbb{R}^{m-n}$ and takes values in the algebra $\mathcal{B}L_2(K)$ of all bounded operators in the space $L_2(K)$

on the cone K. Denote by $\mathcal{L}(K)$ the algebra generated by functions $S^{m-n-1} \ni \theta \mapsto A_0(\theta)$, where the function $A_0(\cdot)$ is related to $A_0 \in \mathcal{A}_0(\mathbb{W})$ by (5.2.4); put $\|A_0(\cdot); \mathcal{L}(K)\| = \sup\{\|A_0(\theta); \mathcal{B}L_2(K)\|; \theta \in S^{m-n-1}\}$.

Proposition 5.2.4 *The algebras $\mathcal{A}_0(\mathbb{W})$ and $\mathcal{L}(K)$ are isomorphic.*

Proof Since

$$\|F^{-1}A_0(\zeta)Fu; L_2(\mathbb{W})\|^2 = \int \|A_0(\zeta)\hat{u}(\cdot, \zeta); L_2(K)\|^2 \, d\zeta,$$

we have

$$\|A_0; \mathcal{B}L_2(\mathbb{W})\| = \sup\{\|A_0(\zeta); \mathcal{B}L_2(K)\|; \zeta \in \mathbb{R}^{m-n}\}.$$

Let $V \subset K$ and let $v : V \to v(V) \subset \mathbb{R}^n$ be a coordinate morphism; remind that the mapping v is homogeneous of degree one. In local coordinates,

$$\begin{aligned}(A_0)_{(v)}(t\zeta)u(x^1) &= (2\pi)^{-n/2} \int e^{ix^1\xi^1} (\tilde{a}_v)_0(\varphi, |x^1|\xi^1, |x^1|t\zeta)\hat{u}(\xi^1)\, d\xi^1 \\ &= (2\pi)^{-n/2} \int e^{itx^1\xi^1} (\tilde{a}_v)_0(\varphi, |x^1|t\xi^1, |x^1|t\zeta)t^n\hat{u}(t\xi^1)\, d\xi^1 \\ &= U_t(A_0)_{(v)}(\zeta)U_t^{-1}u(x^1), \quad (5.2.5)\end{aligned}$$

where $t > 0$ and $U_t u(x^1) = t^{n/2}u(tx^1)$. This implies that $A_0(t\zeta) = U_t A_0(\zeta) U_t^{-1}$ and $\|A_0(\zeta)\| = \|A_0(t\zeta)\|$ for $t > 0$ since the operator U_t is unitary. If $t \to 0$, then $A_0(t\zeta) \to A_0(0)$ in the strong operator topology. Therefore, $\|A_0(0)\| \le \sup\{\|A_0(\zeta)\|; \zeta \in \mathbb{R}^n, \zeta \ne 0\}$. We arrive at the equality

$$\|A_0; \mathcal{B}L_2(\mathbb{W})\| = \sup\{\|A_0(\theta); \mathcal{B}L_2(K)\|; \theta \in S^{m-n-1}\},$$

which means that $\|A_0; \mathcal{A}_0(\mathbb{W})\| = \|A_0(\cdot); \mathcal{L}(K)\|$. □

5.3 Algebras $\mathcal{L}(\theta, \mathbb{R}^n)$ and $\mathcal{L}(0, \mathbb{R}^n)$

Let $\theta \in S^{m-n-1}$ and $A_0(\cdot) \in \mathcal{L}(K)$. Denote by $\mathcal{L}(\theta, K)$ the algebra generated by operators $A_0(\theta)$ in $L_2(K)$. Let also $\mathcal{L}(0, K)$ be the algebra spanned by operators of the form $A_0(0)$. In this section, we are preparing for the study of algebras $\mathcal{L}(\theta, K)$ as $\theta \in S^{m-n-1}$ and $\theta = 0$; here, we deal with the case $K = \mathbb{R}^n$. It turns out that $\mathcal{L}(\theta, \mathbb{R}^n)$ coincides with the algebra $\mathcal{L}(\theta)$ defined in Sect. 2.3.2, and $\mathcal{L}(0, \mathbb{R}^n)$ is isomorphic to the algebra $\mathcal{L}_0(0)$ generated by operators

$$F^{-1}_{\eta\to x} a(\varphi, \omega) F_{y\to\eta}$$

in $L_2(\mathbb{R}^n)$; here, $a \in C^\infty(S^{n-1} \times S^{n-1})$, $\varphi = x/|x|$ and $\omega = \eta/|\eta|$. The proof of these statements is the main result of this section. In what follows, the mentioned algebra $\mathcal{L}(\theta)$ and the algebra $\mathfrak{S}$ in 2.2.1 will be denoted by $\mathcal{L}_0(\theta)$ and $\mathfrak{S}_0$, respectively. Thus, the algebra $\mathcal{L}_0(\theta)$ is generated by operators of the form

$$A(\theta) = F^{-1}_{\eta\to x} \Phi(\varphi, \eta, \theta) F_{y\to\eta} \in \mathcal{B}L_2(\mathbb{R}^n)$$

with a homogeneous function $\xi \mapsto \Phi(\varphi, \xi)$, and $\mathfrak{S}_0$ by operator-valued functions

$$\mathbb{R} \ni \lambda \mapsto E_{\omega\to\varphi}(\lambda)^{-1} a(\varphi, \omega) E_{\psi\to\omega}(\lambda) \in \mathcal{B}L_2(S^{n-1}).$$

5.3.1 A Special Representation of Generators of $\mathcal{L}(0, \mathbb{R}^n)$

Taking into account (5.2.5) and simplifying notations, we will temporarily write generators of the algebra $\mathcal{L}(0, \mathbb{R}^n)$ in the form

$$Au(x) = (2\pi)^{-n/2} \int e^{ix\xi} \tilde{a}(\varphi, |x|\xi) \hat{u}(\xi)\, d\xi. \tag{5.3.1}$$

The nearest goal is to represent the operator (5.3.1) in terms of the Mellin transform. According to Proposition 1.2.2, the following formula for the Fourier transform is valid:

$$\begin{aligned}(Fu)(\xi) &= (M^{-1})_{in/2-\lambda\to\rho} E_{\psi\to\omega}(\lambda) M_{|y|\to\lambda+in/2} u(y) \\ &= (M^{-1})_{in/2-\lambda\to\rho} E_{\psi\to\omega}(\lambda) U(\lambda, \psi),\end{aligned} \tag{5.3.2}$$

where $U(\lambda, \psi) = (Mu)(\lambda + in/2, \psi)$, $\psi = y/|y|$, $\rho = |\xi|$, $\omega = \xi/|\xi|$, and $\lambda \in \mathbb{R}$. The inverse transform admits the representation

$$(F^{-1}v)(x) = M^{-1}_{\mu+in/2\to r} E_{\omega\to\varphi}(\mu)^{-1} M_{\rho\to in/2-\mu} v(\rho\omega), \quad \mu \in \mathbb{R}. \tag{5.3.3}$$

From (5.3.2) and (5.3.3), it follows that

$$Au(x) = (2\pi)^{-3/2} \int_{-\infty}^{+\infty} r^{i\mu-n/2} E_{\omega\to\varphi}(\mu)^{-1}\, d\mu$$

$$\times \int_0^{+\infty} \rho^{i\mu-1} \tilde{a}(\varphi, r\rho\omega)\, d\rho \int_{-\infty}^{+\infty} \rho^{-i\lambda} E_{\psi\to\omega}(\lambda) U(\lambda, \psi)\, d\lambda.$$

We put $a^0(\varphi, \eta) = \lim_{t\to\infty} \tilde{a}(\varphi, t\eta)$ and $\tilde{a}^1 = \tilde{a} - a^0$; the function $\eta \mapsto a^0(\varphi, \eta)$ is zero degree homogeneous. We have $A = A^{(0)} + A^{(1)}$, where

$$A^{(0)}u(x) = (2\pi)^{-1/2} \int_{-\infty}^{\infty} r^{i\lambda - n/2} E_{\omega\to\varphi}(\lambda)^{-1} a^0(\varphi, \omega) E_{\psi\to\omega}(\lambda) U(\lambda, \psi)\, d\lambda, \tag{5.3.4}$$

$$A^{(1)}u(x) = (2\pi)^{-3/2} \int_{-\infty}^{+\infty} r^{i\mu - n/2} E_{\omega\to\varphi}(\mu)^{-1}\, d\mu \tag{5.3.5}$$

$$\times \int_0^{+\infty} \rho^{i\mu - 1} \tilde{a}^1(\varphi, r\rho\omega)\, d\rho \int_{-\infty}^{+\infty} \rho^{-i\lambda} E_{\psi\to\omega}(\lambda) U(\lambda, \psi)\, d\lambda.$$

Introduce the operator

$$\mathfrak{A}^{(0)}(\lambda) = E_{\omega\to\varphi}(\lambda)^{-1} a^0(\varphi, \omega) E_{\psi\to\omega}(\lambda) \tag{5.3.6}$$

and rewrite (5.3.4) in the form

$$(A^{(0)})u(r\varphi) = (2\pi)^{-1/2} \int_{-\infty}^{+\infty} r^{i\lambda - n/2} \mathfrak{A}^{(0)}_{\psi\to\varphi}(\lambda) U(\lambda, \psi)\, d\lambda. \tag{5.3.7}$$

Let us now obtain a similar representation for $A^{(1)}$. To justify the permutation of integrals in (5.3.5), we define the operator $A^{(1)}_\delta$ by (5.3.5) with $\tilde{a}^1(\varphi, r\rho\omega)$ replaced by $\tilde{a}^1_\delta(\varphi, r\rho\omega) = \zeta(r\rho\delta)\zeta(\delta/r\rho)\tilde{a}^1(\varphi, r\rho\omega)$; here $\delta \in (0, 1)$ and $\zeta \in C^\infty(\mathbb{R}_+)$, whereas $\zeta(t) = 1$ for $t \le 1$ and $\zeta(t) = 0$ for $t > 2$. We change the variable $\rho \mapsto \rho/r$ and obtain

$$A^{(1)}_\delta u(x) = (2\pi)^{-3/2} \int_{-\infty}^{+\infty} E_{\omega\to\varphi}(\mu)^{-1}\, d\mu$$

$$\times \int_0^{\infty} d\rho \int_{-\infty}^{+\infty} r^{i\lambda - n/2} \rho^{-i(\lambda-\mu)-1} \tilde{a}^1_\delta(\varphi, \rho\omega) E_{\psi\to\omega}(\lambda) U(\lambda, \psi)\, d\lambda.$$

Since $u \in C^\infty_c(\mathbb{R}^n \setminus 0)$, the function $\lambda \mapsto E(\lambda)U(\lambda, \cdot)$ is rapidly decaying as $\lambda \to \infty$. This allows us to switch two inner integrals, so

$$A^{(1)}_\delta u(x) = \frac{1}{2\pi} \int_{-\infty}^{+\infty} E_{\omega\to\varphi}(\mu)^{-1}\, d\mu \tag{5.3.8}$$

$$\times \int_{-\infty}^{+\infty} r^{i\lambda - n/2} (M\tilde{a}^1_\delta)(\varphi, \lambda - \mu, \omega) E_{\psi\to\omega}(\lambda) U(\lambda, \psi)\, d\lambda,$$

where $(M\tilde{a}^1_\delta)(\varphi, \nu, \omega) = M_{\rho\to\nu} \tilde{a}^1_\delta(\varphi, \rho\omega)$. The function

$$(\lambda, \mu) \mapsto E_{\omega\to\varphi}(\mu)^{-1} (M\tilde{a}^1_\delta)(\varphi, \lambda - \mu, \omega) E_{\psi\to\omega}(\lambda) U(\lambda, \psi)$$

is holomorphic in the domain $\{(\lambda, \mu) : \mathrm{Im}\lambda < n/2, \mathrm{Im}\mu > -n/2\}$ and decays rapidly if $|\lambda| + |\mu| \to \infty$ provided $|\mathrm{Im}\lambda|, |\mathrm{Im}\mu| < h$. Therefore,

$$A_\delta^{(1)} u(x) = (2\pi)^{-1/2} \int_{-\infty}^{+\infty} r^{i\lambda - n/2}\, d\lambda \tag{5.3.9}$$

$$\times \int_{\mathrm{Im}\,\mu = -\tau} E_{\omega \to \varphi}(\mu)^{-1} (M\tilde{a}_\delta^1)(\varphi, \lambda - \mu, \omega) E_{\psi \to \omega}(\lambda) U(\lambda, \psi)\, d\mu$$

for $\tau < n/2$. Let $\delta \to 0$ in (5.3.9). We first find the limit of the left-hand side. The equality

$$A_\delta^{(1)} u(x) = (2\pi)^{-n/2} \int \int e^{i(x-y)\xi} \tilde{a}_\delta^1(\varphi, |x|\xi) \langle \xi \rangle^{-2N} \langle D \rangle^{2N} u(y)\, dy\, d\xi$$

for $N > n/2$ implies that

$$\lim_{\delta \to 0} A_\delta^{(1)} u(x) = (2\pi)^{-n/2} \int e^{ix\xi} \tilde{a}^1(\varphi, |x|\xi) \hat{u}(\xi)\, d\xi = A^{(1)} u(x). \tag{5.3.10}$$

Consider the right-hand side of (5.3.9). It turns out (cf. Lemma 5.3.2 below) that, for $\tau \in (0, 1)$ and any $N \geq 0$, the inequality

$$\begin{aligned} \| E_{\omega \to \varphi}(\mu)^{-1} M(\tilde{a}_\delta^1 - \tilde{a}^1)(\varphi, \lambda - \mu, \omega) E_{\psi \to \omega}(\lambda) U(\lambda, \psi); H^{s+\tau}(S^{n-1}) \| \\ \leq C_N \delta^\varepsilon (1 + |\lambda - \mu|)^{-N} (1 + |\lambda|)^{-N} \end{aligned} \tag{5.3.11}$$

holds, where $\mathrm{Im}\lambda = 0$, $\mathrm{Im}\mu = -\tau$, $\varepsilon \in (0, \min\{\tau, 1 - \tau\})$, and s is an arbitrary non-negative number. Therefore, one can pass to the limit as $\delta \to 0$ under the integral sight in the right-hand side of (5.3.9). Taking into account (5.3.10), we obtain the needed representation

$$(A^{(1)}) u(r\varphi) = (2\pi)^{-1/2} \int_{-\infty}^{+\infty} r^{i\lambda - n/2} \mathfrak{A}_{\psi \to \varphi}^{(1)}(\lambda) U(\lambda, \psi)\, d\lambda \tag{5.3.12}$$

with

$$\mathfrak{A}^{(1)}(\lambda) v(\varphi) = (2\pi)^{-1/2} \int_{\Im\mu = -\tau} E_{\omega \to \varphi}(\mu)^{-1} (M\tilde{a}^1)(\varphi, \lambda - \mu, \omega) E_{\psi \to \omega}(\lambda) v(\psi)\, d\mu, \tag{5.3.13}$$

where τ is an arbitrary number in the interval $(0, \min(1, n/2))$. Together with (5.3.8), this leads to the following statement.

Proposition 5.3.1 *The operator (5.3.1) admits the representation*

$$(Au)(r\varphi) = (2\pi)^{-1/2} \int_{-\infty}^{+\infty} r^{i\lambda - n/2} \mathfrak{A}_{\psi\to\varphi}(\lambda) U(\lambda, \psi)\, d\lambda, \tag{5.3.14}$$

where $\mathfrak{A}(\lambda) = \mathfrak{A}^{(0)}(\lambda) + \mathfrak{A}^{(1)}(\lambda)$, *the operator* $\mathfrak{A}^{(0)}$ *is defined by (5.3.6) and the operator* $\mathfrak{A}^{(1)}$ *by (5.3.13), whereas* $u \in C_c^\infty(\mathbb{R}^n \setminus 0)$, $U(\lambda, \psi) = M_{|y|\to\lambda+in/2}\, u(y)$

It remains to prove the estimate (5.3.11). This proof is split into three lemmas, which will also be of use in what follows. For $\Phi \in C^\infty(S^{n-1} \times S^{n-1})$, introduce the operator

$$\mathfrak{B}_{\psi\to\varphi}(\lambda, \mu) = E_{\omega\to\varphi}(\mu)^{-1} \Phi(\varphi, \omega) E_{\psi\to\omega}(\lambda).$$

Lemma 5.3.2 *Let* $\mathcal{F}$ *be a closed subset contained in a strip* $|\mathrm{Im}\,\lambda| < h$ *and free of the points* $\lambda = \pm i(k + n/2)$ *for* $k = 0, 1, \dots$. *If* $(\lambda, \mu) \in \mathcal{F} \times \mathcal{F}$, $\tau = \mathrm{Im}\,(\lambda - \mu) \geq 0$, *then for any* $s \geq 0$, *the inequalities*

$$\|\mathfrak{B}(\lambda, \mu); H^s(S^{n-1}) \to H^s(S^{n-1})\| \leq C(1 + |\lambda - \mu|)^{s+h}(1 + |\lambda|)^s (1 + |\mu|)^{-\tau} \|\Phi\|_q, \tag{5.3.15}$$

$$\|\mathfrak{B}(\lambda, \mu); H^s(S^{n-1}) \to H^{s+\tau}(S^{n-1})\| \leq C(1 + |\lambda - \mu|)^{s+h}(1 + |\lambda|)^s\, \|\Phi\|_q \tag{5.3.16}$$

hold, where $\|\Phi\|_q = \|\Phi\,; C^q(S^{n-1} \times S^{n-1})\|$ *and* C, q *depend only on* $\mathcal{F}$ *and* s.

Proof From the inequality

$$(1 + |\xi|^2 + |\mu|^2)^t \leq 2^{|t|}(1 + |\lambda - \mu|^2)^{|t|}(1 + |\xi|^2 + |\lambda|^2)^t,$$

due to estimates (1.2.11) and (1.2.12), it follows that

$$\|E(\mu)^{-1} E(\lambda)\,; H^s(\lambda, S^{n-1}) \to H^{s+\tau}(\mu, S^{n-1})\| \leq C_1(\mathcal{F})(1 + |\lambda - \mu|^2)^{|s+\mathrm{Im}\,\lambda|/2} \leq$$
$$\leq C_1(\mathcal{F})(1 + |\lambda - \mu|)^{s+h}$$

for $(\lambda, \mu) \in \mathcal{F} \times \mathcal{F}$. Since $\mathfrak{B}(\lambda, \mu) = E(\mu)^{-1} E(\lambda) \mathfrak{B}(\lambda, \lambda)$ and

$$\|\mathfrak{B}(\lambda, \lambda) w\,; H^s(\lambda, S^{n-1})\| \leq C_2 \|\Phi\|_q\, \|w\,; H^s(\lambda, S^{n-1})\|$$

for some $q = q(s)$), we have

$$\|\mathfrak{B}(\lambda, \mu) w\,; H^{s+\tau}(\mu, S^{n-1})\| \leq C(\mathcal{F})(1 + |\lambda - \mu|)^{s+h}\, \|\Phi\|_q\, \|w\,; H^s(\lambda, S^{n-1})\| \leq$$
$$\leq C(\mathcal{F})(1 + |\lambda - \mu|)^{s+h}(1+|\lambda|)^s\, \|\Phi\|_q\, \|w\,; H^s(S^{n-1})\|.$$

The statement of the lemma now follows from the obvious estimates

$$\|v\,;\,H^{s+\tau}(S^{n-1})\| \le \|v\,;\,H^{s+\tau}(\mu, S^{n-1})\|,$$
$$\|v\,;\,H^{s}(S^{n-1})\| \le (1+|\mu|^2)^{-\tau/2}\|v\,;\,H^{s+\tau}(\mu, S^{n-1})\|$$

applied to $v = \mathfrak{B}(\lambda, \mu)w$. □

Lemma 5.3.3 *Let $\tilde{a}^1_\delta$ and $\tilde{a}^1$ be the same as in (5.3.11) and let $\varepsilon \in (0, 1/2)$. Then, for all non-negative integers N and q, the inequality*

$$\sup\{|\nu|^N \|M(\tilde{a}^1_\delta - \tilde{a}^1)(\cdot, \nu, \cdot); C^q(S^{n-1} \times S^{n-1})\|; \varepsilon < \operatorname{Im}\nu < 1-\varepsilon\} \le c\delta^\varepsilon$$

holds with a positive constant $c = c(N, q, \varepsilon)$. Moreover,

$$\sup\{|\nu|^N \|M\tilde{a}^1(\cdot, \nu, \cdot); C^q(S^{n-1} \times S^{n-1})\|; \varepsilon < \operatorname{Im}\nu < 1-\varepsilon\} < \infty.$$

Proof From definitions, it directly follows that

$$\partial^k_\rho \tilde{a}^1(\varphi, \rho\omega) = O((1+\rho)^{-k-1}) \tag{5.3.17}$$

for k=0, 1, ... Moreover, $\partial^k_\rho(\zeta(\delta\rho)\zeta(\delta/\rho)) = O(\rho^{-k})$ uniformly with respect to $\delta \in (0, 1)$. We have

$$M(\tilde{a}^1_\delta - \tilde{a}^1)(\varphi, \nu, \omega)$$
$$= (2\pi)^{-1/2}(i\nu\ldots(i\nu - N+1))^{-1}\int_0^\infty \rho^{-i\nu+N-1}\partial^N_\rho(\tilde{a}^1_\delta - \tilde{a}^1)(\varphi, \rho\omega)\,d\rho.$$

The integrand equals zero as $\rho \in (\delta, \delta^{-1})$ and is estimated by $\rho^{\operatorname{Im}\nu-1}(1+\rho)^{-1}$. Therefore,

$$|M(\tilde{a}^1_\delta - \tilde{a}^1)(\varphi, \nu, \omega)| \le c(N)\delta^\varepsilon|\nu|^{-N}, \quad \operatorname{Im}\nu \in (\varepsilon, 1-\varepsilon).$$

The derivatives $\partial^\alpha_\varphi\partial^\beta_\omega M(\tilde{a}^1_\delta(\varphi, \nu, \omega) - \tilde{a}^1(\varphi, \nu, \omega)$ also admit similar estimates, since for all α, β the function $\partial^\alpha_\varphi\partial^\beta_\omega\tilde{a}^1(\varphi, \rho\omega)$ satisfies condition (5.3.17). This implies the first inequality of the lemma. The second inequality follows from the fact that the corresponding estimate is obvious for the function $\tilde{a}^1_\delta$. □

Lemma 5.3.4 *For any $\tau \in (0, \min(1, n/2))$, $s \ge 0$, and $N = 0, 1, \ldots$, the following estimate is valid:*

$$\|E_{\omega\to\varphi}(\mu)^{-1}M(\tilde{a}^1_\delta - \tilde{a}^1)(\varphi, \lambda-\mu, \omega)E_{\psi\to\omega}(\lambda)U(\lambda, \psi); H^{s+\tau}(S^{n-1}) \to L_2(S^{n-1})\|$$
$$\le C(N, \tau)\delta^\varepsilon(1+|\lambda-\mu|)^{-N}(1+|\lambda|)^{-N},$$

where $U(\lambda, \psi) = M_{|y| \to \lambda + in/2}\, u(y)$, $u \in C_c^\infty(\mathbb{R}^n \setminus 0)$, ε *is a positive number,* λ *is real,* $\operatorname{Im}\mu = -\tau$.

Proof It is sufficient to compare (5.3.16) and the first inequality in Lemma 5.3.3 and to consider that the function $U(\lambda, \psi)$ decreases rapidly as $\lambda \to \infty$. □

5.3.2 Coincidence of Algebras $\mathcal{L}(\theta, \mathbb{R}^n)$ and $\mathcal{L}_0(\theta)$

We return to Proposition 5.3.1. Formula (5.3.14) can be rewritten in the form $A = M^{-1}\mathfrak{A}M$, where M is the Mellin transform. Denote by $\mathfrak{S}$ the algebra generated by operator-valued functions $\mathbb{R} \ni \lambda \mapsto \mathfrak{A}(\lambda) = \mathfrak{A}^{(0)}(\lambda) + \mathfrak{A}^{(1)}(\lambda)$ with the norm

$$\|\mathfrak{A}; \mathfrak{S}\| = \sup\{\|\mathfrak{A}(\lambda); \mathcal{B}L_2(S^{n-1})\|,\ \lambda \in \mathbb{R}\}.$$

Since the Mellin transform is a unitary operator from $L_2(\mathbb{R}^n)$ onto $L_2(\mathbb{R} \times S^{n-1})$, the mapping $A \mapsto \mathfrak{A}$ defined on the set of operators of the form (5.3.1) extends to an isomorphism of the algebra $\mathcal{L}(0, \mathbb{R}^n)$ onto the algebra $\mathfrak{S}$. Remind that $\mathfrak{S}_0$ means the algebra spanned by operator-valued functions of the form

$$\mathbb{R} \ni \lambda \mapsto E_{\omega \to \varphi}(\lambda)^{-1} a(\varphi, \omega) E_{\psi \to \omega}(\lambda),$$

where $a \in C^\infty(S^{n-1} \times S^{n-1})$. (In Sect. 2.2.1, we used the notation $\mathfrak{S}$ instead of $\mathfrak{S}_0$.)

Proposition 5.3.5 *The algebras* $\mathfrak{S}$ *and* $\mathfrak{S}_0$ *coincide.*

Proof Let $\mathfrak{A} = \mathfrak{A}^{(0)} + \mathfrak{A}^{(1)}$ be an arbitrary generator of $\mathfrak{S}$. According to Lemma 5.3.2 and to the second inequality in Lemma 5.3.3, the norm of

$$\mathfrak{B}(\lambda, \mu) = E(\mu)^{-1}(M\tilde{a}^1)(\varphi\,, \lambda - \mu\,, \omega)E(\lambda)\,, \quad \operatorname{Im}\lambda = 0\,,\ \ \operatorname{Im}\mu = -\tau\,,$$

for any $N \in \mathbb{N}$ is subject to the estimates

$$\|\mathfrak{B}(\lambda, \mu)\,; L^2(S^{n-1}) \to H^\tau(S^{n-1})\| \le C_N(1 + |\lambda - \mu|)^{-N}\,,$$

$$\|\mathfrak{B}(\lambda, \mu)\,; L^2(S^{n-1}) \to L^2(S^{n-1})\| \le C'_N(1 + |\lambda - \mu|)^{-N}(1 + |\lambda|)^{-\tau}\,.$$

This implies that the function

$$\mathbb{R} \ni \lambda \mapsto \mathfrak{A}^{(1)}(\lambda) = \frac{1}{\sqrt{2\pi}} \int_{\operatorname{Im}\mu = -\tau} \mathfrak{B}(\lambda, \mu)\, d\mu$$

belongs to the algebra $C_0(\mathbb{R}) \otimes \mathcal{K}L_2(S^{n-1})$. This algebra is contained in $\mathfrak{S}_0$; therefore, $\mathfrak{A}^{(1)} \in \mathfrak{S}$. Moreover, $\mathfrak{A}^{(0)} \in \mathfrak{S}_0$. So, $\mathfrak{A} = \mathfrak{A}^{(0)} + \mathfrak{A}^{(1)} \in \mathfrak{S}_0$ and hence $\mathfrak{S} \subset \mathfrak{S}_0$.

We now show that $\mathfrak{S}_0 \subset \mathfrak{S}$. Denote by

$$\mathfrak{A}^{(0)}(\cdot) = E_{\omega\to\varphi}(\cdot)^{-1} a(\varphi, \omega) E_{\psi\to\omega}(\cdot)$$

an arbitrary generator of $\mathfrak{S}_0$, and let $\zeta \in C^\infty(\mathbb{R})$, $\zeta(t) = 1$ for $t < 1/2$ and $\zeta(t) = 0$ for $t > 1$. For $\delta \in (0, 1)$, we put $a_\delta(x, \xi) = (1 - \zeta(|x||\xi|/\delta))a(\varphi, \omega)$ and consider the Ψdo $\operatorname{Op} a_\delta \in \mathcal{L}(0, \mathbb{R}^n)$. Under the isomorphism $\mathcal{L}(0, R^n) \cong \mathfrak{S}$, to this Ψdo there corresponds an operator-valued function $\mathfrak{A}_\delta = \mathfrak{A}^{(0)} - \mathfrak{A}_\delta^{(1)}$, where for $\lambda \in \mathbb{R}$, the operator $\mathfrak{A}_\delta^{(1)}(\lambda) : C^\infty(S^{n-1}) \to C^\infty(S^{n-1})$ is defined by

$$\mathfrak{A}_\delta^{(1)}(\lambda)w = \frac{1}{\sqrt{2\pi}} \int_{\operatorname{Im}\mu=-\tau} \mathfrak{B}_\delta(\lambda, \mu) w \, d\mu, \tag{5.3.18}$$

while

$$\mathfrak{B}_\delta(\lambda, \mu) = E(\mu)^{-1}(M\tilde{a}_\delta^1)(\varphi, \lambda - \mu, \omega)E(\lambda),$$
$$\tilde{a}_\delta^1(\varphi, \xi) = \zeta(|\xi|/\delta)a(\varphi, \omega).$$

The inclusion $\mathfrak{S}_0 \subset \mathfrak{S}$ will be proven if we show that

$$\lim_{\delta\to 0} \sup\{\|\mathfrak{A}_\delta^{(1)}(\lambda); \mathcal{B}L_2(S^{n-1})\|; \lambda \in \mathbb{R}\} = 0.$$

The formula

$$\partial_\varphi^\alpha \partial_\omega^\beta (M\tilde{a}_\delta^1)(\varphi, \nu, \omega) =$$
$$= (2\pi)^{-1/2}[i\nu \dots (i\nu - k + 1)]^{-1} \int_0^\infty \rho^{-i\nu+k-1} \partial_\rho^k \partial_\varphi^\alpha \partial_\omega^\beta \tilde{a}_\delta^1(\varphi, \rho\omega) \, d\rho,$$

where $\operatorname{Im}\nu = -\tau$, implies, due to the equality $\tilde{a}_\delta^1(\varphi, \rho\omega) = 0$ for $\rho > \delta$, that

$$|\partial_\varphi^\alpha \partial_\omega^\beta (M\tilde{a}_\delta^1)(\varphi, \lambda, \omega)| \le C_k|\lambda - \mu|^{-k}\delta^\tau, \quad k \in \mathbb{Z}_+. \tag{5.3.19}$$

Furthermore, according to Lemma 5.3.2, we have

$$\|\mathfrak{B}_\delta(\lambda, \mu); \mathcal{B}L_2(S^{n-1})\| \le C(1+|\lambda-\mu|)^\tau (1+|\mu|)^{-\tau} \|M\tilde{a}_\delta^1(\cdot, \lambda-\mu, \cdot); C^q(S^{n-1} \times S^{n-1})\|,$$

where q is a sufficiently large number. Comparing this formula with (5.3.19), we obtain

$$\|\mathfrak{B}_\delta(\lambda,\mu)\,;\mathcal{B}L_2(S^{n-1})\| \le C_k'(1+|\lambda-\mu|)^{-k}\delta^\tau\,, \quad k\in\mathbb{Z}_+\,.$$

From here and (5.3.18), it follows that $\|\mathfrak{A}_\delta^{(1)}(\lambda);\,\mathcal{B}L_2(S^{n-1})\| \le C\delta^\tau$. □

Proposition 5.3.6 *The algebra $\mathcal{L}(\theta,\mathbb{R}^n)$ coincides with the algebra $\mathcal{L}_0(\theta)$ for $\theta=0$ and for all $\theta\in S^{m-n-1}$.*

Proof Since $\mathcal{L}(0,\mathbb{R}^n)$ and $\mathcal{L}_0(0)$ are isomorphic to $\mathfrak{S}$ and $\mathfrak{S}_0$, respectively, Proposition 5.3.5 implies the coincidence of the algebras $\mathcal{L}(0,\mathbb{R}^n)$ and $\mathcal{L}_0(0)$. Suppose now that $\theta\in S^{m-n-1}$. We restrict ourselves to the proof of the inclusion $\mathcal{L}_0(\theta)\subset\mathcal{L}(\theta,\mathbb{R}^n)$; in essence, only this inclusion is used below (for proof of the inverse inclusion see [38]). Let $\operatorname{Op} a$ be a Ψdo in $\mathbb{R}^m_n$ with a homogeneous symbol a,

$$(\operatorname{Op} a)v(x) = (2\pi)^{-m/2}\int e^{ix\xi}a(\varphi,\xi)\hat{v}(\xi)\,d\xi,$$

$a(\varphi,t\xi)=a(\varphi,\xi)$ as $t>0$, $\varphi=x^1/|x^1|$, and let $\zeta\in C^\infty(\mathbb{R}_+)$, $\zeta(t)=1$ for $t<1/2$, $\zeta(t)=0$ for $t>1$. We put $\tilde{a}_\delta(\varphi,\xi)=\zeta(|\xi^1|/\delta)\zeta(|\xi^2|/\delta)a(\varphi,\omega)$, $\delta>0$ and introduce the operator

$$(\operatorname{Op}(a-a_\delta))v(x) = (2\pi)^{-m/2}\int e^{ix\xi}(a(\varphi,\xi)-\tilde{a}_\delta(\varphi,|x^1|\xi))\hat{v}(\xi)\,d\xi.$$

To the operator $\operatorname{Op} a_\delta$ there corresponds the function $S^{m-n-1}\ni\theta\mapsto(\operatorname{Op} a_\delta)(\theta)\in\mathcal{B}L_2(\mathbb{R}^n)$, where

$$(\operatorname{Op} a_\delta)(\theta)u(x^1) = (2\pi)^{-n/2}\int e^{ix^1\xi^1}\tilde{a}_\delta(\varphi,|x^1|\xi^1,|x^1|\theta)\hat{u}(\xi^1)\,d\xi$$

$$= \zeta(|x^1|/\delta)(2\pi)^{-n/2}\int e^{ix^1\xi^1}\zeta(|x^1||\xi^1|/\delta)a(\varphi,\xi^1,\theta)\hat{u}(\xi^1)\,d\xi^1.$$

The following equality is valid:

$$\|\operatorname{Op} a_\delta\,;\mathcal{B}L^2(\mathbb{R}^m)\| = \sup\{\|(\operatorname{Op} a_\delta)(\theta)\,;\mathcal{B}L^2(\mathbb{R}^n)\|\,;\ \theta\in S^{m-n-1}\}. \tag{5.3.20}$$

Since $\operatorname{Op}(a-a_\delta)\in\Psi^0(\mathbb{R}^m_n)$, to prove the inclusion $\mathcal{L}_0(\theta)\subset\mathcal{L}(\theta,\mathbb{R}^n)$, it suffices to prove that $\|\operatorname{Op} a_\delta;\,\mathcal{B}L_2(\mathbb{R}^m)\|\to 0$ as $\delta\to 0$.

Let us fix $\theta\in S^{m-n-1}$ and let $\tilde{b}_\delta^{(1)}(x^1,\xi^1)=\zeta(|\xi^1|/\delta)(a(\varphi,\xi^1,|x^1|\theta)-a(\varphi,\xi^1,0))$, $\tilde{b}_\delta^{(2)}=\zeta(|\xi^1|/\delta)a(\varphi,\xi^1,0)$. Then,

$$(\operatorname{Op} a_\delta)(\theta)u(x^1) = \zeta(|x^1|/\delta)(\operatorname{Op} b_\delta^{(1)})(\theta)u(x^1)+\zeta(|x^1|/\delta)(\operatorname{Op} b_\delta^{(2)})u(x^1) \tag{5.3.21}$$

for $u \in C_c^\infty(\mathbb{R}^n \setminus 0)$. From the proof of Proposition 5.3.5, it follows that $\|\operatorname{Op} b_\delta^{(2)}; \mathcal{B}L_2(\mathbb{R}^n)\| \to 0$ as $\delta \to 0$. Furthermore, the function $\tilde{b}_\delta^{(1)}$ satisfies the estimate

$$|\partial_\rho^k \tilde{b}_\delta^{(1)}(r\varphi, \rho\omega)| \le C_k \rho^{-k} r(r^2+\rho^2)^{-1/2}, \quad k \in \mathbb{Z}_+, \tag{5.3.22}$$

where $\rho = |\xi^1|$ and $\omega = \xi^1/|\xi^1|$. Integrating by parts in

$$(M\tilde{b}_\delta^{(1)})(r\varphi, \nu, \omega) = \frac{1}{\sqrt{2\pi}} \int_0^\infty \rho^{-i\nu-1} \tilde{b}_\delta^{(1)}(r\varphi, \rho\omega)\, d\rho, \quad \operatorname{Im}\nu > 0,$$

and assuming that $\tau = \operatorname{Im}\nu$, we deduce from (5.3.22) that

$$|(M\tilde{b}_\delta^{(1)})(r\varphi, \nu, \omega)| \le C_k' |\nu|^{-k} \int_0^\infty \rho^{\tau-1} r(r^2+\rho^2)^{-1/2}\, d\rho =$$

$$= C_k' |\nu|^{-k} r^\tau \int_0^\infty \rho^{\tau-1}(1+\rho^2)^{-1/2}\, d\rho = C_k'' |\nu|^{-k} r^\tau.$$

The derivatives $\partial_\varphi^\alpha \partial_\omega^\beta (M\tilde{b}_\delta^{(1)})(r\varphi, \nu, \omega)$ obey the same inequalities. Due to Proposition 5.3.1 we have

$$(\operatorname{Op} b_\delta^{(1)})u(r\varphi) = \frac{1}{\sqrt{2\pi}} \int_{\operatorname{Im}\lambda=0} r^{i\lambda-n/2}\, d\lambda \tag{5.3.23}$$

$$\times \int_{\operatorname{Im}\mu=-\tau} E(\mu)^{-1}(Mb_\delta^{(1)})(r\varphi, \lambda-\mu, \omega) E(\lambda) U(\lambda, \psi)\, d\mu.$$

The (positive) number τ in this inequality satisfies $\tau < n/2$. If $n \ge 2$, then τ can be taken in $(1/2, 1)$. Then (see proof of Proposition 5.3.5),

$$\|\zeta(|\cdot|/\delta)(\operatorname{Op} b_\delta^{(1)})u\|^2 \le C \int_0^\delta r^{n-1}\, dr \Big(\int_{\operatorname{Im}\lambda=0} r^{\tau-n/2}(1+|\lambda|)^{-\tau}$$

$$\times \|U(\lambda, \cdot)\|\, d\lambda \int_{\operatorname{Im}\mu=-\tau} |\lambda-\mu|^{-k}\, d\mu \Big)^2$$

$$\le C' \int_0^\delta r^{2\tau-1} r\, dr \int_{\operatorname{Im}\lambda=0} (1+|\lambda|)^{-2\tau}\, d\lambda \int_{\operatorname{Im}\lambda=0} \|U(\lambda, \cdot)\|^2\, d\lambda = C'\delta^{2\tau}\|u\|^2.$$

In view of (5.3.20) of (5.3.21), it follows from here that $\lim_{\delta\to 0} \|\operatorname{Op} a_\delta; \mathcal{B}L_2(\mathbb{R}^m)\| = 0$. Thus, the inclusion $\mathcal{L}_0(\theta) \subset \mathcal{L}(\theta, \mathbb{R}^n)$ is proven for $n \ge 2$. For $n = 1$, the operator-valued function $\mu \mapsto E(\mu)^{-1}$ has a pole at $\mu = -i/2$. Therefore, for $\tau \in (1/2, 1)$, the right-hand side of (5.3.23) contains the additional summand

$$-(2\pi)^{-3/2} \int_{\operatorname{Im}\lambda=0} r^{i\lambda-1/2}\Big(\int_{S^0} (M\tilde{b}_\delta^{(1)})(r\varphi, \lambda + i/2, \omega)\, d\omega\Big) E(\lambda) U(\lambda, \psi)\, d\lambda,$$

which is estimated as before. □

5.4 Localization in $\mathcal{L}(\theta, K)$

Remind that the algebra $\mathcal{L}(\theta, K)$ is generated by operators $A_0(\theta)$ in $L_2(K)$ (cf. formulas (5.2.4) and (5.2.5)). To simplify notations, we write generators $A(\theta)$ of $\mathcal{L}(\theta, K)$ in local coordinates in the form

$$A_{(\upsilon)}(\theta)u(x) = (2\pi)^{-n/2} \int e^{ix\xi} (\tilde{a}_\upsilon)_0(\varphi, |x|\xi, |x|\theta)\hat{u}(\xi)\, d\xi; \tag{5.4.1}$$

here, we write x and ξ instead of x^1 and ξ^1.

Proposition 5.4.1 *The algebra $\mathcal{L}(\theta, K)$ is irreducible, whereas $\mathcal{K}L_2(K) \subset \mathcal{L}(\theta, K)$.*

Proof Let D be the (smooth) base of the cone K, $D = K \cap S^{N-1}$, and let $\{V_j\}$ be an atlas on K, i.e., $V_j = V_j' \times \mathbb{R}_+$, where $\{V_j'\}$ is an atlas on D. Introduce a partition of unity $\{\eta_j\}$ on K subject to the covering $\{V_j\}$ and consisting of homogeneous functions of degree 0. According to Theorem 4.2.13, every element $A(\theta) \in \mathcal{L}(\theta, K)$ can be written in the form

$$A(\theta) = \sum v_{ij}^* A_{ij}(\theta)(v_{ij}^{-1})^* \tag{5.4.2}$$

with $A_{ij}(\theta) = (v_{ij}^{-1})^*(\eta_i A(\theta)\eta_j)v_{ij}^*$, while $A_{ij}(\theta) \in \mathcal{L}(\theta, \mathbb{R}^n)$. The converse is also true: if the last inclusion holds for all $A_{ij}(\theta)$, then the operator (5.4.2) belongs to $\mathcal{L}(\theta, K)$.

Owing to Propositions 2.3.5, 2.3.9, and 5.3.6, the algebra $\mathcal{L}(\theta, \mathbb{R}^n)$ is irreducible and contains the ideal $\mathcal{K}L_2(\mathbb{R}^n)$. Therefore, any operator of the form $(v_{ij}^{-1})^*(\eta_i Q \eta_j)v_{ij}^*$ for $Q \in \mathcal{K}L_2(K)$ belongs to the algebra $\mathcal{L}(\theta, \mathbb{R}^n)$. Hence, $Q \in \mathcal{L}(\theta, K)$ for all $Q \in \mathcal{K}L_2(K)$. □

Denote by $\bar{K}$ the compactification of the cone K obtained by adding the cone vertex and the "infinitely remote copy" D_∞ of the base D (i. e. the set of the "ends" of the cone generatrices). Introduce the algebra $C(\bar{K})$ of continuous functions on this compact. We apply Proposition 1.3.26 for localization in the algebra $\mathcal{L}(\theta, K)$ with $C(\bar{K})$. According to Proposition 5.4.1, $\mathcal{K}L_2(K) \subset \mathcal{L}(\theta, K)$. From Proposition 2.3.10, we obtain the inclusions

$\chi A(\theta)\zeta \in \mathcal{K}L_2(K)$ for each operator $A(\theta) \in \mathcal{L}(\theta, K)$ and for arbitrary functions $\chi, \zeta \in C(\bar{K})$ satisfying $\chi\zeta = 0$. Thus, the requirements of Proposition 1.3.26 are fulfilled.

For any $z \in \bar{K}$, we define some algebra $l(\theta)_z$; below, we will show that $l(\theta)_z$ is isomorphic to the local algebra $\mathcal{L}(\theta, K)_z$. Below, $A(\theta)$ denotes operators obtained from operators of class $\Psi_0^0(\mathcal{M})$. Such operators can be taken as generators of the algebra $\mathcal{L}(\theta, K)$; they admit a representation of the form (5.4.1).

(i) Points of the set K. For any $z \in K$, let us introduce the algebra $l(\theta)_z := C(S^*(K)_z)$ of continuous functions on a fiber of the bundle $S^*(K)$ of cotangent spheres. Assume that φ is the direction from vertex 0 to point z. (Thus, for $z \in K$, the algebras $l_z(\theta)$ are independent of θ. However, it is convenient to keep this parameter to the unity of the notations.) According to Definition 5.1.1, there exists a limit

$$(a^0)_\upsilon(\varphi, \xi, \theta) = \lim_{t\to\infty} (\tilde{a}_\upsilon)_0(\varphi, |z|t\xi, |z|t\theta), \tag{5.4.3}$$

which is a zero degree homogeneous function of the variables (ξ, θ). Putting $\theta = 0$, we obtain the function $\xi \mapsto a^0(\varphi, \xi, 0)$ on the fiber $S^*(K)_z$ and introduce the mapping

$$p_z(\theta) : A(\theta) \mapsto \sigma^0(\varphi, \cdot, 0) \in l(\theta)_z. \tag{5.4.4}$$

(ii) Points of the set D_∞. Let φ be the direction from vertex 0 to $z \in D_\infty$. For $\theta \neq 0$, the function $\xi \mapsto a^0(\varphi, \xi, \theta)$ appearing in (5.4.3) is well-defined and continuous on the fiber $T^*(\bar{K} \setminus 0)_z$ of the cotangent bundle; remind that the coordinate transforms on K are homogeneous of degree 1. This function extends to be continuous on the compact set $\bar{T}^*(\bar{K} \setminus 0)_z$ obtained by gluing to the fiber the infinitely remote sphere S^{n-1}. We put $l(\theta)_z := C(\bar{T}^*(\bar{K} \setminus 0)_z)$ and introduce the mapping

$$p_z(\theta) : A(\theta) \mapsto a^0(\varphi, \cdot, \theta) \in l(\theta)_z. \tag{5.4.5}$$

(iii) The vertex of K. To the point $z = 0$, we relate the algebra $l(\theta)_0 := \mathcal{L}(0, K)$. This algebra is generated by operators $A(0)$ admitting in local coordinates the representation of the form

$$A(0)_{(\upsilon)}u(x) = (2\pi)^{-n/2} \int e^{ix\xi} (\tilde{a}_\upsilon)_0(\varphi, |x|\xi, 0)\hat{u}(\xi)\, d\xi \tag{5.4.6}$$

(compare with (5.4.1)). By $p_0(\theta)$, we denote the mapping

$$p_0(\theta) : A(\theta) \mapsto A(0) \in l(\theta)_0. \tag{5.4.7}$$

Therefore, if points $z \in K$ are placed on the same generatrix of K, then the mappings $p_z(\theta)$ do not depend neither on z nor on θ; dependence on θ is present in $l_z(\theta)$ for $z \in D_\infty$. For z in $\bar{K} \setminus 0$, the algebras $l_z(\theta)$ are commutative. Actually, the algebra $l_0(\theta)$ does not depend on θ and is commutative.

To prove the following Proposition 5.4.2, one can use the argument from the proof of Proposition 2.3.11 with obvious changes.

Proposition 5.4.2 *Let $\mathcal{L}(\theta, K)_z$ for $z \in \bar{K}$ be a local algebra obtained by localization in $\mathcal{L}(\theta, K)$ with the algebra $C(\bar{K})$. Then, the localizing mapping $p_z(\theta)$ defined by (5.4.4), (5.4.5), and (5.4.7) extends to an isomorphism $\mathcal{L}(\theta, K)_z \simeq l(\theta)_z$.*

5.5 Localization in $\mathcal{L}(0, K)$

Assume, as in Sect. 5.2, that $\mathcal{A}$ stands for the C^*-algebra generated in $L_2(\mathcal{M})$ by all proper Ψdo in the class $\Psi_0^0(\mathcal{M})$ on a w-manifold $\mathcal{M}$. In Propositions 5.2.2 and 5.2.3, local algebras that arise by the localization in $\mathcal{A}$ are listed. This list and Proposition 5.2.4 show that the study of the local algebras actually reduces to the study of the algebras $\mathcal{L}(\theta, K)$ (cf. notations at the beginning of Sect. 5.3), $\theta \in S^{m-n-1}$. After localization in $\mathcal{L}(\theta, K)$ (Proposition 5.4.2), the local algebra $\mathcal{L}(0, K)$ arises; we turn to the study of this algebra. Taking the Mellin transform $M_{r \to \lambda}$ along the generatrices of the cone K as an intertwining operator, we pass from $\mathcal{L}(0, K)$ to the isomorphic algebra $\mathfrak{S}(D)$; elements of $\mathfrak{S}(D)$ are some functions $\mathbb{R} \ni \lambda \mapsto \mathfrak{A}(\lambda) \in \mathcal{B}L_2(D)$, where $D = K \cap S^{N-1}$ is the base of the cone K. We apply to $\mathfrak{S}(D)$ the localization principle formulated in Proposition 1.3.22. The role of J is played by the ideal $C_0(\mathbb{R}) \otimes \mathcal{K}L_2(D)$; here, $C_0(\mathbb{R})$ is the set of continuous functions vanishing at the infinity, and $\mathcal{K}L_2(D)$ is the ideal of compact operators in $L_2(D)$. As a commutative algebra $\mathcal{C}$, we take the algebra $C(D)$ of continuous functions on D. In this section, we describe the corresponding local algebras.

Let $A(0)$ be a Ψdo in the algebra $\mathcal{L}(0, K)$ (i.e., an operator obtained by localization from a Ψdo in $\mathcal{A}$). We have the equality $A(0) = \sum v_{ij}^* A_{ij}(0)(v_{ij}^{-1})^*$ similar to (5.4.2), whereas $A_{ij}(0) = (v_{ij}^{-1})^*(\eta_i A(0)\eta_j)v_{ij}^*$. In view of (5.3.14), for any operator $A_{ij}(0)$, there corresponds a function $\lambda \mapsto \mathfrak{A}_{ij}(\lambda)$. Denote by D_{ij} the intersection of the domain of v_{ij} with the sphere S^{N-1}. Let $v_{ij}^0 : D_{ij} \to S^{n-1}$ be the mapping generating the diffeomorphism v_{ij}. We introduce the operator

$$\mathfrak{A}_D(\lambda) = \sum (v_{ij}^0)^* \mathfrak{A}_{ij}(\lambda)((v_{ij}^0)^{-1})^* : L_2(D) \to L_2(D). \tag{5.5.1}$$

From (5.3.14), it follows that

$$A(0)u = (2\pi)^{-1/2} \int_{-\infty}^{+\infty} r^{i\lambda - n/2} \mathfrak{A}_D(\lambda) U(\lambda, \cdot)\, d\lambda. \tag{5.5.2}$$

Due to the Parseval equality for the Mellin transform, we have

$$\|A(0); \mathcal{B}L_2(K))\| = \sup\{\|\mathfrak{A}_D(\lambda); \mathcal{B}L_2(D)\|; \lambda \in \mathbb{R}\}.$$

Assume that the algebra $\mathfrak{S}(D)$ is generated by functions $\lambda \mapsto \mathfrak{A}_D(\lambda)$ defined by (5.5.1) with the norm $\|\mathfrak{A}_D; \mathfrak{S}(D)\| = \sup\{\|\mathfrak{A}_D(\lambda); \mathcal{B}L_2(D)\|; \lambda \in \mathbb{R}\}$. We arrive at the following statement.

Proposition 5.5.1 *The algebras $\mathcal{L}(0, K)$ and $\mathfrak{S}(D)$ are isomorphic.*

Let us now make sure that the localization principle in Proposition 1.3.22 is applicable to the algebra $\mathfrak{S}(D)$. If a function $\lambda \mapsto \mathfrak{A}_D(\lambda)$ is defined by (5.5.1), where $u_{ij} \in C_0(\mathbb{R}) \otimes \mathcal{K}L_2(S^{n-1})$, then $\mathfrak{A}_D \in \mathfrak{S}(D)$. This implies the inclusion $J := C_0(\mathbb{R}) \otimes \mathcal{K}L_2(D)) \subset \mathfrak{S}(D)$. The localizing algebra $C(D)$ is a subalgebra of $\mathfrak{S}(D)$; the unity of $\mathfrak{S}(D)$ belongs to $C(D)$. For $\mathfrak{A} \in \mathfrak{S}(D)$ and $a \in C(D)$, the commutator $[\mathfrak{A}, a]$ falls into J. Therefore, conditions (i) and (ii) of Proposition 1.3.22 are fulfilled. Condition (iii) is also fulfilled; this can be easily deduced from the fact that any irreducible representation of the ideal J is equivalent to a representation of the form $\pi(\lambda) : \mathfrak{A} \mapsto \mathfrak{A}(\lambda)$.

Denote by J_φ the ideal generated in the algebra $\mathfrak{S}(D)$ by functions in $C(D)$ equal to zero at $\varphi \in D$. The local algebra $\mathfrak{S}(D)/J_\varphi$ is denoted by $\mathfrak{S}(D)_\varphi$. Remind that elements $\mathfrak{A}$ of $\mathfrak{S}(D)$ are related to operators in $\mathcal{L}(0, K)$ by (5.3.14). For $A(0) \in \mathcal{L}(0, K)$, the principal symbol a^0 is defined, $a^0(\varphi, \cdot, 0) \in C(S^*(K)_\varphi)$, where as usual, $C(S^*(K)_\varphi)$ is the fiber at $\varphi \in D \subset K$ of the cospherical bundle $C(S^*(K))$. The same function a^0 is also called the principal symbol of the element $\mathfrak{A} \in \mathfrak{S}$ connected with $A(0)$ by (5.3.14). The next assertion follows, in fact, from Proposition 2.2.10.

Proposition 5.5.2 *Let $\mathfrak{A} \in \mathfrak{S}(D)$ and a^0 be the principal symbol of $\mathfrak{A}$. The mapping*

$$\mathfrak{S}(D) \ni \mathfrak{A} \mapsto a^0(\varphi, \cdot, 0) \in C(S^*(K)_\varphi)$$

generates an isomorphism $\mathfrak{S}(D)_\varphi \cong C(S^*(K)_\varphi)$.

5.6 Invariant Description of Local Algebras

Remind that the localization procedure used in Sects. 5.2, 5.4, and 5.5 consisted of several stages. First, the localization was applied to the algebra $\mathcal{A}$ generated in $L_2(\mathcal{M})$ by operators in $\Psi_0^0(\mathcal{M})$. Then, two more stages of localization were carried out in local algebras that appeared in the previous step. As a result, local algebras of three types arise: algebras of continuous (scalar) functions on cotangent spheres and non-commutative algebras $\mathcal{L}(K, \theta)$ and $\mathfrak{S}(K)$. The description of $\mathcal{L}(K, \theta)$ and $\mathfrak{S}(K)$ depended on the choice of local coordinates. We now demonstrate how to make definitions of these algebras "invariant" under the change of variables.

Let T be an n-dimensional edge of the manifold $\mathcal{M}$, $0 \le n \le m-1$, where $m = \dim \mathcal{M}$. Assume also that $\{O_\alpha\}$ is a finite collection of neighborhoods in $\mathcal{M}$ covering T and that $\kappa_\alpha : O_\alpha \to \kappa_\alpha(O_\alpha)$ is a homeomorphism onto an open subset of the product $\bar{K} \times \mathbb{R}^{m-n}$. Since, by definition, any edge T is connected, we can assume that the conical manifold K is independent of α. For neighborhoods O_α and O_β such that $O_\alpha \cap O_\beta \neq \emptyset$, the mapping $\kappa_{\alpha\beta} = \kappa_\beta \circ \kappa_\alpha^{-1} : O_\alpha \cap O_\beta \to O_\alpha \cap O_\beta$ is an admissible diffeomorphism. Let us choose $z \in T \cap O_\alpha \cap O_\beta$ and suppose for simplicity that $\kappa_\alpha(z) = \kappa_\beta(z) = 0$. Denote by $\pi^{(1)}$ and $\pi^{(2)}$ the projections $\bar{K} \times \mathbb{R}^{m-n} \to \bar{K}$ and $\bar{K} \times \mathbb{R}^{m-n} \to \mathbb{R}^{m-n}$, let $\kappa_{\alpha\beta}^{(1)} = \pi^{(1)} \circ \kappa_{\alpha\beta}$ and $\kappa_{\alpha\beta}^{(2)} = \pi^{(2)} \circ \kappa_{\alpha\beta}$, and define

$$k_{\alpha\beta}^{(1)}(x^1) = \lim_{t\to+0} t^{-1}\kappa_{\alpha\beta}^{(1)}(tx^1, 0),\ \ x^1 \in K,$$

$$k_{\alpha\beta}^{(2)}(x^2) = \lim_{t\to+0} t^{-1}\kappa_{\alpha\beta}^{(2)}(0, tx^2),\ \ x^2 \in \mathbb{R}^{m-n}.$$

Then, $k_{\alpha\beta}^{(1)}$ is an admissible diffeomorphism $\bar{K} \to \bar{K}$ homogeneous of degree 1, and $k_{\alpha\beta}^{(2)}$ is a linear isomorphism.

Introduce a bundle $\mathcal{E} = (E, T, \bar{K})$ with base T and fiber $\bar{K}$ using the mappings $(Id_{\alpha\beta}, k_{\alpha\beta}^{(1)}) : (O_\alpha \cap O_\beta \cap T) \times \bar{K} \to (O_\alpha \cap O_\beta \cap T) \times \bar{K}$ as transition functions. Let

$$h_\alpha : (O_\alpha \cap T) \times \bar{K}) \to E|(O_\alpha \cap T) \tag{5.6.1}$$

be the natural trivialization, and χ_α such a partition of unity on T that $\operatorname{supp}\chi_\alpha \subset O_\alpha \cap T$. For points on the same generatrix of the cone, addition and multiplication by non-negative numbers are defined. Therefore, the following mapping makes sense:

$$\tau : T \times \bar{K} \to E : (z, x) \mapsto \sum_i \chi_\alpha(z) h_\alpha(z, x).$$

Since τ is an isomorphism of bundles, we conclude that the bundle $\mathcal{E}$ is trivial. We endow the fiber $E(z)$ with the measure $\mu(z)$ transferred from $\bar{K}$ using the mapping $\tau(z) = \tau|\{z\} \times \bar{K}$; the cone $\bar{K}$ is assumed to be endowed with the Euclidean measure.

Remind that under localization at $z^0 \in T$, we must pass from an operator $A \in \mathcal{A}$ to an operator

$$(A_0 u)(x) = \lim_{t\to 0}(U_t(\kappa^{-1})^*(\chi A \chi)\kappa^* U_t^{-1} u)(x)$$

on the wedge $\mathcal{W} = \bar{K} \times R^{m-n}$, where χ is a cut-off function, $\chi(z^0) = 1$, and U_t is the unitary operator: $u(\cdot) \mapsto t^{m/2}u(t\cdot)$ in $L_2(\mathcal{W})$, while $t > 0$ and the point z^0 is taken as the origin.

Suppose that $z^0 \in T \cap O_\alpha \cap O_\beta$ and $\kappa_\alpha(z^0) = \kappa_\beta(z^0) = 0$. Assuming $A^{(\gamma)} = (\kappa_\gamma^{-1})^*(\chi A \chi)\kappa_\gamma^*$ for $\gamma = \alpha, \beta$, we obtain $A^{(\alpha)} = \kappa_{\alpha\beta}^* A^{(\beta)} \kappa_{\beta\alpha}^*$ for $\kappa_{\alpha\beta} = \kappa_\beta \circ \kappa_\alpha^{-1}$. Introduce the operator $A_0^{(\gamma)} = \lim_{t\to 0} U_t A^{(\gamma)} U_t^{-1}$. It is clear that

$$A_0^{(\alpha)} = \lim_{t\to 0}(U_t \kappa_{\alpha\beta}^* U_t^{-1} U_t A^{(\beta)} U^{-1} U_t \kappa_{\beta\alpha}^* U_t^{-1}).$$

Since

$$(U_t \kappa_{\alpha\beta}^* U_t^{-1} f)(x) = t^{m/2}(\kappa_{\alpha\beta}^* U_t^{-1} f)(tx) = t^{m/2}(U_t^{-1} f)(\kappa_{\alpha\beta}(tx)) = f(\kappa_{\alpha\beta}(tx)/t),$$

we have

$$A_0^{(\alpha)} = k_{\alpha\beta}^* A_0^{(\beta)} k_{\beta\alpha}^* \ \text{ for } \ k_{\alpha\beta}(x) = \lim_{t\to 0} \kappa_{\alpha\beta}(tx)/t.$$

Points $x \in \bar{K} \times \mathbb{R}^{m-n}$ will be written in the form $x = (x^1, x^2)$, where $x^1 \in \bar{K}$ and $x^2 \in \mathbb{R}^{m-n}$. We assume that $((k_{\alpha\beta}^{(1)})^* u)(x^1, x^2) = u(k_{\alpha\beta}^{(1)} x^1, x^2)$ and $((k_{\alpha\beta}^{(2)})^* u)(x^1, x^2) = u(x^1, k_{\alpha\beta}^{(2)} x^2)$. Introduce the notation $\hat{u}(\cdot, \xi^2) = F_{x^2 \to \xi^2} u(\cdot, x^2)$, where F is the Fourier transform on $\mathbb{R}^q$, and represent the operator $A_0^{(\gamma)}$, $\gamma = \alpha, \beta$, in the form

$$(A_0^{(\gamma)} u)(\cdot, x^2) = \int \exp(i x^2 \xi^2) A_0^{(\gamma)}(\xi^2) \hat{u}(\cdot, \xi^2)\, d\xi^2 \tag{5.6.2}$$

(cp. (5.2.4)). Then,

$$\begin{aligned} &(k_{\alpha\beta}^* A_0^{(\beta)} k_{\beta\alpha}^* u)(\cdot, x^2) \qquad (5.6.3)\\ &= \int \exp\{i(k_{\alpha\beta}^{(2)} x^2)\xi^2\}(k_{\alpha\beta}^{(1)})^* A_0^{(\beta)}(\xi^2)(k_{\beta\alpha}^{(1)})^*((k_{\beta\alpha}^{(2)})^* u)\hat{}(\cdot, \xi^2)\, d\xi^2 \\ &= \int \exp(i x^2 \eta^2)(k_{\alpha\beta}^{(1)})^* A_0^{(\beta)}(({}^t k_{\alpha\beta}^{(2)})^{-1}\eta^2)(k_{\beta\alpha}^{(1)})^* \hat{u}(\cdot, \eta^2)\, d\eta^2. \end{aligned}$$

Using formula (5.6.2) for $A_0^{(\alpha)}$ and taking into account (5.6.3), we obtain

$$A_0^{(\alpha)}(\eta^2) = (k_{\alpha\beta}^{(1)})^* A_0^{(\beta)}(({}^t k_{\alpha\beta}^{(2)})^{-1}\eta^2)(k_{\beta\alpha}^{(1)})^*. \tag{5.6.4}$$

Denote by $\mathfrak{T}(T)$ the tangent bundle over T and by $\mathfrak{T}(T)_z$ the fiber over a point $z \in T$. Let $\mathfrak{t}_\alpha : \mathfrak{T}(T)_{z^0} \to \mathbb{R}^q$ be the derivative of the mapping $\kappa_\alpha|T$ at z^0, so that $k_{\alpha\beta}^{(2)} = \mathfrak{t}_\beta \mathfrak{t}_\alpha^{-1}$. Put

$$A_0(z^0; \zeta) = (h_\alpha^{-1})^* A_0^{(\alpha)}({}^t\mathfrak{t}_\alpha^{-1}\zeta) h_\alpha^*$$

for $\zeta \in \mathfrak{T}^*(T)_{z^0}$. In view of (5.6.4), the operator

$$A_0(z^0;\zeta) : L_2(E(z^0);\mu(z^0)) \to L_2(E(z^0);\mu(z^0)) \tag{5.6.5}$$

is well-defined for $\zeta \in \mathfrak{T}^*(T)(z^0)$.

Denote by $\mathcal{A}(z^0;\zeta)$ the algebra generated by operators (5.6.5) for a fixed ζ. In the proof of Proposition 5.2.4, it was stated that if ζ_1 and ζ_2 belong to the same ray of the fiber $\mathfrak{T}^*(T)_{z^0}$, then, for the algebras $\mathcal{A}(z^0;\zeta_1)$ and $\mathcal{A}(z^0;\zeta_2)$, there exists a unitary intertwining operator. Since we make no distinction between equivalent representations of $\mathcal{A}$, to each ray ρ in the fiber $\mathfrak{T}^*(T)_{z^0}$, we have to relate only one algebra $\mathcal{A}(z^0;\zeta)$ with an arbitrarily fixed $\zeta \in \rho$. To make the description simpler, let us introduce a Riemannian metric on the edge T and relate to a ray ρ the algebra $\mathcal{A}(z^0;\theta)$, where $\theta = \rho \cap S^*(T)_{z^0}$ and $S^*(T)$ is the cospherical bundle over T. Thus, for an edge T, $0 < \dim T < \dim \mathcal{M}$, and for points $z^0 \in T$ and $\theta \in S^*(T)_{z^0}$, the local algebra $\mathcal{A}(z^0;\theta)$ is the algebra spanned by operators (5.6.5) for $\zeta = \theta$.

We turn to the algebra of operator-valued functions $\mathbb{R} \ni \lambda \mapsto \mathfrak{A}(z^0;\lambda)$. Assume that the bundle $\mathcal{E}$ is equipped with a Riemannian metric and introduce the Mellin transform

$$(Mu)(\lambda + in/2,\varphi) := \tilde{u}(\lambda + in/2,\varphi) := \frac{1}{\sqrt{2\pi}} \int_{-\infty}^{+\infty} r^{-i(\lambda+in/2)-1} u(r\varphi)\,dr,$$

for a function u on $E(z^0)$; here $r = |x|$ and $\varphi = x/|x|$ for $x \in E(z^0) \setminus 0$. Define the operator $\mathfrak{A}(z^0;\cdot) = MA_0(z^0;0)M^{-1}$, which replaces a function $\mathbb{R} \ni \lambda \mapsto \tilde{u}(\lambda + in/2,\cdot)$ by $\mathbb{R} \ni \lambda \mapsto \mathfrak{A}(z^0;\lambda)\tilde{u}(\lambda + in/2,\cdot)$. For $z^0 \in O_\alpha \cap O_\beta \neq \emptyset$, we put $\mathfrak{A}^{(\gamma)}(z^0;\cdot) = M^{(\gamma)} A_0^{(\gamma)}(z^0;0)(M^{(\gamma)})^{-1}$ for $\gamma = \alpha, \beta$, where

$$(M^{(\gamma)}u)(\lambda + in/2,\varphi_\gamma) := \tilde{u}(\lambda + in/2,\varphi_\gamma) := \frac{1}{\sqrt{2\pi}} \int_{-\infty}^{+\infty} r_\gamma^{-i(\lambda+in/2)-1} u(r_\gamma\varphi_\gamma)\,dr_\gamma,$$

$r_\gamma = |x_\gamma|$, and $\varphi_\gamma = x_\gamma/|x_\gamma|$. We relate the operators $\mathfrak{A}(\lambda)$ and $\mathfrak{A}^{(\alpha)}(\lambda)$. Using the trivialization h_α from (5.6.1), we obtain

$$\begin{aligned} \mathfrak{A} &= MA_0(0)M^{-1} = M(h_\alpha^{-1})^* A_0^{(\alpha)}(0)h_\alpha^* M^{-1} \\ &= M(h_\alpha^{-1})^*(M^{(\alpha)})^{-1} M^{(\alpha)} A_0^{(\alpha)}(0)(M^{(\alpha)})^{-1} M^{(\alpha)} h_\alpha^* M^{-1} \\ &= M(h_\alpha^{-1})^*(M^{(\alpha)})^{-1} \mathfrak{A}^{(\alpha)}(M^{(\alpha)})^{-1} M^{(\alpha)} h_\alpha^* M^{-1}. \end{aligned} \tag{5.6.6}$$

Let us calculate $(M^{(\alpha)}h_\alpha^* M^{-1}\tilde{u})(\lambda + in/2, \varphi_\alpha)$. We have

$$(h_\alpha^* M^{-1}\tilde{u})(x_\alpha) = (M^{-1}\tilde{u}(h_\alpha(x_\alpha))$$
$$= \frac{1}{\sqrt{2\pi}} \int_{-\infty}^{+\infty} |h_\alpha(x_\alpha)|^{i(\lambda+in/2)}\tilde{u}(\lambda + in/2, h_\alpha(x_\alpha)/|h_\alpha(x_\alpha)|)\, d\lambda.$$

Therefore,

$$(M^{(\alpha)}h_\alpha^* M^{-1}\tilde{u})(\lambda + in/2, \varphi_\alpha) \tag{5.6.7}$$
$$= \frac{1}{2\pi} \int_0^{+\infty} |x_\alpha|^{-i(\lambda+in/2)-1}\, d|x_\alpha| \int_{-\infty}^{+\infty} |x_\alpha|^{i(\lambda+in/2)} |h_\alpha(x_\alpha)/|x_\alpha||^{i(\lambda+in/2)}$$
$$\times \tilde{u}(\lambda + in/2, h_\alpha(x_\alpha)/|h_\alpha(x_\alpha)|)\, d\lambda$$
$$= |h_\alpha(\varphi_\alpha)|^{i(\lambda+in/2)}\tilde{u}(\lambda + in/2, h_\alpha(\varphi_\alpha)/|h_\alpha(\varphi_\alpha)|)$$

(since the mapping h_α is homogeneous of degree 1). From (5.6.7), it follows that

$$(M(h_\alpha^{-1})^*(M^{(\alpha)})^{-1}\tilde{v})(\lambda + in/2, \varphi) = |h_\alpha^{-1}(\varphi)|^{i(\lambda+in/2)}\tilde{v}(\lambda + in/2, h_\alpha^{-1}(\varphi)/|h_\alpha^{-1}(\varphi)|). \tag{5.6.8}$$

Denote by $((g_\alpha^{-1})^*\tilde{v})(\lambda+in/2, \varphi)$ the right-hand side of (5.6.8) and by $(g_\alpha^*\tilde{u})(\lambda+in/2, \varphi_\alpha)$ the corresponding expression in (5.6.7). Comparing (5.6.6)–(5.6.8), we obtain

$$\mathfrak{A}(\lambda) = (g_\alpha^{-1})^*\mathfrak{A}^{(\alpha)}(\lambda)g_\alpha^*.$$

The operator $\mathfrak{A}(z^0; \lambda)$ is well-defined for functions on the set $D(z^0) = \{x \in E(z^0) : |x| = 1\}$ and implements a continuous mapping $L_2(D(z^0); \nu(z^0)) \to L_2(D(z^0); \nu(z^0))$, where $\nu(z^0)$ is a measure induced on $D(z^0)$ by the measure $\mu(z^0)$ on the cone $E(z^0)$. By $\mathfrak{S}(D(z^0))$, we mean the algebra generated by functions of the form $\mathbb{R} \ni \lambda \mapsto \mathfrak{A}(z^0; \lambda)$ with the norm

$$\|\mathfrak{A}(z^0; \cdot); \mathfrak{S}(D(z^0))\| = \sup\{\|\mathfrak{A}(z^0; \lambda); \mathcal{B}L_2(D(z^0); \nu(z^0))\|, \lambda \in \mathbb{R}\}. \tag{5.6.9}$$

Thus, all local algebras that appeared in Sect. 5.2, 5.4, and 5.5 admit an invariant description.

5.7 The Spectrum of C^*-Algebra of Pseudodifferential Operators on Manifold with Edges

Introduce a set whose points parametrize one-dimensional representations. Denote by $\mathfrak{M}$ the space of maximal ideals of $\overline{C}(\mathcal{M}_0)$ (cf. definition of this algebra at the beginning of Sect. 5.2). The set $\mathcal{M}_0$ is homeomorphically embedded into $\mathfrak{M}$. The passage from $\mathcal{M}_0$ to $\mathfrak{M}$ can be described as gluing to $\mathcal{M}_0$ the boundary of a tubular neighborhood of each edge. The set $\mathfrak{M}$, in a natural way, is endowed with a structure of a smooth manifold with boundary. Principal symbols of operators in $\Psi_0^0(\mathcal{M})$ extend to be continuous functions on the cospherical bundle $S^*(\mathfrak{M})$.

Let T be a stratum of $\mathcal{M}$, $0 \le \dim T = n \le m-1$, $\dim \mathcal{M} = m$, and $K_T \times \mathbb{R}^{m-n}$ a local model of $\mathcal{M}$ in the vicinity of T; here, K_T is the cone with the base D_T. As in Sect. 5.6, we introduce locally trivial bundles $\mathcal{E}(T)$ and $\mathcal{D}(T)$ with the base space T. A fiber $E(x)$ of $\mathcal{E}$ over x is a cone diffeomorphic to K_T and generated by rays tangent to $\mathcal{M}$ at x and orthogonal to T. The fiber $D(x)$ is the base of the cone $E(x)$.

For an operator A in $\Psi_0^0(\mathcal{M})$, we introduce mappings

$$\pi(x,\omega) : A \mapsto a^0(x,\omega), \tag{5.7.1}$$

where $(x,\omega) \in S^*(\mathcal{M}_0)$, $\omega \in S^*(\mathcal{M}_0)_x$, and a^0 is the principal symbol of A;

$$\pi(x,\varphi,\omega) : A \mapsto a^0(x,\varphi,\omega), \tag{5.7.2}$$

where $x \in T$, $\omega \in S^*(\mathfrak{M})_{(x,\varphi)}$ (points (x,φ,ω) can be identified with $\omega \in S^*(E(x))_\varphi$, where $S^*(E(x))_\varphi$ is the fiber of $S^*(E(x))$ over $\varphi \in D(x)$);

$$\pi(x,\theta) : A \mapsto A(x;\theta) \in \mathcal{A}(x,\theta) \subset \mathcal{B}L_2(E(x);\mu(x)), \tag{5.7.3}$$

where $(x,\theta) \in S^*(T)$, $x \in T$, $\theta \in S^*(T)_x$, while it is assumed that $\dim T > 0$; for $\dim T = 0$ the mappings (5.7.3) are not defined. As $A(x;\theta)$, we take the operator from formula (5.6.5), and $\mathcal{A}(x;\theta)$ means the algebra defined after this formula,

$$\pi(x,\lambda) : A \mapsto \mathfrak{A}(x;\lambda) \in \mathcal{B}L_2(D(x);\nu(x)), \tag{5.7.4}$$

where $x \in T$, $0 \le \dim T \le m-1$, $\lambda \in \mathbb{R}$ and the function $\lambda \mapsto \mathfrak{A}(x;\lambda)$ is defined by (5.6.9). Note that for $\dim T < m-1$, the operator $\mathfrak{A}(x;\lambda)$ acts in an infinite-dimensional Hilbert space. If $\dim T = m-1$, then the cone K_T is just a finite collection of rays, and all the operators $\mathfrak{A}(x;\lambda)$, for $x \in T$ and $\lambda \in \mathbb{R}$, act in a space of finite dimension (its dimension is equal to the number of rays in the cone K_T).

Concluding the discussion in Sects. 5.2–5.6, we obtain the following result.

Theorem 5.7.1 *The mappings (5.7.1)–(5.7.4) extend to representations of $\mathcal{A}$. These representations are irreducible and pairwise non-equivalent. Any irreducible representation of $\mathcal{A}$ is equivalent either to one of the representations (5.7.1)–(5.7.4) or to the identity representation.*

One can describe the topology on the spectrum $\widehat{\mathcal{A}}$ taking as a "hint" Theorem 2.3.2 and its proof; this is left to the interested reader.

We now present a solving composition series. Denote by Θ the ideal in $\mathcal{A}$ equal to the overlap of kernels of all representations of the form (5.7.3) for all edges of positive dimensions, $\Theta = \bigcap \ker \pi(x, \theta)$. Put $\Lambda = \bigcap \ker \pi(x, \lambda)$; all representations of the form (5.7.4) for all edges participate in this intersection. As before, $\operatorname{com} \mathcal{A}$ denotes the ideal generated by commutators of elements in $\mathcal{A}$; remind that $\operatorname{com} \mathcal{A}$ is the intersection of kernels of all one-dimensional representations. Introduce the following ideals:

$$I_0 = \Theta \cap \Lambda \cap \operatorname{com} \mathcal{A}, \;\; I_1 = \Lambda \cap \operatorname{com} \mathcal{A}, \;\; I_2 = \operatorname{com} \mathcal{A}.$$

Theorem 5.7.2 *Assume that the set of edges of dimension ≥ 1 is non-empty, and there are no edges of dimension* $\dim \mathcal{M} - 1$. *Then, the composition series*

$$0 \subset I_0 \subset I_1 \subset I_2 \subset \mathcal{A}$$

is the shortest solving series; hence, the length of $\mathcal{A}$ is equal to 3. Whereas,

$$I_0 = \mathcal{K}L_2(\mathcal{M}), \quad (5.7.5)$$

$$I_1/I_0 \simeq \bigoplus_{\{T:\dim T \geq 1\}} (C(S^*(T) \otimes \mathcal{K}L_2(K_T)) \simeq (\bigoplus_{\{T:\dim T \geq 1\}} C(S^*(T))) \otimes \mathcal{K}H, \quad (5.7.6)$$

$$I_2/I_1 \simeq \bigoplus_T (C(T \times \mathbb{R}) \otimes \mathcal{K}L_2(D_T)) \simeq (\bigoplus_T C(T \times \mathbb{R})) \otimes \mathcal{K}H, \quad (5.7.7)$$

$$\mathcal{A}/I_2 \simeq C(S^*(\mathfrak{M})), \quad (5.7.8)$$

where H is an infinite-dimensional separable Hilbert space.

Let us outline the proof. To establish the first (from the left) two isomorphisms (5.7.6) and (5.7.7), one can apply the scheme of the proof of Theorem 2.3.3. The second isomorphism (5.7.6) follows from the fact that, for any T, the space $L_2(K_T)$ is infinite-dimensional and separable. If $\dim T < m - 1$, then $L_2(D_T)$ is also infinite-dimensional and separable. Hence, there holds the second isomorphism in (5.7.7). □

The restrictions on the dimensions of edges in the theorem are not caused by the merits of the case and needed only to simplify the description. If we refuse these restrictions, then the length of the algebra may change. For example, if the manifold has only zero-dimensional edges (conical points), then the length of $\mathcal{A}$ equals 2 (cp. Theorem 2.2.19).

For an $(m-1)$-dimensional edge T, the cone K_T consists of a finite number of rays and the dimension of the representations $\pi(x, \lambda)$ for $x \in T$ is equal to the number of rays in the cone. Therefore, increasing the number of $(m-1)$-dimensional edges, to which there correspond different numbers of rays, one can increase the length of the algebra.

Bibliographical Sketch

Chapter 1. Section 1.1 is a relatively short and elementary introduction to the theory of pseudodifferential operators. Any book of [15, 39, 42, 43] provides a deeper exposition of this theory.

The results of Sect. 1.2 belong to B. A. Plamenevskii; proofs of the propositions formulated in this section can be found in [21].

The notion of solvable algebra was introduced by A. S. Dynin [6]. Examples in [6] are based on papers by L. Boutet de Monvel [1], R. G. Douglas and R. Howe [5], I. Ts. Gohberg and N. Ya. Krupnik [11]. Algebras of Wiener–Hopf operators in "piecewise smooth" cones and algebras of Toeplitz operators on bounded symmetric domains were studied by A. S. Dynin, P. S. Muhly, J. N. Renault, H. Upmeier, V. N. Senichkin, etc. (irreducible representations, spectral topology, solvability); in this regard, see the survey [27]. In particular, H. Upmeier used solving series to obtain formulas for the index of Toeplitz and Wiener–Hopf operators [44, 45]. Papers by C. O. Cordes and his followers (see the monograph [2] and the references therein) are related to this subject.

The sufficient triviality condition for the fields of elementary algebras is taken from the paper [28]. The information on maximal radical series is borrowed from [13]. The localization principle in similar formulations was used by R. G. Douglas [4], I. B. Simonenko [40, 41], and A. S. Dynin [7].

Chapter 2. S. G. Mikhlin introduced the notion of symbol for a singular integral operator on $\mathbb{R}^n$ and proved that, if the symbol is nonzero everywhere, then the operator is Fredholm. Irreducible representation of the algebra generated by ΨDO with smooth symbols on $\mathbb{R}^n$ was described by I. Ts. Gohberg. These results were generalized for manifolds by R. T. Seeley. We present proofs of the mentioned results different from the original ones.

The spectrum of the algebra $\mathcal{A}$ (see the description of Chap. 2) and one of the algebra $\mathfrak{S}$ generated by meromorphic ΨDO were studied in [23] (all equivalence classes of irreducible representations are listed, and the spectral topology is described). In this book, another scheme of investigation is used. The major difference from [23] is that the localization principle from Sect. 1.3 is consistently applied. When implementing this

© The Author(s), under exclusive license to Springer Nature Switzerland AG 2023

B. Plamenevskii, O. Sarafanov, *Solvable Algebras of Pseudodifferential Operators*, Pseudo-Differential Operators 15, https://doi.org/10.1007/978-3-031-28398-7

scheme, we use some techniques from [26]. Let us note that for one-dimensional singular integral operators on a contour, some infinite-dimensional representations become two-dimensional. The operators implementing these representations turn out to be unitary equivalent to the operator symbols introduced in [11].

In this chapter, some results from [25] are presented; we give here new proofs based on the localization principle.

Chapter 3. The theorem containing the list of all (equivalence classes of) irreducible representations of the algebra $\mathcal{A}$ considered in Chap. 3 is proved in [22] (earlier this theorem was formulated in [26]). The description of the topology on the spectrum $\widehat{\mathcal{A}}$ is taken from [26], and the construction of the solving composition series is taken from [22]; the paper [28] contains an outline of such a construction, while the properties of the solving series are not verified there. Another approach to describing algebras of ΨDO with singularities using the Levi-Civita connections was outlined by A. S. Dynin in [8].

Chapter 4 is based on the paper by V. N. Senichkin [37]. In connection with the theorem on the boundedness of ΨDO in weighted spaces, we mention another approach discussed in book [21] and papers [24, 31]; in particular, these works contain another approach to definition of bounded non-proper ΨDO in weighted spaces with weight power outside the Stein interval.

Chapter 5. The results of the paper [38] are exposed; proofs are partially reconsidered due to the use of the localization principle.

ΨDO on manifolds with conical singularities were studied in [21]. In [29,30], ΨDO are defined for a wider class of "stratified" piecewise smooth manifolds than that in Chaps. 4 and 5 (informally, manifolds with intersecting edges of different dimensions). In [29], the approach exposed in Chap. 4 is generalized, and in [30] all (up to equivalence) irreducible representations of the corresponding algebra of ΨDO are found. The results of [29,30] are not included in the book.

Let us mention some studies of C^*-algebras generated by ΨDO on manifolds with boundary. One of the chapters in [21] is devoted to the study of the spectrum of algebras of ΨDO on smooth manifolds with boundary; symbols ("coefficients") allow isolated singularities. The dependence of the spectrum on singularities and on the choice of weighted function spaces is determined. In the paper by A. Yu. Kokotov [16] the spectrum of C^*-algebras generated by pseudodifferential operators in a polyhedron is described. O. V. Sarafanov [32,33] considered pseudodifferential boundary value problems for operators from Chaps. 4 and 5 on manifolds with smooth closed edges on the boundary. In [33], operators composing a boundary value problem are introduced, and in [32], the spectrum of C^*-algebras generated by boundary value problems is studied.

Other approaches to ΨDO with singularities and other references are given in the books [17, 35, 36].

This short sketch is not intended to be complete; its purpose is only to show the origin of the presented results and to point out some related works.

References

1. Boutet de Monvel, L. (1971). Boundary problems for pseudodifferential operators. *Acta Math, 126*, 11–51.
2. Cordes, H. O. (1987). *Spectral theory of linear differential operators and comparison algebras*. London Mathematical Society Lecture Note Series (Vol. 76).
3. Dixmier, J. (1977). *C*-algebras*, Amsterdam: North-Holland Publishing Company.
4. Douglas, R. G. (1972). *Banach algebra technics in operator theory*. Pure and Applied Mathematics, (Vol. 49). New York, London: Academic Press.
5. Douglas, R. G., & Howe, R. (1971). On the C*-algebras on the quarter plane. *Transactions of the American Mathematical Society, 158*, 203–217.
6. Dynin, A. S. (1978). Inversion problem for singular integral operators. *Proceedings. National Academy of Sciences. United States of America, 75*, 4668–4670.
7. Dynin, A. S. (1986). Multivariable Wiener-Hopf operators I. Representation. *Integral Equations and Operator Theory, 9*, 537–556.
8. Dynin, A. S. (1987). Multivariable Wiener-Hopf operators II. Spectral topology and solvability. *Integral Equations and Operator Theory, 10*, 554–576.
9. Friedrichs, K. O. (1970). *Pseudodifferential operators: An introduction*. Lecture Notes, Courant Institute of Mathematical Sciences.
10. Gakhov, F. D. (1966). *Boundary value problems*. International Series of Monographs on Pure and Applied Mathematics (Vol. 85). Pergamon Press.
11. Gohberg, I. T., & Krupnik, N. Y. (1970). Algebra generated by one-dimensional singular integral operators with piecewise continuous coefficients. *Functional Analysis and its Applications, 4*(3), 193–201.
12. Gohberg, I. T., & Krupnik, N. Y. (1992). *One-dimensional linear singular integral equations*. Operator Theory: Advances and Applications. Basel: Birkhäuser Verlag.
13. Handelman, D., & Yin, H.-S. (1988). Toeplitz algebras and rotational automorphisms associated to polydiscs. *American Journal of Mathematics, 110*(5), 887–920.
14. Hörmander L. (1965). Pseudodifferential operators. *Communications on Pure and Applied Mathematics, 18*, 501–517.
15. Hörmander, L. (2007). *The analysis of linear partial differential operators III. Pseudo-Differential operators*. Classics in Mathematics. Berlin: Springer-Verlag.
16. Kokotov, A. Yu. (1992). Representation of C^*-algebras generated by pseudodifferential operators in a polyhedron. *St. Petersburg Mathematical Journal, 3*(3), 541–562.
17. Melrose R. B. (1993). *The Atiyah-Patodi-Singer index theorem*. Research Notes in Mathematics. A K Peters/CRC Press.

© The Author(s), under exclusive license to Springer Nature Switzerland AG 2023

B. Plamenevskii, O. Sarafanov, *Solvable Algebras of Pseudodifferential Operators*, Pseudo-Differential Operators 15, https://doi.org/10.1007/978-3-031-28398-7

18. Mikhlin, S. G. (1965). *Multidimensional singular integrals and integral equations*. International Series of Monographs on Pure and Applied Mathematics (Vol. 83), Pergamon Press.
19. Murphy G. (1990). *C*-algebras and operator theory*. Academic Press.
20. Muskhelishvili, N. I. (1958). *Singular Integral Equations: Boundary problems of functions theory and their applications to mathematical physics*. Dordrecht: Springer.
21. Plamenevskii, B. A. (1989). *Algebras of pseudodifferential operators*. Mathematics and its Applications, Soviet Series (Vol. 43). Dordrecht: Kluwer Academic Publishers.
22. Plamenevskii, B. A. (2010). Solvability of the algebra of pseudodifferential operators with piecewise smooth coefficients on a smooth manifold. *St. Petersburg Mathematical Journal, 21*(2), 317–351.
23. Plamenevskii, B. A., & Senichkin, V. N. (1984). On the spectrum of C^*-algebras generated by pseudodifferential operators with discontinuous symbols. *Mathematics of the USSR-Izvestiya, 23*(3), 525–544.
24. Plamenevskii, B. A., & Senichkin, V. N. (1987). Representations of C^*-algebras generated by pseudodifferential operators in weighted spaces. *Problems in Mathematical Physics, 12*, 165–189. (In Russian).
25. Plamenevskii, B. A., & Senichkin, V. N. (1988). On representations of an algebra of pseudodifferential operators with multidimensional discontinuities in the symbols. *Mathematics of the USSR-Izvestiya, 31*(1), 143–169.
26. Plamenevskii, B. A., & Senichkin, V. N. (1990). The spectrum of an algebra of pseudodifferential operators with piecewise smooth symbols. *Mathematics of the USSR-Izvestiya, 34*(1), 147–179.
27. Plamenevskii, B. A., & Senichkin, V. N. (1995). Solvable operator algebras. *St. Petersburg Mathematical Journal, 6*(5), 895–968.
28. Plamenevskii, B. A., & Senichkin, V. N. (1996). *On composition series in algebras of pseudodifferential operators and in algebras of Wiener-Hopf operators*. Mathematical Topics (Vol. 11, pp. 373–404). Advances in Partial Differential Equations. Berlin: Akademie Verlag.
29. Plamenevskii, B. A., & Senichkin, V. N. (2000). On a class of pseudodifferential operators in $\mathbb{R}^m$ and on stratified manifolds. *Sbornik ,191*(5), 725–757.
30. Plamenevskii, B. A., & Senichkin, V. N. (2002). Representations of C^*-algebras of pseudodifferential operators on piecewise smooth manifolds. *St. Petersburg Mathematical Journal, 13*(6), 993–1032.
31. Plamenevskii, B. A., & Tashchiyan, G. M. (1993). A convolution operator in weighted spaces. *Journal of Mathematical Sciences, 64*, 1363–1381.
32. Sarafanov, O. V. (2004). The spectrum of the C^*-algebra of pseudodifferential boundary value problems on a manifold with smooth edges. *Journal of Mathematical Sciences, 120*, 1195–1239.
33. Sarafanov, O. V. (2005). Calculus of pseudodifferential boundary value problems on manifolds with smooth edges. *American Mathematical Society Translations, Series 2, 214*, 183–236.
34. Schochet, C. (1981). Topological methods for C*-algebras I: Spectral sequences. *Pacific Journal of Mathematics, 96*(1), 193–211.
35. Schulze, B.-W. (1991). *Pseudodifferential operators on manifolds with singularities*. Studies in Mathematics and its Applications (Vol. 24). Amsterdam: North Holland.
36. Schulze, B.-W. (1994). *Pseudo-differential boundary value problems, conical singularities, and asymptotics*. Berlin: Akademie Verlag.
37. Senichkin, V. N. (1995). Pseudodifferential operators on manifolds with edges. *Journal of Mathematical Sciences, 73*, 711–747.
38. Senichkin, V. N. (1997). The spectrum of the algebra of pseudodifferential operators on a manifold with smooth edges. *St. Petersburg Mathematical Journal, 8*(6), 985–1013.
39. Shubin, M. A. (2001). *Pseudodifferential operators and spectral theory*. Berlin: Springer-Verlag.

40. Simonenko, I. B. (1965). A new general method for studying linear operator equations of the type of singular integral equations, I, II. *Izvestiya Akademii Nauk SSSR. Seriya Matematicheskaya, 29*(3,4), 567–586, 757–782. (In Russian).
41. Simonenko, I. B. (1967). Operators of convolution type in cones. *Mathematics of the USSR-Sbornik, 3*(2), 279–293.
42. Taylor, M. (1981). *Pseudodifferential operators*. Princeton, NJ: Princeton University Press.
43. Treves, J.-F. (1980). *Introduction to pseudodifferential and Fourier integral operators* (Vols. 1–2). University Series in Mathematics. New York: Springer.
44. Upmeier, H. (1987). Index theory for Toeplitz operators on bounded symmetric domains. *Bulletin of the American Mathematical Society, 16*, 109–112.
45. Upmeier, H. (1988). Index theory for multivariable Wiener-Hopf operators. *Journal fur die Reine und Angewandte Mathematik, 384*, 57–79.

Printed in the United States
by Baker & Taylor Publisher Services